Asper Nation

ASPER NATION

Canada's Most Dangerous Media Company

Marc Edge

NEW STAR BOOKS

VANCOUVER

2007

NEW STAR BOOKS LTD.

107 — 3477 Commercial Street | Vancouver, BC V5N 4E8 | CANADA
1574 Gulf Rd., #1517 | Point Roberts, WA 98281 | USA
www.NewStarBooks.com | info@NewStarBooks.com

Publication of this work is made possible by the support of the Canada Council, the Government of Canada through the Department of Canadian Heritage Book Publishing Industry Development Program, the British Columbia Arts Council, and the Province of British Columbia through the Book Publishing Tax Credit.

Printed and bound in Canada by Marquis Printing, Cap-St-Ignace, QC
First printing, October 2007

LIBRARY AND ARCHIVES CANADA CATALOGUING IN PUBLICATION

Edge, Marc, 1954–

Asper nation : Canada's most dangerous media company / Marc Edge.
Includes bibliographical references and index.
ISBN 978-1-55420-032-0

    1. CanWest Global Communications Corp. — History.   2. Asper, I.H.,
1932–2003.   I. Title.
HD2810.12.C378D34 2007   384.5506'571   C2007–903983–9

For the Clarks – Lynda, Al, Laura, Spencer, and Chloe –
and especially their hot tub, without which
this book could never have been written.

CONTENTS

Preface   ix

Introduction : The Asper Disaster   3

Chapter 1: Citizen Asper   9

Chapter 2: CanWest Rising   27

Chapter 3: Going Global   50

Chapter 4: The Black Plague   70

Chapter 5: Media Power   86

Chapter 6: Convergence   109

Chapter 7: The Gazette Intifada   131

Chapter 8: L'Affaire Russell Mills   152

Chapter 9: Dishonest Reporting   170

Chapter 10: Like Father, Like Children   190

Chapter 11: Media Bias   208

Chapter 12: The Press We Deserve   228

Conclusion: The Pictures in Our Heads   246

Appendix: The Asper Agenda   265

Notes   268

Bibliography   310

Index   320

T his book follows from my first – *Pacific Press: The Unauthorized Story of Vancouver's Newspaper Monopoly* – which was published in 2001. Because it was my doctoral dissertation, I had to end that narrative sufficiently far in the past for it to qualify as an academic history. The end point I chose pushed the limits at only ten years earlier, and since 1991 momentous events had transformed the *Vancouver Sun* and my alma mater, *The Province*. *Pacific Press* had changed hands twice with Conrad Black's 1996 takeover of parent Southam Newspapers, which he sold four years later to CanWest Global Communications. I was only able to touch on those developments briefly in an updated preface to my dissertation. Many of my former colleagues at the *Province* confided that journalism at the Vancouver dailies had deteriorated sharply as a result. They urged me to write a sequel that would chronicle the regime of Black and his Vancouver-based lieutenant, David Radler. With my hands full writing the history of *Pacific Press*, I was only able to briefly research current events by chronicling some apparent political favoritism during the 2000 federal election.[1]

The following spring, David Asper's infamous *National Post* column defending prime minister Jean Chrétien from his accusers was published. The spectacle of a newspaper owner publicly scolding his own journalists for daring to delve into a scandal surrounding the prime minister was daunting indeed for press freedom. I had already accepted a position teaching at a university in Singapore, however, where I would spend three years. From that distance it was difficult to keep up with events back home. I managed to read reports, depressing as they were, on the Internet. Byline strikes broke out across the country and columnists quit in droves to protest the draconian editorial practices of the Aspers. Amidst the furor over the firing of Russell Mills in mid-2002, I returned home on leave fully intending to go sailing and to stay out of the controversy. Instead, I began research into a money trail that led from the Aspers to journalism schools that should have housed their toughest critics. In fact, as the beneficiaries of CanWest funding, some faculty members served as their staunchest supporters. I never did get out sailing that summer.

A year teaching in Texas followed, during which I did some research into Black's imploding newspaper empire. He and Radler had personally pocketed millions in questionable "non-compete" payments as part of the sale of Southam to CanWest, among other deals. The newspaper business in Canada, I well knew from my research, was non-competitive enough without owners paying each other to ensure their monopoly profits and political influence. A workshop on corporate governance of media companies hosted by media economist Robert Picard in Sweden was the perfect venue for presenting that research. My paper chronicled Black's battle with the independent directors of Southam, who did what they could to block his takeover only to be branded an "obdurate rump." The irony came in the subsequent collapse of Black's house of cards. Hollinger International was the antithesis of good corporate governance, with Black's wife Barbara Amiel and his political cronies, such as Henry Kissinger and Richard Perle, installed as directors. Perle, a former U.S. assistant secretary of defence, sat on Hollinger's three-member executive committee but admitted to not even reading documents Black and Radler gave him to sign. Instead Perle spent most of his time masterminding the invasion of Iraq as head of the Pentagon's volunteer Defense Policy Board.[2]

Thus, while I had no idea as yet that I would be asked to write a book about the intertwined tale of Southam, Black and the Aspers, I had already done much of the research. It was also published in such academic journals as *Textual Studies in Canada*, for which I chronicled Senator Keith Davey's conclusion that Canadians get "the press we deserve" for failing to demand better. *The International Journal on Media Management* in Switzerland carried my account of stock market effects on Southam ownership.[3] *Journalism & Mass Communication Educator* ran my exposé of Canadian journalism schools in 2004. The late Margaret "Peggy" Blanchard published my book review of *Hidden Agendas* in *Journalism and Mass Communication Quarterly*. Its blaming of media bias on individual journalists, while absolving media owners, just didn't fit with my understanding of the subject. Once I examined the research more closely, I understood why.

David Spencer of the University of Western Ontario invited me to present my research on the 2003 Lincoln report on Canadian broadcastingto the 2005 Canadian Communication Association conference, which his school hosted, and he even billeted me during my stay. He also published my paper earlier this year as "Convergence and the 'Black News Hole'" in his new online peer-reviewed *Canadian Journal of Media Studies*.[4] I am fortunate to have Dr. Spencer as a mentor. When I returned to Canada full-time that fall to teach for a year at BC's newest university, Thompson Rivers, I began to learn more about the Asper family's media machinations. Preparation for my lectures in the News and Media Business course I taught showed the appalling level to which journalism had fallen in Canada. That sorry state, it seemed from my research, was thanks largely to Winnipeg's first family of converged media. So when New Star Books publisher Rolf Maurer asked me in early 2006 if I was interested in writing a book about CanWest, I leapt at the chance. Such an important book had to be written, I decided, even if it meant taking a year off from my budding second career in teaching. The opportunity to write this book was one I could not pass up, and I thank Rolf for it.

Finally, a bit of serendipity resulted in some valuable comparative research being conducted concurrently for Blackwell's forthcoming *International Encyclopedia of Communication*. When asked if I would be interested in writing its entry on cross-ownership of media, I gladly agreed because I realized how relevant such research would be to this book. Examining how such laws have been changing in countries around the world — and not changing in others, notably the US — has helped add valuable insight. It explained how removing a ban on cross ownership in Canada decades ago left the floodgates open for the tsunami of convergence to transform the country's media landscape in 2000.

When I submitted a draft manuscript in late 2006, Rolf had several valuable suggestions for improvement. In addition to telling the Asper story, I had decided to explore some issues I felt helped put it in context, such as media bias and convergence. Not only was Rolf all in favor of illuminating those aspects, he decided even more "theory" was needed to explain why such tight media control should concern Canadians. I thus added the Media Power chapter at his suggestion. It includes almost everything I have learned on the subject since I began studying it a decade ago. When I tell other scholars my publisher actually asked for more theory, not less, they shake their heads in amazement.

One problem I encountered in writing this book was that the story kept changing even as I was telling it. I was able to work some breaking aspects into the narrative, such as CanWest's takeovers of Alliance Atlantis and the *New Republic*. Later developments failed to make it into the book, such as CanWest's May 2007 sale (for CDN$320 million) of its New Zealand operations and its June 2007 decision not to sell Network TEN in Australia. CanWest had also set the wheels in motion to buy back the shares it had sold in creating an income trust. David Asper was bidding to buy the Winnipeg Blue Bombers after 77 years of community ownership, and seeking $80 million in government funding for a new stadium. Revelations from Conrad Black's Chicago fraud trial proved useful, but his July 2007 conviction came just as we entered production, with sentencing and appeals to follow.

CRTC approval of the contentious Alliance Atlantis deal was still pending, as was its review of media diversity announced for mid-September 2007.[5] The hearings under new CRTC chair Konrad von Finckenstein promised to pit Prime Minister Stephen Harper's deregulationist agenda against a growing grassroots movement for media reform. The clamor for change came after the CRTC eliminated limits on television advertising that spring, thus granting one of CanWest's fondest wishes. "We want the market to make decisions, not the CRTC," announced von Finckenstein, adopting the stance of the minority Conservative government.[6] That prompted calls for von Finckenstein's resignation and for Canadians to protest the CRTC's laissez-faire approach to media regulation.[7] As a result of its interim nature, therefore, this history of CanWest must be considered necessarily incomplete. It does aim, however, to provide some needed context for the CRTC's diversity review and for the ensuing CanWest Global licence renewal hearings in 2008.

I met Izzy Asper once, briefly, in the mid-1980s at the Law Courts in downtown Vancouver, which I covered for the *Province*. His nicotine habit led to him being banished outside, as a ban on smoking had recently been imposed. Through the glass walls of Arthur Erickson's leaky landmark, I can still see him guffawing outside with Phil Needham, my competition from the *Vancouver Sun*, who was also a smoker. That was fine as far as I was concerned; as a reporter for the tabloid *Province* I was more interested in juicy criminal trials than in Asper's legal battle for control of CKVU.

One day he intercepted me and introduced himself. I recall how smoothly charming and sartorially splendid he was. He ingratiated himself with me by relating how he had been a longtime newspaper columnist. I was impressed with the man, but my interest in his court case did not increase. Now, of course, I wish I had taken the opportunity to get to know him a bit better. Everyone who knew him seemed to agree he was endearing and personable. Except for his business partners, of course, who universally found him insufferable, and for the journalists ultimately confronted with his notions of ownership privilege.

This book is less an indictment of Izzy Asper and his heirs than of a system that allowed them to gain control over so much of Canada's news media and use it to promote an ideological agenda. There will always be people like the Aspers — successful in business and eager to use their fortunes to influence public opinion. If Izzy Asper hadn't been the one to demonstrate the peril of permitting such concentrated and now converged media ownership, it would have been another tycoon.

Some will undoubtedly find that in chronicling the media biases of Izzy Asper and his heirs I betray a bias of my own. That is inevitable, as every writer necessarily comes with a bias. Mine is that of a disillusioned former journalist who spent two decades growing increasingly frustrated by a craft to which I had long aspired. Since then, I have dedicated myself to studying the news business to help improve it through research and informed criticism. Most reviews of *Pacific Press* seemed to agree it went a good way towards that end. One, however, pointed to my bias. "His history is not a neutral analysis, as is obvious from its self-proclaimed status as an 'unauthorized' history," wrote Stephen Ward of the University of British Columbia. "A more balanced view on the legacy of Pacific Press will have to await other histories with other perspectives."[8]

Professor Ward found me remiss for failing to interview the current managers of Pacific Press, the most senior of whom were among his faculty colleagues. My inclusion of "considerable strong opinion and evaluation," he seemed to think, rendered my research somehow suspect. My training in historical research methods under Pat Washburn at Ohio University, however, impressed on me the need to illuminate the "why" behind what merely happened. What possible insight could the current managers of Pacific Press have provided — other than spin — into events decades earlier? I found it much more useful to interview the managers and editors of the era I was researching.

The main thing I learned in researching this book is the importance of perspective. Perceptions of media bias seem to say as much about the reader as they do about the writer. We all have our own preconceptions of the world that influence the way we perceive bias. The stronger and more extreme our prejudices, the more biased we will consider reports that differ from our way of looking at things. Professor Ward was recently named to replace Donna Logan as director of UBC's journalism school. (Without the need for a competition, incidentally.) As such, I wouldn't expect him to share my perspective, as the pro-corporate bias of that school has been apparent since its inception.

Whether or not a biography or a corporate history is authorized by its subject does not necessarily speak to its neutrality, or its quality. Charles Bruce's 1968 authorized history *News and the Southams*, for example, examined both sides of important issues. Allan Levine's 2002 commissioned history of CanWest, *From Winnipeg to the World*, was told by comparison from a decidedly Asper perspective. As this book went to press, Levine was collaborating with Peter C. Newman on an Asper biography that may or may not be authorized by his heirs. "Peter would never do an authorized biography," his agent insisted to the *Globe and Mail* in 2006. "Let's be really, really clear about that."[9] Morley Walker, books editor of the *Winnipeg Free Press*, wasn't so sure. His colleague Gordon Sinclair Jr. was also writing a biography of Asper, but his was unauthorized. Sinclair had received a "sizeable" advance from McClelland & Stewart in 2004 for *Izzy Asper: The Maverick, the Mogul and All That Jazz*. It was planned for publication in May 2007, and its cover was even posted on Amazon.ca. Sinclair's research, however, was stymied by the Aspers, according to Walker. "Early on, while trying to set up interviews with members of Izzy's Winnipeg mafia, he heard the same refrain," Walker reported. " 'Sorry, the family has asked us not to speak to you. They have someone else in mind to write the book.' " That turned out to be Newman, whose fawning profile of Asper in his book *Titans* made him the family's biographer of choice. "Newman was in town last month conducting interviews at the Ramada Inn — with many of the people who wouldn't go on record with Gord," reported Walker in July 2006. "Asper's widow, Babs, gave Newman an estimated 10 boxes of material to aid his research." The project was being kept "so hush-hush," quipped Walker, "you'd think they were negotiating a hostile takeover of Bell Globemedia."

> Normally, when a star author signs a book deal, the news is trumpeted by the publisher. But if Newman doesn't have a contract, who is paying for his time? Could it be CanWest Global itself? This is the inescapable conclusion, although it's unthinkable that a journalist of Newman's stature would risk his reputation by taking money from the family of the subject he's writing about. And would this mean they would vet the material? That's not biography writing. That's vanity publishing. On the other hand, Newman has made no bones over the years about his need for money to pay off his ex-wives.[10]

The intrigue increased when McClelland & Stewart cancelled Sinclair's contract in late 2006 after he missed his deadline. As a result, concluded the *Globe and Mail*, Newman "should face an uncluttered market."[11] Well, we hate to clutter things up, but I guess our little project brewing way out here on the west coast flew somewhat under the radar. Of course, the idea was Rolf's. I just typed the thing out. Editor Carellin Brooks helped make sense of it. Proofreader Stefania Alexandru caught things we both missed. Any remaining errors and/or omissions are, of course, my responsibility alone. I am especially grateful for the valuable input of readers John Miller, Bob Hackett, David Spencer, Albert Rose, Dan Riffe, Dane Claussen, and Jennifer Kirkey.

This book is neither authorized nor a biography, instead qualifying more as a critical corporate media history. It is much more a work of scholarship than of journalism. It relies almost entirely on secondary sources, containing little original research. So much has already been written and reported about CanWest that what I felt was more needed was an analysis to help put it all into context. That's what this book hopes to do. In no way does it claim to be a comprehensive history. More decades will have to pass for that to even be attempted, after which some of the story's principals will hopefully be at liberty to contribute their insights. Journalism has often been called the "first draft of history." This qualifies only as a rough second draft. To paraphrase Stephen Ward, a more balanced view on the legacy of CanWest Global Communications will have to await other histories with other perspectives.

Asper Nation

The Asper Disaster

W hen Conrad Black sold the Southam newspaper chain to Izzy Asper's CanWest Global Communications in July of 2000, many Canadians expressed relief. Concentration of newspaper ownership by the five largest chains had risen under Black's domination from 73 percent in 1996 to 93 percent in 1999.[1] The neo-conservative press baron consistently injected political partisanship into the newspapers he acquired worldwide starting in the 1980s. His mid-'90s takeover of Southam, Canada's oldest and largest news media company, led many to predict that right-wing propaganda would soon come disguised as news. Then Black launched the *National Post* in 1998, and it changed journalism in Canada forever. Reporting from a decidedly partisan perspective, it eschewed traditional journalistic notions of neutrality for thinly-veiled advocacy of conservative causes.

From its inception, the *National Post* openly sought to "unite the right" of Canada's fractured conservatives in an attempt to displace the ruling Liberal party. Many thus saw Black's unprecedented control of the country's press as dangerous politically, and it brought renewed calls for long-sought limits on media ownership. When his stewardship of Southam proved short-lived after Black quickly "flipped" the newspaper chain to CanWest, much journalistic anxiety abated. Concentration of ownership fell to 78 percent by the five largest chains.[2] The reduction, however, was at the expense of a new "convergence" between newspaper and television ownership.

Any relief was short-lived when it soon became apparent that Southam's new owner was just as ardently activist as Black. Worse, Izzy Asper and his heirs were more influential under convergence than he ever was. They seemed even more intent than Black on using the Southam dailies to sway public opinion. It was well known that Asper supported the Liberal party and had been leader of its Manitoba branch in the early 1970s. It soon became apparent, however, that his ideology was much closer to Black's neo-conservatism than most realized. Black noted in a published parting shot to Canadians that Asper was "a St. Laurent-Howe Liberal, as I am." The reference was to the pro-business post-World War II government of Liberal prime minister Louis St. Laurent. One of Canada's leading conservative think tanks was fittingly named after his "minister of everything", C.D. Howe. "When he was leader of the Manitoba Liberal party, Izzy Asper advocated a flat tax and workfare," noted Black. "*National Post* has no more appreciative reader than he."[3] Asper family actions would soon result in two federal inquiries into Canada's media, but nothing would come from either.

Asper had been a Winnipeg tax lawyer, a nationally-syndicated tax columnist, and a politician before founding the country's third television network in the mid-1970s. He had done more than just advocate for a flat tax as a writer. He authored a best-selling 1970 book aimed, according to Larry Zolf, at "showing the rich how they could destroy the welfare state and get even richer, and all on tax reform." The CBC commentator, who grew up with Asper, noted he had "a government minimalist policy for everything and had his own academics to pronounce him right."[4] Asper's book, *The Benson Iceberg*, was critical of tax reforms proposed by Liberal finance minister Edgar Benson. It played a major role in thwarting Benson's bid to shift the tax burden from Canada's working class onto corporations. Asper then entered politics as Manitoba Liberal leader, but he failed to bring his party to power. The fiscally conservative policies Asper espoused as a longtime party insider, however, helped prompt a rightward turn by the Liberals, according to Zolf. In the *National Post*, Asper acquired the perfect vehicle to promote his views on tax policy, which most of all included tax cuts for rich Canadians such as himself.

The Southam newspaper empire, which dated to the 19th Century, included the largest circulation dailies in Ottawa, English Montreal, Regina, Edmonton, Calgary, and Vancouver. It published the dominant daily in the two largest cities in each of the three westernmost provinces. In unprecedented local control, the Aspers owned both dailies, most of the community newspapers, and the dominant television station in Vancouver.[5] One result was open media support for British Columbia's neo-conservative Liberal

Party, including a $35,000 campaign contribution from CanWest. Black could initially only bring himself to sell half of the *National Post*, but within a year he let CanWest have the other half as well. Asper proved adept at using the *Post* to promote his views nation-wide, and soon his heirs joined in as well. Most of their agenda jibed nicely with that of the upstart daily, except for its continual criti-cism of Liberal prime minister Jean Chrétien, who was an Asper family friend. *National Post* journalists quickly learned that criti-cizing the prime minister or delving too closely into a scandal sur-rounding Chrétien was frowned on. The message was delivered not only in Asper-authored columns, but also by dropping some of Chrétien's sharpest critics from the newspaper's payroll.

Soon Izzy Asper's favourite issues, which included the Middle East conflict and the CBC, were being covered by Southam newspa-pers as never before. Asper claimed media coverage of the Middle East, particularly by the CBC, was biased against Israel. Criticism of the public broadcaster, which Asper also complained was unfair competition for his Global TV stations, became a regular feature in the *National Post*. Newspaper owners in Canada had seldom exer-cised their influence so openly. Even Black, who was well known for voicing his opinions in print, knew better than to be seen interfer-ing overtly in journalistic independence. The Aspers were from the entertainment medium of television, however, and they had little understanding of journalism ethics.

CanWest Global, known to some as the "Love Boat network" for its incessant Hollywood reruns, had become Canada's most prof-itable broadcaster. The profits resulted largely from Asper's abil-ity to manipulate the country's lax broadcasting regulations to his advantage. The mounting proceeds went to expanding the com-pany internationally in the 1990s, with operations in Australia, New Zealand, South America, and the UK. They also went to buy-ing Southam and providing Black and other company insiders with controversial "non-compete" payments. The resulting "conver-gence" of newspapers and television worried many Canadians, who felt it allowed too much political power to be held by too few media owners. While laws in other countries, including the US, prohib-ited joint ownership of newspapers and television stations, a ban in Canada had been lifted decades earlier. Consumer advocates insisted that the CRTC impose strict licence conditions on CanWest Global and CTV, which had partnered with the *Globe and Mail*. The networks, however, refused a CRTC demand for a "firewall" of separation between their print and television newsrooms.

When David Asper, Izzy's eldest son, wrote a 2001 *National Post* column defending Chrétien from his critics, most saw political favouritism at work. All three federal opposition parties demanded an inquiry into concentration of media ownership. Instead the matter was shuffled off to a committee that had already been formed to examine broadcasting policy.[6] The Aspers appeared undaunted by the inquiry as they stepped up the nationwide dissemination of their views. In late 2001, CanWest decreed that regular national editorials written at company headquarters in Winnipeg would run in its newspapers across Canada. The edict ran contrary to the long-standing Southam policy of allowing publishers the editorial independence to reflect the views of their own communities. Protests by journalists erupted across the country as the episode escalated into one of the sorriest chapters in Canadian journalism history. Reporters at the *Montreal Gazette* went on "byline strike" by withdrawing their names in protest from atop stories. The uprising was suppressed when CanWest threatened them with firing. "The slightest misstep, the memo warned — even 'gossiping' — could lead to dismissal," reported the *Toronto Star*.[7] David Asper called CanWest's critics "riff raff" and "bleeding hearts." He mocked them by paraphrasing a song by the band REM. "It's the end of the world as they know it," he said, "and I feel fine."[8]

The response brought even more protests from journalists. Halifax *Daily News* columnist Stephen Kimber quit in January of 2002 after his criticism of the Aspers was killed. The *Daily News* editor resigned after admitting interference in content from CanWest headquarters. Reporters at the *Regina Leader-Post* went on byline strike after an article on a speech critical of CanWest's national editorials was altered by editors. The journalists were suspended by CanWest, which banned such protests at its newspapers. Dozens of former Southam editors and publishers took out a full-page ad in newspapers across the country, but CanWest refused to print it. The erstwhile executives called on Ottawa to step in if the Aspers did not relent in their editorial strong-arm tactics. Soon Canadians who welcomed Black's sale of the Southam dailies realized that Asper ownership had been a change for the worse.

The controversy exploded into a full-scale disaster for press freedom in the summer of 2002. After the *Ottawa Citizen* ran an editorial calling for Chrétien's resignation, David Asper fired longtime publisher Russell Mills. Opposition politicians renewed their calls for an inquiry into the press in Canada. The Vienna-based International Press Institute called the Mills firing "an attack on press freedom by an unholy coalition between politics and big business."[9] Debate in Parliament was dominated for days by accusations that Chrétien had ordered the firing personally. CanWest quietly abandoned its policy of imposing national editorials on its newspapers, but skeptics warned that Asper family influence would only become less obvious.[10]

Then in early 2003, the *Globe and Mail* reported what it described as "this country's most aggressive attempt to centralize editorial operations across a newspaper chain." It revealed an internal CanWest memo setting out plans for a central news desk at company headquarters in Winnipeg. Some saw it as an attempt to covertly influence public opinion by the shaping the news instead of openly advocating in editorials.[11] Perhaps symbolically, Southam Newspapers also got a new name — CanWest Publications. The following month, Senate hearings on the news media were announced. After he died in 2003, Izzy Asper's three children inherited CanWest and continued to support their father's favourite causes, with one important exception. Instead of supporting the Liberal party, they openly backed the rejuvenated Conservatives of Stephen Harper, who came to power in 2006 with a minority government.

Harper had been personally endorsed by David Asper, and CanWest's relationship with the new ruling party in Ottawa was uncomfortably close for some critics. Bev Oda, a former CanWest executive, was named Heritage Minister with responsibility for media regulation. Derek Burney, a longtime Tory who headed Harper's transition team to power, was named chairman of CanWest's board of directors. A senior Global Television executive even ran as a Conservative candidate in Toronto. The Harper government and the Aspers engaged in an unseemly honeymoon of mutual back scratching. When the Senate inquiry into Canada's news media issued a report with only mild proposals for reform, even those were rejected out of hand by Oda. A new round of corporate media consolidation saw CanWest acquire Alliance Atlantis, one of Canada's largest media companies. The takeover was accomplished only with massive American investment, disregarding the country's limits on foreign ownership. Most expected federal regulators to look the other way, however, under a CanWest-friendly Conservative government. Meanwhile, CanWest beefed up its own news service with dozens of new hires in advance of its promised pullout from the Canadian Press news co-operative in mid-2007.

This book looks at the origins of Asper family influence on Canadian media and politics. It examines Izzy Asper's writings on taxation policy and his campaign positions as Manitoba Liberal leader for some needed perspective on CanWest's media dominance. It scrutinizes the business practices of CanWest Global Communications both in Canada and around the world. It chronicles the intertwined tales of Southam, Black, and the Aspers to better understand how Canada came to suffer some of the world's tightest media control. It concludes that the country's current concentration of media ownership has resulted in an increasingly centralized perspective — an Asper Nation, if you will.

CHAPTER 1
Citizen Asper

I srael Harold "Izzy" Asper was many things at once, a fountain of energy who combined his passions as liberally as he mixed his martinis. He could be ruthless in business and relentless in court, yet charming socially and generous to a fault in his devotion to philanthropy. The chain-smoking, piano-playing jazz aficionado embarked on successive careers, which often overlapped. He was by turns a lawyer, a tax consultant, a newspaper columnist, a best-selling author, a university lecturer, a politician, and finally a businessman. Commerce was a field he came to only in his forties, but it was where he had his greatest success, making deals and fighting boardroom battles. "Business was much more suited to Asper's unvarnished style," noted a 1996 *Maclean's* profile. "He is egotistical but unpretentious, and often unguarded. He likes to amuse and to stir the pot."[1] On the campaign trail with Asper in 1972, *Globe and Mail* reporter Martin O'Malley described him in terms that might have endeared the candidate to some. Others might have found the description derogatory.

> Brash. Cocky. Impudent. He is 39, drives a Firebird convertible, wears snappy tweed suits and keeps his hair moderately past his collar. He once earned $200,000 in a peak year as a tax lawyer and consultant, and now he is slugging it out, dipping into his first electoral swim after being criticized for shying away from three previous by-elections since he was elected Liberal leader in October, 1970.[2]

The *Globe and Mail*'s Edward Greenspon saw Asper in 1987 as "a raspy-voiced chain-smoker whose fast-talking and outgoing style turn as many people off as on."[3] Greenspon was obviously turned on by the gregarious Winnipegger. He described him in a magazine cover story the next year as "overloaded with energy, charm and brains."[4] Trevor Cole labelled him "a work of entrepreneurial art" in

1991. "When Izzy fixes on a goal, he is like a four-year-old's windup toy racer, moving relentlessly forward, bouncing off obstacles and roaring back, until he achieves it." According to Cole, Asper was "driven by the legacy of a workaholic father . . . who was never satisfied with himself or his sons." The result was a "pitiless" work ethic. "He will work until the dark and corrugated lids of his eyes leave him slits to see through and his voice seems to rise from the centre of the earth," wrote Cole. "Then he'll sleep for a day or more."[5]

One friend and business associate, who wished to remain anonymous for obvious reasons, called Asper "a Machiavellian genius," and "the most aggressive businessman I know."[6] Even before his two defining business deals, Asper warranted inclusion in Peter C. Newman's 1998 re-examination of the Canadian establishment. The dean of Canadian business journalism deemed Asper "virtually immune to criticism" and found that he tackled problems "with the grace of a tank."[7] Modesty, noted Newman, was "not one of Asper's dominant traits."[8]

> Asper is one of a kind. . . . A Titan of his own making, he is solidly of the New Canadian Establishment, but not of it. A graduate of his unique school of meritocracy, he tries hard to ignore most establishment rites, and whatever establishment still exists in Manitoba returns the favor, by trying to ignore him.[9]

Yet in later life, after he had risen to the top of the media business in Canada, others found Asper an unlikely candidate for power and influence. Ed Pearce described him as "relatively short and overweight, with a florid face and gray, thinning hair," in 2001. "He is a heavy smoker who looks uncomfortable in a suit and has an obvious disdain for ties. In short, Izzy Asper does not fit the preconceived image of an international media mogul."[10] Gordon Pitts portrayed him in his 2002 book *Kings of Convergence* as a man of contradictions — worldly yet firmly grounded by his Manitoba roots. "He is very smart but defensive, carrying a two-by-four on his shoulder about being a Westerner and, some say, a Jewish outsider."[11]

Rural beginnings

Asper was born in 1932 in Minnedosa, Manitoba, a town of 2,000 on the Little Saskatchewan River 200 kilometres west of Winnipeg, where his father ran the Lyric theatre. Leon Ausereper trained as a violinist at the Odessa Conservatory of Music but fled the Ukraine in the early 1920s following the Bolshevik revolution. He changed his name to suit a new country and got into the theatre business as an orchestra conductor in Regina during the days of silent pictures. Movie sound made musicians redundant, so he became an owner instead. The hard times of the Depression paradoxically proved a boon for the movie business, as people sought refuge in the new media miracle. Izzy Asper literally spent his childhood in the theatre, taking tickets before shows and scraping gum off the seats afterward. From Minnedosa, the family moved in 1941 to nearby Neepawa, a larger town of 3,500. There a second theatre, the Roxy, was added to the budding media empire of Leon Asper. Soon the family business had grown into a thriving chain of movie houses with the addition of two more theatres in Winnipeg. It was profitable enough to move the family to a comfortable home in the city's affluent River Heights neighbourhood. There Izzy Asper would plant himself in the frozen tundra, stubbornly refusing to be moved by opportunities in eastern Canada and elsewhere.

It was in Winnipeg that Asper first showed a flair for media, founding a newspaper in his senior year at Kelvin High School in 1949. Unfortunately, the *Kelvin Gazette* was shut down after only one issue when Asper ran afoul of the local media regulator. "In one report of a Grade 12 party, references were made to labels on whiskey bottles," Asper explained. "The principal thought the reference was to alcohol consumed at the party. He closed the paper and I was suspended from school for about a week."[12] Larry Zolf, who shot hoops with a teenaged Asper at the Young Men's Hebrew Association, described him on the basketball court as an "aggressive fast shooter." They spent summers together at B'nai Brith Camp in Sandy Hook, where the older Asper once saved him from drowning. "Izzy rescued me from a watery grave," recalled Zolf, "pulling me out by the hair." Camp counselors Nathan "Toozy" Divinsky and Allan Gotlieb, a future ambassador to the US, were preoccupied playing chess, according to Zolf. The libertarian Divinsky would become a University of BC math professor and future prime minister Kim Campbell's first husband. According to Zolf, Divinsky preached the right-wing gospel of Ayn Rand to the impressionable young campers.

> Toozy got nowhere with me, but did make an impression on Izzy, who liked Ayn Rand's ideas of rugged individualism and

unfettered freedom and certainly preferred those ideas to my North End view of street-socialist thuggery for the greatest good for the greatest number.[13]

At the University of Manitoba, Asper studied law and excelled in debate, one year being crowned not just campus champion, but best in western Canada.[14] He also resumed his aborted career in journalism, writing a column on music for the *Manitoban* student newspaper. He dabbled in political writing, but there he would again run afoul of the administration. "I remember being summoned to the president's office when I advocated a student exchange with the Iron Curtain countries," recalled Asper. "This was . . . in the heyday of McCarthyism when everyone saw a Communist saboteur behind every bush."[15] Asper did more than just write about politics, putting his debating skills to use by running for office. His campaign in the faculty of arts was a harbinger of things to come, according to a profile years later in the *Toronto Star*. "His campaign slogans suggested some of the flair for self-promotion that would later become a trademark. 'Izzy clever? Izzy ever! Izzy Asper,' ran one slogan. 'Arts got a headache?' asked another. 'Get Asper-in.'"[16]

Tax columnist

After graduating in 1957 with a Master's degree in law, Asper began practicing in Winnipeg as a specialist in taxation. He became a leading expert in the arcane world of estate planning, tax holidays, and corporate income lumping. He married his sweetheart, Ruth "Babs" Bernstein, and they started a family. David was born in 1958, followed by Gail in 1960, and finally by Leonard in 1964. As the holder of a graduate degree, Asper even found time to lecture at his alma mater. Soon, however, Asper's ambition and creativity strained against the constraints of his dull, gray area of law. He started writing a column on taxes for the *Winnipeg Free Press*, and his facility for turning complex and often boring issues into interesting reading was apparent. The *Free Press* was the flagship of FP Publications, which in the mid-1960s was Canada's largest newspaper chain. Soon Asper's column was syndicated in ten FP dailies across the country, including the *Globe and Mail* and the *Vancouver Sun*. Writing under the byline I.H. Asper, he did more than dispense advice and dissect regulations. He also commented on the wisdom of various tax provisions, often in language that reflected his libertarian influence. "The father of the present Canadian tax system wasn't Adam Smith, or John Maynard Keynes, or even John Galbraith," quipped Asper in 1971. "It was Robin Hood."

> It took Karl Marx, the founder of communism, to set out the specifics of an ideal system for the Communist state. In the Communist Manifesto, Marx recommended a heavy progressive or graduated tax as an excellent method of transferring wealth from the aggressive rich to the non-aggressive masses. Canada has adopted the Marx tax system.[17]

Pointing out that almost 70 percent of income tax was collected from only a quarter of taxpayers, Asper noted that the others had their government services subsidized. The political power of the majority, he reasoned, ensured that governments would continue to increase social services — and taxes — in order to get re-elected. "The present system is inconsistent with democratic principle," he argued, "in that the majority can legislate the taxes that will be paid by everybody but them." Luckily, there was a simple and fair solution to this problem, according to Asper. "A flat tax would do more than remove this danger," he wrote. "First, it would restore the incentive to work harder. Next, the middle and upper income group would have enough money left after taxes to acquire greater ownership of Canadian industry. . . . The brain drain might also be stopped. Also, the incentive to avoid taxes would be remarkably reduced."[18]

The tax burden

The income tax system in Canada was the subject of heated debate in the late 1960s. A Royal Commission on Taxation headed by accountant Kenneth Carter reported in six volumes in 1967, finding that Canada's working poor paid more than their fair share of taxes while the wealthy exploited numerous loopholes and often paid none. It proposed to tax income from the sale of assets, or "capital gains," the same as earned income. As Carter observed, "a buck is a buck."[19] The proposed reforms would have increased the corporate tax burden in Canada by about 27 percent, noted Linda McQuaig. Opposition from the business community, however, killed most of them.[20] A White Paper on taxation with proposals for legislation introduced by finance minister Edgar Benson in November of 1969 was a substantial retreat from Carter's reforms. Even its watered-down provisions, McQuaig noted, enraged many businessmen.[21]

Leading the charge against shifting the tax burden onto the wealthy was Izzy Asper, who attacked the White Paper regularly in his column, in speeches, at legal conferences, and at Liberal party meetings. It was serendipity, however, that led Asper to author his seminal critique of Benson. He came down with mononucleosis and was bedridden for several months, which gave him the time and opportunity to write *The Benson Iceberg*. It turned out to be a best-sellet. The title was a reference to Asper's theory that 90 percent of the White Paper's consequences were unseen. He claimed its provisions, if adopted, would "alter the Canadian social and economic order."[22] Asper argued that the proposed tax changes were only a means to a larger end. "The end is the reshaping of society."[23] The White Paper proposals, he pointed out, raised fundamental questions, including: "Should every Canadian be his brother's keeper, or just his helper?"[24] Introduction of a capital gains tax, Asper argued, could have unforeseen consequences. In his tax practice, Asper said he had seen three projects cancelled in the months after the White Paper was introduced. "The investors concluded that in the light of the current tax proposals, the rewards for success, after taking into account the new taxes, would not be large enough to warrant the risk."[25] Asper made his own position perfectly clear, in contrast to that of those he saw behind the movement for change.

> I believe that the present and even greater social objectives can be furthered in a free enterprise economy in which private capital is maintained and is not dissipated through excessive estate taxation and capital taxation. The authors of the White Paper belong to the ever growing number of economists and academics who believe that money is more productive in the hands of government than in the private sector.[26]

Legislation introduced in 1971, according to McQuaig, was "a pale version of the White Paper, itself a pale version of the Carter report. Business had won."[27] Rather than shift the tax burden onto the rich, it left many of the loopholes intact. It even eliminated the estate tax that had been a significant check on the increasing concentration of wealth in Canadian society.[28] Only half of capital gains would be taxed, and capital losses would be tax deductible. The top tax bracket was lowered from 80 percent to 61.1 percent.[29] Instead of leading to reform of the income tax system to make it more fair to a majority of Canadians, the end result was almost exactly the opposite. "Under Finance Minister John Turner, the government introduced several new tax measures that greatly enriched corporate tax breaks," noted McQuaig. "By the time Turner quit the Cabinet in 1975, his new corporate tax breaks were saving companies close to $1 billion a year — on top of the tax savings they were already enjoying before Turner's stint at Finance."[30]

Leap into politics

By then, Asper was himself a Liberal politician, having jumped into politics shortly after the publication of his book. He continued to write his influential national column on taxation. He declared his candidacy for the vacant leadership of the Manitoba Liberal party in October of 1970. He told the *Winnipeg Free Press* he decided to "answer the call" after being pressured to seek the position.[31] He used his arguments against the White Paper as a springboard for his candidacy, since the NDP government of Premier Ed Schreyer had embraced its proposals.[32] His appeal was firmly to the middle ground as an alternative to Schreyer's socialism. His brand of Liberalism was fiscally conservative, in contrast to what he saw as a spendthrift NDP.

Asper's three-week campaign for the Liberal leadership stressed the image of a "dynamic young leader and a family man," noted the *Winnipeg Free Press*. It proved a successful appeal to a younger generation of Liberals that had been re-energized federally by the magnetic appeal of Pierre Trudeau. Asper's 11-year-old son David took an active part in whipping up support on the convention floor. The hit version of Asper's campaign slogan "Let the Sunshine In" blared over loudspeakers.[33] His appeal to delegates combined traditional Liberal ideals of social justice with his own brand of fiscal responsibility. He warned them the party might disintegrate provincially, leaving a polarization between left and right. "If Manitoba is exposed to a long reign of socialism, it will become an economic wasteland," Asper told the convention, "which will take a succeeding government a decade to correct."

> One does not have to be a socialist to have a social conscience, or a Tory to cherish free enterprise. To me, the Liberals straddle the middle. We seek both social and financial justice. We must

continually search for new answers to problems. We must provide equal opportunity for all to succeed in life. But, unlike the
socialist approach, we must not dictate what that life is.[34]

The main plank of Asper's platform was a new deal for Manitoba and the west, and he put the inequality borne by the region in
the strongest terms possible. "We are suffering from the criminal
injustice of 100 years of a Confederation economically structured
to make us a colony of Eastern Canada," he told the convention.
After out-polling his nearest challenger more than 2–1, Asper took
the stage for his acceptance speech. He declared he would lead the
party to a "new and true liberalism that will no longer permit ero-
sion of rights through totalitarian statism or government indifference."[35] In keeping with his carefully-crafted image, however, Asper
first introduced to delegates the future first family of converged
Canadian media. "This is Leonard . . . that's Gail . . . and David . . .
and this is Babs, my wife."[36] The political love-in even extended to
the editorial page of the *Winnipeg Free Press*. The venerable daily's
fawning endorsement must have left a lasting impression on Asper
of the power of editorial laudation. "Mr. Asper brings to our problems a mind familiar with the search for creative solutions and
the imagination and will to apply them," enthused the *Free Press*.
"What he has to say in the months ahead may change the mood and
the complexion of Manitoba politics."[37] The new Liberal leader had
made a considerable financial sacrifice to enter political life, the
newspaper noted.

> I.H. Asper chose a branch of law as his avocation and has turned
> to public service at the expense of a large and responsible prac-
> tice, out of a pressing sense of public obligation. . . . He has cho-
> sen the personally most expensive way to serve his province
> and has done it not after his practice had reached its limit but
> while his deviation from it must still be a costly venture for him.
> For that alone he deserves the congratulations and thanks of
> his fellow Manitobans.[38]

Uphill battle

The only problem for Asper was that the Liberal party in Manitoba
had been decimated in the 1969 provincial election, winning only
four of the legislature's 57 seats. That was the minimum required

for official party status in the province, but the Liberals soon lost that designation when one of their MLAs was appointed to the Senate. Schreyer called a by-election in early 1971 and challenged Asper to run, but the cagey lawyer demurred. It turned out to be a wise decision, as the NDP won the seat easily. As *Globe and Mail* reporter Don Newman noted, if Asper guessed wrong about his chances of winning and lost, his political career could be cut short. "If he fails to be elected, his carefully tailored and expensively priced image as the logical alternative to the Schreyer brand of democratic socialism will be dashed."[39] The reference was to Asper's reliance on the political polling and image-polishing techniques of Martin Goldfarb. The Toronto consultant helped pioneer the use of focus groups and "motivational research" in Canadian political campaigning in the late 1960s.[40] From his initial appearances in business suits with white collars and thick, black-rimmed glasses, Asper began sporting wider sideburns and wearing trendier leisure suits. Doffing his glasses, he began to look less like the tax lawyer he was and more like the actor Al Pacino.

The lack of elected office did not stop Asper from weighing in on the issues. He began staking out a political ground well to the right fiscally of most Liberals, and even of many Conservatives. When a Liberal policy conference endorsed a guaranteed annual income to eliminate poverty in 1970, Asper opposed strengthening the social safety net. He instead called for a resolution to dismantle programs such as old age security and the family allowance.[41] In early 1971, he placed his own blueprint for social reconstruction before Manitoba Liberals. It was based on the Depression-era "New Deal" policies of US president Franklin D. Roosevelt. Asper's plan included a job bank for the unemployed and those on welfare, sometimes called "workfare." Straight welfare payments, Asper argued, "destroy dignity and are obscene." In periods of high unemployment, Asper proposed that workfare jobs be created in tourism, mining, and agriculture.[42]

Asper was not shy about playing politics with his column, but he objected when others in the media did so. When a Senate committee released a report critical of Benson's White Paper proposals, the CBC noted the senators each sat on an average of five corporate boards. Asper scolded the government broadcaster, and it would not be the last time. "This is the stuff of politics, not objective comment," he wrote. "It serves no useful purpose to dismiss the Senate report by labeling it the tool of big business."[43] When Asper weighed in on the growth of Manitoba's civil service, however, the whistle was blown on his dual identity. The NDP, Asper claimed, had increased the provincial payroll to more than twice Ontario's, on a per-capita basis. Manitoba's minister of finance, however, pointed out that the growth of Manitoba's civil service had actually slowed in the past year. The province had 1.2 civil servants for every 100 residents, noted Saul Cherniak in a letter to the editor, compared to 1.0 in Ontario. But Cherniak saved his most cutting criticism for Asper's undeclared conflict of interest. After that, Asper was identified at the end of his columns as not only a Winnipeg tax lawyer, but also as leader of the Manitoba Liberal Party.

> Many of your readers may not be aware that your columnist on tax matters ... wears a second hat as Leader of the Liberal Party in Manitoba. The confusion regarding the dual identity of Mr. Asper may be due to the fact that he is not a member of the Manitoba Legislature — or that his party holds only 3 seats out of 57.[44]

Running for office

Asper had another chance to take a seat in the Manitoba legislature when a by-election was called in the fall of 1971, this time in his home town of Minnedosa.[45] Asper reportedly spent a month in the riding trying to drum up support, but decided against running when Goldfarb's polls showed he couldn't win.[46] Finally, in the spring of 1972, after leading the Manitoba Liberals from the legislature's public gallery for eighteen months, Asper ran in a by-election in Winnipeg. The highly ethnic working class riding of Wolseley had been a Conservative stronghold for more than twenty years, but Asper expressed optimism. He campaigned as a champion of western rights, proposing constitutional changes to provide equal opportunity for development in Canada's forgotten half.

Federal tariffs designed to protect eastern Canadian industries cost each Manitoba family an average of $700 a year, Asper argued. For example, he said, a colour television set made in Ontario cost $100 more in Winnipeg than it did in the US, just 100 kilometres to the south. Freight rates also put western Canadian industries at a disadvantage, according to Asper. It cost twice as much to ship goods from Winnipeg to Vancouver, he pointed out, as it did from Toronto to Vancouver, about twice the distance. The disparity dated to the "National Policy" of the 1878 government of Sir John A. Macdonald that protected infant central Canadian industries from cheap US imports. In the 1970s, however, the tariffs amounted to nothing less than "economic colonialism," according to Asper.[47]

Asper had taken up the cause of western alienation early on as leader of the Manitoba Liberals. He told Nick Hills of Southam News that his desire to fight Ottawa was so strong that "if Louis Riel were alive today, I would be in the trenches with him."[48] In January of 1971 he delivered a speech in Toronto that issued a dire warning. "There is a new breed of man in Western Canada," he told the Canadian Club. "A man who has grown impatient [and] a new society which will no longer tolerate regional economic disparity."[49] The recent formation of the Western Canadian Party, Asper said, should be taken as a wake-up call similar to that delivered by Quebec separatists in the 1960s. "It is the new West, not the old West, which demands a new relationship within the political and economic decision-making structure of Canada," he railed. "While the East worries about American domination, the West simply asserts that it will not accept a Canada governed predominantly from Toronto, Montreal, and Ottawa."[50]

Asper's speech was reprinted in the next day's *Globe and Mail* and outlined in detail his litany of grievances on behalf of western Canadians. "In many senses, current Western discontent begins in fear: fear that our people will somehow be at a disadvantage in a bilingual Canada; fear that our own multicultural society will be submerged by strident Eastern bilingualism."[51] The Manitoban's pleas evoked some comment in the Eastern press. "It's a familiar complaint," wrote columnist George Bain, "but Mr. Asper gave it an added kick."[52] Asper's by-election campaign was not all east-bashing, noted the *Globe and Mail*. He had some innovative ideas for improving the economic fortunes of Winnipeg and Manitoba by trading on their wintry charms.

> One of Mr. Asper's proposals to put life back into Manitoba is to accentuate the negative and make Manitoba "the snow capital of North America." He insists the winters are no worse than in Ottawa or Montreal but he thinks people might be attracted to

snow instead of repelled by it. "Do you know there are people in Los Angeles who've never seen snow? We could start a spend-Christmas-in-Winnipeg campaign."[53]

Election victory

On by-election night, Asper easily outpolled his rivals and regained official party status for the Liberals, along with their caucus room and parking spot at the legislature. His defeated Conservative opponent issued a challenge to the new MLA, warning that a provincial election could produce a different result in Wolseley. "Mr. Asper will have to put up or shut up," said Ernest Enns. "He's had more exposure the past two years, and got a large sympathy vote from the people who want to give him a chance."[54] Whatever else might result from his foray into politics, there was little chance that Asper would shut up about the raw deal received by the west from the rest of Canada.

While he had to wait until July to take his seat, Asper practically leapt to his feet to be heard once he was ensconced in the legislature. "While most newly-elected members traditionally wait at least several days before making their maiden speeches," noted the *Globe and Mail*, "Mr. Asper addressed the assembly only five minutes after being introduced."[55] The strict rules of debate in the legislature, however, were not suited to Asper's spontaneous and volatile speaking style. He was cited for violations three times in his first half hour as a member. Outside the House, he was able to make waves without being ruled out of order. In early 1973, he threw down a gauntlet to Ottawa, declaring: "As far as I'm concerned the union has broken up."[56]

In a wide-ranging and remarkable interview with the *Free Press*, Asper threatened to act as nothing less than a "kamizake." He vowed, in effect, to commit political suicide in a showdown with the federal government over western grievances. "He would be prepared to be a sort of martyr for the West — to resort to 'intemperate acts' alien to traditional federal-provincial diplomacy," warned the *Free Press*. "If he were to become premier, the province would be in for a period in which things could be 'agonized, anguished, bumpy.'"[57] Asper urged that laws be passed in areas of provincial jurisdiction to fight unjust federal tariffs. For example, he suggested a new health regulation require that all shirts sold in Manitoba be made there to guard against rashes. In echoing the growing call of Western separatists, Asper proposed that the Prairie provinces form a united front against Ottawa. "The first thing I would do if I were elected is I would form an immediate public, written and declared alliance with Saskatchewan," he said. The idea, noted reporter Tim Traynor, ignored the political reality that the NDP was in power in the neighboring province. Families in Saskatchewan, Asper claimed, paid up to $1,000 more a year for goods than those in eastern Canada due to high tariffs and unjust freight rates. Not all of Asper's complaints on behalf of western Canadians were economic, however. Asper also faulted the federal government broadcaster for marginalizing westerners culturally. The CBC, Asper declared, was a "dangerous and insulting vehicle." Despite its purpose of promoting Canadian unity, Asper complained the CBC gave little information about the west. "I didn't see the Dauphin Ukrainian Festival covered on the national CBC," he said, noting that 70,000 people flocked from across North America for the occasion. "But, by God, let 300 people get together in St. Jean-Baptiste and throw needles at a dart board or something and that'll be a 20-minute special." In what might have been his first major assault on the CBC, Asper spoke directly through Traynor to the mandarins in Ottawa. It was a statement that showed Asper was well aware of the power of images and stereotypes to marginalize a people culturally. It also revealed the depths of his ferocity in the rhetorical arts.

> And when you do portray us Westerners, you insult us to the core because you put on whiny mouth-organ background kind of music, playing kind of 'aw-shucks' prairies music and you show waving wheat fields and you show a very deeply-pensive, grained face and the wind blowing against the [grain] elevator, and that, you tell Canada, is the West. You are our enemy, CBC.[58]

Corporate welfare bums

Asper was quickly making enemies of his own, not the least of whom was Liberal prime minister Pierre Trudeau. Asper's complaints, noted Nick Hills, "caused in Ottawa — particularly in the prime minister's office — a cold, brooding anger in return."[59] Trudeau's government took several steps to redress western grievances before calling an election in the fall of 1972. It moved the federal mint to

Winnipeg and headquartered the new Canada Development Corporation in Vancouver. Asper seemed to moderate his criticism of the federal government during the election campaign. He also used his column to oppose the campaign waged by federal NDP leader David Lewis against "corporate welfare bums." Lewis pointed out that giant Imperial Oil had paid only $290 million income tax on earnings of $1.5 billion over the previous six years. That 19 percent rate, Lewis noted, was the same as paid by a family of four with a breadwinner earning $11,000. The reason Imperial Oil paid such a low tax rate, Lewis claimed, was due to tax breaks, deferred taxes, and government grants. Asper quickly jumped to the defence of Imperial Oil and recent tax measures introduced by federal finance minister John Turner. In a column eighteen days before the election, Asper scolded Lewis for getting his facts wrong.

> Either he has been guilty of deliberately attempting to mislead the Canadian public by presenting distorted and selectively concocted data, or if his own sincerity and integrity are not found wanting, he is guilty of having engaged and listened to tax advisers who apparently do not understand the Canadian tax system and have fed him incorrect information.[60]

Asper pointed out that Imperial Oil also paid oil and gas exploration royalties, along with other taxes to provincial and municipal governments. That brought its total tax burden in 1971 to 58 percent. It was an argument that ignored the fact that a family of four also faced tax burdens beyond federal income tax. Lewis disputed Asper's logic, and his integrity, in a letter to the editor two days later. "It is no surprise to me that Mr. Asper should come to the defence of the corporate tax rip-off," Lewis wrote. "Mr. Asper is not only leader of the Manitoba Liberal Party. He is also one of the tax lawyers well versed in the tax concessions and loopholes available to the corporations. Hence both in his politics and in his profession he defends the corporate tax advantages at the expense of the ordinary taxpayer."[61]

The 1972 federal election proved a setback for Trudeau's Liberals, as they were returned with only a minority government. Left holding just seven seats west of Ontario, they ended up with only a two-seat edge over the Progressive Conservatives of Robert Stanfield. That gave Lewis and his NDP the balance of power. Trudeau moved even further to appease the west by bruiting a new National Policy that would count it in for a change. "I'm satisfied this is not going to be a window-dressing operation," declared Asper after spending two days in Ottawa for talks with the prime minister.[62] A conference of Liberal leaders was planned for Vancouver in June of 1973, by which time Asper expressed optimism he would be premier of Manitoba. "A restructuring of Confederation as it affects the West is really what we're talking about — economically, socially and constitutionally," he said.[63]

General election

Asper's future in politics, however, would depend on how he fared in the Manitoba election, which was called for June 28. The odds were heavily against him, as the Liberals held only four of the legislature's 57 seats. Hopes within the party were high for winning more seats in in 1973, however, as they had attracted almost a quarter of the popular vote in 1969. Asper ran on a platform of tax cuts and western rights he called "the incentive society." He convinced some leading businessmen and political figures to run under the slogan "Self Control not State Control." Asper hit the campaign trail hard, preferring spontaneous visits to factories and beer parlours over staged rallies and meetings. He accused Conservative leader Sidney Spivak of using a "politics of fear" against the NDP to polarize the electorate, leaving no middle ground for the Liberals. Asper demonized the NDP just as eagerly, however. "Make no mistake about it," he warned, "state control ultimately means state ownership."[64]

Asper waged his election campaign like an all-out war. Lloyd Axworthy, one of the party's few MLAs, noted his leader's zeal on the campaign trail. While most candidates would glad-hand captive voters waiting at bus stops, Asper took the opportunity for exposure to another level. "He would get on the bus, shake every hand and get off at the next stop and go back and do it again," Axworthy recalled.[65] Asper's personal style, however, hurt the party's chances to finish anything better than third, according to Frances Russell. "Although they are fielding the best stable of candidates of all three parties," she noted in the *Globe and Mail*, "the outspokenness and greenness of their leader ... has cost them credibility."[66] Asper's opposition to a $3 billion hydroelectric project in northern Manitoba caused one of his candidates to resign and two more to run as independents.[67] Despite being raised in small towns, Asper's image as a big-city lawyer also worked against him, as it reportedly demolished the party's base in rural Manitoba.[68]

Asper's economic arguments were also steadily refuted by Schreyer. The Liberal leader pointed out that the province had the highest rate of provincial income tax, at 42.5 percent of the federal rate. The premier countered by explaining that health care was heavily subsidized in Manitoba. Taken together, it had one of the lowest rates of provincial income tax and medicare premiums. Asper pointed to numerous examples of welfare being collected by able-bodied recipients. Schreyer produced figures that showed Manitoba had one of the lowest levels of employable welfare recipients in Canada, at 10.2 percent.[69] According to the *Globe and Mail*, Schreyer took "particular delight in using every opportunity to demolish the personal credibility of Mr. Asper."[70] The premier had taken a dislike early on to the Liberal leader, referring to him at one point as a "disgusting little shyster" with a "very big and quick mouth."[71] Asper's daughter Gail came home from school crying one day because classmates were calling her an American. "Why were they calling her an American?" Asper recalled. "Because Schreyer had called me a Philadelphia lawyer." The assessment was actually quite accurate, as Asper would eventually make a career of exploiting legal technicalities, but not as a lawyer.[72]

Political defeat

On election night, the NDP was returned to power with 31 seats for an increased majority while Asper's Liberals placed a distant third. They added one seat by winning five ridings, but they received only 19 percent of the popular vote. "The Liberal party got caught in a draft of polarization," Asper explained. "But we withstood the blizzard. We were not wiped out, as some predicted we would be."[73] Asper finished in a virtual tie in Wolseley, being counted the winner by five votes on election night, then losing in a recount by one ballot. Asper retook the riding by three votes in a judicial recount a month later. The matter was not settled until September, when the Manitoba Court of Appeal upheld Asper's victory.[74] It didn't take long, however, for Asper to realize he was not cut out for campaigning. He even admitted to a reporter that he basically hated politics.[75] Years later, he would explain why to Peter C. Newman. "Everybody else wants to talk about paving Main Street," Asper said, "while I wanted to discuss climbing great political mountains."[76] Another observer noted that Asper didn't quite fit the mold that was increasingly required of successful politicians. "His character, ironically, didn't suit the TV age," wrote John Stackhouse in 1990. "He talked too quickly for rural Manitobans, and his ideas, such as reducing provincial trade barriers, were too radical."[77]

There was another barrier to Asper being elected, and like the proverbial elephant it went largely unacknowledged. The prejudice was noted briefly, however, in the *Globe and Mail*. "It's not one of the things that's mentioned on public platforms," wrote Egon Frech, "but, behind the scenes, one hears Liberal talk about the fact that Mr. Asper is a Jew, and a big-city lawyer to boot."[78] Anti-Semitism was a fact of 1970s life in Winnipeg, noted Newman, where a small but vital Jewish community thrived. "Winnipegers knew only too well how to apply cold showers to douse the ambitions of upstart Jew boys," wrote Newman.[79] Asper had experienced discrimination throughout his life and well understood the exclusions he faced. "It was just a given that Eaton's didn't hire Jews for summer jobs, and neither did the banks or insurance companies," he told Newman. "Local universities at the time had strict quotas, and there were no Jews allowed into the Manitoba Club."[80] His older brother Aubrey, who went into teaching, recalled the prejudice they endured while growing up. "[It] included name calling — 'dirty Jew,' and that kind of thing, references to our parents and to money, because we were in business." But while he learned to turn the other cheek, Aubrey said his younger brother never did. "Izzy was more combative than I was. He got into fights. He took the bait when he was baited. . . . He had a lot of nerve even then."[81]

What next?

Following the election, Asper attempted several times to step down as Liberal leader, according to the *Free Press*, only to be persuaded to stay on. He called politics a "con" but remained as party head until the spring session ended in 1974. After that, Asper declared himself a "free agent."[82] Asper announced his resignation as leader that August, claiming he had been misunderstood when he referred to politics as a con. "Certainly I have learned that the profession of politics is the most noble, the most selfless and the most outstanding calling one can assume."[83] On stepping down, the *Free Press* lauded Asper for having "rebuilt and rejuvenated lagging Liberal forces."[84] Asper denied any interest, however, in a group applying for a licence to operate a new television station in Winnipeg.

> He declined to talk about his future, saying only he had received a number of offers for "very exciting possibilities." As for his reported connection with Canwest Broadcasting Ltd. of Winnipeg, which hopes to establish a third English-language television [station] in the city. Mr. Asper said, "They are clients of mine — that's it."[85]

While he stayed on as a member of the legislature until early 1975, the *Free Press* continued to press Asper on his plans. In noting that his Wellington Crescent mansion was up for sale, the newspaper reported Asper's denial that he intended to move to Toronto as an executive at Global Television. "I have no intentions to go into business," he said.[86] Asper may have been a washout as party leader, but as a failed Liberal politician he was in line for a plum patronage appointment — to the Air Canada board of directors.[87] The post came with a generous stipend for attending meetings, plus free first-class air travel for Asper and his family. When the Liberals fell from power in the mid-1980s, however, the perk was rescinded by Brian Mulroney's Conservatives, who appointed their own people.[88]

The connections Asper made during his brief political career helped him wield influence in Liberal party back rooms for decades. "Asper stayed an unelected Liberal," noted *Maclean's* in its 1996 profile, "close to Turner as a sometimes ruthless strategist and admiring of Pierre Trudeau, straining, when he could, to raise the voice of the West in whatever configuration of Confederation was being plotted."[89]

CHAPTER 2
CanWest Rising

After his party's defeat at the polls, Izzy Asper cast about for a more suitable — and profitable — career. He could have returned to law, but his practice had atrophied from years of neglect during his foray into politics. Besides, Asper longed for the larger stage he had been denied as a politician. A career in business seemed ideal. It would allow Asper to combine his legal and tax expertise with the knowledge of government he had gained from his time in politics. He already had experience as a capitalist, having founded the Manitoba Distillery in his home town of Minnedosa in 1965. Asper had big plans for the small company, installing himself as CEO and his young tax law protégé, Gerald Schwartz, as president. As if to foreshadow Asper's subsequent partnerships, however, the business was soon sold when investors got a buyout offer they couldn't refuse. Schwartz left to do a Harvard MBA and then honed his takeover skills on Wall Street, but their first business experience together would come in handy.

There was only one problem with Asper's hopes for becoming a capitalist — he was broke. His six-figure income as a tax lawyer had turned into $14,000 a year in politics, draining his savings through five years of personal deficit financing. He still had his lavish home, on which he could borrow, but otherwise Asper's career in business would have to be launched using OPM — other people's money. The ardent critic of public assistance would even avail his new business liberally of federal government financing. At one point, Ottawa would be invested in almost half of Asper's entrepreneurial enterprise through its new Canada Development Corporation.

Asper's quest for business opportunities began while he was still a Manitoba MLA, and his political connections proved useful in getting him started. His executive assistant, Peter Liba, drew the Liberal leader's attention to a call issued by Ottawa in 1973 for licence applications to operate independent television stations. The federal government's aim was to supplement the programming offered by the CBC and CTV networks. It was also prompted in part by the fact that US television stations close to the border were siphoning off advertising dollars from Canadian broadcasters. Southern Ontario merchants were buying ads on TV stations in Buffalo, New York, which carried about 40 percent Canadian advertising. On the west coast, KVOS-TV in Bellingham, Washington, was the most profitable station of its size in the US. It sold 90 percent of its advertising through an office in Vancouver.

Asper and Liba decided to form a company to bid for the Winnipeg licence, but they had little capital of their own to invest. "We started off making long lists of people who had money," Liba told CanWest historian Allan Levine.[1] Their major investor was Paul Morton, who like Asper hailed from a local Jewish theatre-owning family. Seymour Epstein, a broadcast engineer and former CRTC policy official, was hired and became a minor investor. Their application was submitted in Morton's name. According to the *Globe and Mail*, that was because it would have caused "a political stink for Liberal Ottawa to be seen handing the station to one of the party's provincial leaders."[2] Their main competition for the Winnipeg licence came from Craig Broadcasting, which operated radio and TV stations in nearby Brandon, Manitoba. Despite having no previous broadcasting experience, Asper and company were named the winning bid. According to Gordon Pitts, rejected applicant Stuart Craig was "convinced to his dying day that he had been snookered by the politically connected Asper." The future media mogul, after all, was "still leader of the Liberal Party in Manitoba — even while he was seeking a broadcast licence."[3] The next hurdle — coming up with a television station — was cleared ingeniously. The strategy used would provide the template for much of CanWest's subsequent success. Instead of creating a Canadian television station from scratch, they bought one in the US and moved it north.

The cross-border television competition in the Winnipeg market came from KCND-TV in Pembina, North Dakota, 100 kilometres to the south. CanWest planned to set up its transmitter between Winnipeg and Pembina, blocking the KCND signal. That greatly lessened KCND's value, which prompted its owner to sell for a bargain price. Over the 1975 Labour Day long weekend CanWest hired a fleet of trucks and shipped the station across the border lock, stock, and transmitter. It was reassembled inside a vacant Winnipeg super-market and CanWest reversed the first two letters of the station's call sign to conform to CRTC protocol. Vancouver broadcasting critic Herschel Hardin considered the CRTC's licensing of CKND a mistake from the beginning because it was "against everything the commission was supposed to stand for."[4] Instead of staving off creeping American influence on Canadian cultural life, it opened the border to a torrent of Hollywood programming, according to Hardin.

> The CKND licence should never have been handed out. ... It had none of the promises, fantasy as they were, that had induced the commission to licence Global. When the CanWest application was filed early in 1974, it referred glowingly to the unquali-fied success of Global; by the time the application was heard only a few months later, Global had gone belly up.[5]

Rescuing Global

Global Television had started up in 1974 as a regional network of six stations in southern Ontario broadcasting from Windsor to Ottawa. Radio pioneer Al Bruner chose the name, according to Peter Des-barats, because he envisioned it some day spreading worldwide. "He would talk about broadcasting to India with the same enthu-siasm that had hypnotized the CRTC into giving him what should have been the most lucrative TV licence ever granted," recalled Desbarats, a Global news anchor. The CRTC granted Bruner "an incredibly generous franchise," according to Desbarats, "that gave him access to the largest and wealthiest television audience in Canada." It failed, he claimed, because Bruner and Frank Buckley, the cough medicine mogul who handled Global's finances, were "better visionaries than executives." Their biggest mistake, accord-ing to Desbarats, was the rich diet of indigenous programming they promised the CRTC, which proved unpalatable to viewers. "Overly ambitious Canadian productions sank the network before it was fairly launched," he claimed. "Global became the first Canadian TV franchise to go bankrupt."[6]

According to Edward Greenspon, Global failed due to a series of mishaps that resulted in it becoming "the biggest embarrassment in Canadian broadcasting history." Its first mistake was in choosing January 6, 1974 to go on the air. "Since the television season starts in September," noted Greenspon in *Report on Business* magazine, "most US programs had already been bought up and most of the advertising budgets committed."[7] Bad luck added to Global's woes. An Arab oil embargo against the US, imposed due to its support for Israel in the Yom Kippur War, led to an energy crisis there. The country switched unseasonably to daylight savings time to save power. The time change was made the very day Global began broadcasting. It put Ontario temporarily an hour behind the time in eastern US states, ruining some of Global's best-laid plans. "Global had devised a schedule that would put its foreign programs on the air at the same time they were showing on the US network," noted Greenspon. "Under Canadian broadcast regulations, that meant cable systems would have to give preference to Global's signal." To make things worse, Global's late-night variety show, which was scheduled to start 30 minutes before NBC's popular *Tonight Show Starring Johnny Carson*, instead started a half hour after *Carson*.

> Billboards promoting Global's programs couldn't be changed. Viewers switched on expecting comedy and got drama or vice versa. They quickly went back to the stations they knew. Even before winter melted away, the runoff from Global had reached $50,000 a day and the bank was unwilling to extend more credit.[8]

Global soon became insolvent after investors Odeon Theatres and Maclean Hunter pulled the plug only three months into operations. Asper and Morton helped rescue Global, partnering with Toronto radio veteran Allan Slaight in an $11.2 million bailout.[9] Their takeover bid, noted Greenspon, "won regulatory approval in record time."[10] According to Ryerson University communication professor Matthew Fraser, the CRTC "gratefully" approved Asper's partnership in Global. "At first blush, the rescue seemed like an ideal solution," observed Fraser. "The CRTC failed to understand, however, that Izzy Asper was above all a shrewd tax lawyer. Patriotism came after profits."[11] The Global partnership was problematic from the start due to Asper's controlling nature, according to Greenspon. A 1976 restaurant argument led Slaight to invoke the "shotgun" clause of their partnership agreement to end the association. He pulled the trigger with a $6.8-million offer for CanWest's half just before Christmas, thinking the holiday season would make it difficult to raise the money to match his offer. Asper and Morton, however, had the last laugh when they did just that.[12]

Asper was not about to repeat Global's mistake of spending heavily on Canadian programming. Instead, CanWest favoured tried and tested Hollywood fare, running cheap favourites such as *The Honeymooners, I Love Lucy* and *Gilligan's Island*. The Canadian content requirements set down by the CRTC for third television stations were not as strict as the 60 percent required of the CBC and CTV networks. By the time CKND's licence came up for renewal in 1977, however, local content made up only 20 percent of its programming, compared to the 50 percent it had promised. "Except for sports and the regular news slots, CKND had junked most of its provincial local programs, many of which had not got on the air at all," noted Hardin. "CKND was also packing more American content into the high-revenue fall period and balancing it off with more Canadian content in the low-viewing summer period — an old trick."[13]

'A licence to print money'

CanWest had hit on the formula that would soon make it the most profitable television company in Canada and turn it into an international giant. They would prove the truth of Roy Thomson's observation about the television business as owner of the first private TV broadcaster in Scotland. It was, said Lord Thomson of Fleet, "just like having a licence to print your own money." The CanWest Global formula was to pack as much Hollywood programming into its schedule as it could get away with, rather than invest in costly original content. The mounting profits were used to add more Canadian stations, expand overseas, and ultimately acquire Southam.

CanWest Global's financial success came in large part as a result of two broadcasting rules made by the federal government in the mid-1970s. The first was the CRTC's 1973 "simultaneous substitution" rule that Global had hoped to cash in on. It allowed Canadian television stations to cut into US network shows carried at the same time on local cable systems and substitute their own commercials. The other was Bill C-58, which discouraged Canadian businesses from advertising on US stations by deeming the cost not tax deductible. The bill was designed to protect the Canadian magazine industry, which was suffering at the hands of US competition. By the time it passed in 1976, Bill C-58 was expanded to cover broadcast media as well. That led to a cross-border business war that ran until the 1988 Free Trade Agreement.

This "television border war" saw the US Congress retaliate by passing similar legislation that prohibited American companies from claiming as a business expense the cost of attending conventions held in Canada. More than 100 conventions planned for Canada that year were cancelled as a result. The value of Canadian advertising flowing south to American broadcasters in 1977 was cut in half, to $9 million. The US television networks launched legal action against the CRTC and Toronto's Rogers Cablevision, charging them with piracy. Despite high-level government talks, the dispute could not be resolved diplomatically and it went to court in 1977. The Supreme Court of Canada upheld the CRTC's right to allow interception of cross-border broadcasting signals, but the ruling proved a mixed blessing. While a victory for Canadian sovereignty and a boon economically, the result was a cultural "Trojan horse," according to professor of communication Barry Berlin. The bulk of the revenue that stayed in Canada was unfortunately used to purchase more US programming. "Ironically," noted Berlin, of Canisius College in border town Buffalo, "the problem of the extensive amount of American fare on Canadian television worsened thereby."[14]

CanWest Capital Corp.

Shortly after CKND went on the air in 1975, Asper finished what he called his "Magnum Opus."[15] His blueprint for a Canadian merchant bank, or diversified holding company, had been encouraged from New York by Gerald Schwartz, his former law partner. "He pushed and prodded me into our joint venture in starting CanWest Capital," recalled Asper. It would take almost two years, however, to put in place the $20 million in investment capital Asper sought. He contributed $1 million only by mortgaging his home, while Schwartz committed $250,000. Their hunt for investors began fruitlessly in western Canada, then moved to the eastern banks. Turned down by all but the Toronto-Dominion, which came in for 10 percent, Asper appealed to Montreal businessman Paul Demarais. Western Canadian investment was essential under Asper's plan to obtain government support. Demarais invested the required amount through his Winnipeg-based insurance company, Great West Life. CanWest's most important investor, however, would be the federal government through its new Canada Development Corporation (CDC).

The CDC had been launched in 1971 to assist Canadian entrepreneurs with venture capital. It was headquartered in Vancouver in no small part due to Asper's advocacy for the Western provinces while he was Liberal leader in Manitoba. A 1975 share offering raised $100 million in capital, and Asper eyed a piece of that for CanWest. "They decided they liked the concept so much, they upped their proposed stake from 25% to 35%," he recalled, adding that CDC officials even promised CanWest another 10 percent if needed.[16] Asper would take all of that — and more. Not everyone in Ottawa was in favour of such a large investment being made in Asper's business, no matter how politically well connected he was. Conservative finance critic Sinclair Stevens even raised the matter in the House of Commons. The federal government's initial $7 million investment would include an annual management fee to Asper of $100,000, Stevens charged. Canadian Investment Partners Ltd., as he mistakenly referred to it, planned to take control of Global Television, putting Ottawa in competition with its own CBC, according to Stevens. Asper denied the CDC had yet invested in his company, which had not even been formed. "There is virtually no commitment of any kind," he said.[17] Negotiations with CDC lawyers, Asper recalled, were difficult because some "didn't want a deal at all."[18] Marathon bargaining sessions in the spring of 1977, however, finally brought CanWest Capital Corp. to life.

Soon CanWest made a wide range of investments, but despite the federal involvement many were not even in Canada. A foray into pay TV through San Francisco-based subsidiary USTV saw transmission towers erected in cities across the US. A takeover of Ohio-based fertilizer company Na-Churs International was made through a leveraged buyout. CanWest also bought Monarch Life Assurance of Winnipeg in 1978 and picked up Crown Trust from Conrad Black a year later. The Macleod-Stedman hardware store chain was added in 1980, as was Miami-based Aristar financial group. Asper's holdings in CKND and Global Television, which increased to 60 percent, were also folded into the growing CanWest portfolio.[19]

Heart-breaking breakup

The recession of the early 1980s, which was fuelled by a rise in interest rates to 20 percent, put the brakes on CanWest's rapid growth. Asper's enterprise was profitable, but many others were not, prompting some CanWest investors to urge selling divisions for needed cash. "CanWest was not planning to be profitable before its 10th year, but made money in 1981, its fifth year," reported the *Globe and Mail*. "In 1982, 'a very, very profitable' year, profit was three times the 1981 level, but Mr. Asper would not reveal the actual level."[20] Offers rolled in for CanWest's operations, and anxious board members jumped at the chance to sell, considerably to Asper's chagrin. USTV and Crown Trust were the first to go, with Monarch Life close behind, over Asper's objections. Its sale in 1983 for $68 million, on an investment of $6 million four years earlier, was supported by both Schwartz and the CDC, ruining Asper's relationship with both. The CDC, which had grown into a multinational conglomerate and one of Canada's largest companies with assets of $7.6 billion, had lost more than $500 million over the previous two years.[21] While CanWest was growing in value, it was not producing income, prompting investors to cash out. By 1983, the CDC had accumulated 49 percent of the company, and it joined the exodus from Asper's creation.[22] Soon the CDC itself would fall victim to a change in the political winds, as the Conservative government of Brian Mulroney privatized the company in the mid-1980s.[23]

For Asper, the divestitures were nothing less than a personal tragedy. The sale of Monarch Life, according to Schwartz, caused the close friends to split. "Izzy probably felt that I had betrayed him by agreeing to sell Monarch," Schwartz told CanWest historian Levine.[24] Asper kept the television assets of CanWest and Schwartz took the rest, moving to Toronto and building his Onex Corp. into a specialist in leveraged buyouts. "I believe in planting trees, growing them, and then eating the apples," Asper told the *Globe and Mail*. "Gerry believes in growing trees, selling them, and then looking for other trees."[25] According to Schwartz, however, what soured his relationship with Asper was the earlier decision to divest USTV. Pay TV was "the whole disaster at CanWest," according to Schwartz. "The leveraged buyouts made all the money and a lot of it went down that sinkhole." He finally sided with other board members who clamoured to sell USTV after $40 million in losses.[26]

The turmoil took a toll on Asper's health, and he suffered a heart attack in October of 1983. He was put on the waiting list for quadruple bypass surgery. While he could have had the procedure done quickly at the Mayo Clinic just south of Winnipeg in Minnesota, Asper stubbornly refused. "I'm a Canadian, damn it," he told Peter C. Newman. "I trust the system, so I had my surgery in Winnipeg, even though I had to wait around for six weeks, doing nothing."[27]

Big profits from TV

The television side of CanWest that Asper kept, in partnership with Morton and Epstein, proved lucrative. Ottawa's simultaneous substitution rule was the key to making it Canada's most profitable broadcaster. By 1984, a federal Task Force on Broadcasting Policy estimated the rule had provided between $36 million and $42 million in revenue annually for Canadian broadcasters.[28] By 1997, that figure was estimated to have risen to $100 million.[29] According to Matthew Fraser, simulcasting was nothing short of a "bonanza" for Canadian broadcasters but a cultural net loss for viewers.

> The simulcasting twist meant that Canadian TV stations now had a powerful incentive to fill up their prime-time schedules with popular American shows procured at a marginal cost and air them against US border station slots. The result was the rapid colonization of Canadian prime-time schedules by simulcasted American shows.[30]

CanWest led the simulcasting charge, buying large amounts of US network programming for broadcast not only in Canada but soon in other countries as well. In addition to CKND and Global's string of southern Ontario rebroadcasters, CanWest's Canadian holdings soon included new outlets in Regina and Saskatoon, which were licenced by the CRTC in the mid-1980s. CanWest also acquired existing stations in Vancouver and Halifax in the late 1980s. The takeover of CKVU in Vancouver came after a lengthy legal battle that started when Asper sued the station's local ownership in 1984 for refusing to sell it to him. Asper had lent them $12 million in 1979 to help thwart a takeover bid by Charles Allard of Edmonton.

In return, he got an option to buy their station if they couldn't pay him back. They couldn't, but they wouldn't sell either, so Asper took them to court. The dispute ran for three years and reportedly cost Asper more in legal fees than CKVU was worth before it was finally his.[31]

Scheduling popular American programs at the same time as US networks and selling their own advertising for both soon saw Can-West turning customers away. The CTV and CBC networks also aired US programming, but they were at a disadvantage competing for it because of their stricter Canadian content requirements. CTV was a network of local affiliates operated by different owners, so it had only 40 hours of national programming to fill per week. Can-West Global had a full 126 hours a week to work with, which soon made it the largest importer of American television programming in the world.

Not a network

CanWest also did not have the expense of transmitting its signal to remote locations across Canada because it was not a national network. Until the late 1990s, it lacked stations in several Canadian provinces, including Alberta and Quebec. As a result, Can-West avoided the costly re-broadcasting expenses incurred by CTV. By confining itself to the more lucrative larger markets, CanWest could skim the cream of advertising dollars without the expense of providing service to all. As far as the CRTC was concerned, Can-West was not a network but instead a "system." It was a distinction exploited for decades by Asper, which brought constant complaints from CTV executives. One of the few scholarly studies of CanWest Global concluded it was "invisible to researchers" because it did not fit the dominant network form. CanWest was nonetheless "changing the nature of television in Canada" by the early 1990s, according to Paul Taylor of the University of Washington. Its success, noted Taylor, was due to the "unique and carefully crafted regulatory position" devised by its owners.

> The CanWest Global System (CGS) is part of a fundamental shift in Canadian television yet it has been largely ignored in scholarly discussions of the country's broadcasting system and its future. ... CGS has no national network obligations

because each owned and operated station is licensed as an independent entity. This degree of carefully constructed and fiercely defended regulatory freedom has allowed CGS to become the most profitable television broadcasting entity in Canada.[32]

Taking advantage of the fact that it was not deemed a network, CanWest Global invested only $44 million in Canadian content for the 1990–91 programming season. That was half of CTV's $88 million Cancon commitment. "As a result," noted Taylor, "American programming dominates the prime-time schedules of CGS stations, which tend to take full advantage of simultaneous substitution regulations to maximize audience size and revenues."[33] Airing more American content, which could often be purchased for 10 percent of its production cost, made CanWest more profitable than the larger CTV. By counting its divisions separately, the CRTC even ranked CanWest Global as the first and second most profitable broadcaster in Canada.[34]

A report by the accounting firm Coopers & Lybrand set out the vast difference in profitability between airing US programming and original Canadian content. A popular American program bought for $80,000 an episode might attract $200,000 in advertising revenue, for a profit of $120,000. A Canadian drama might cost $200,000 an episode to produce but attract only $125,000 in advertising for a loss of $75,000. The difference, or "opportunity cost," between airing a cheap, popular US show and an expensive Canadian production was therefore about $195,000 per episode. Over a season of 22–26 episodes, that amounted to about $5 million.[35]

Despite its business success, CanWest Global was a fractious enterprise. Partnerships were not suited to Asper's controlling style, and almost every joint venture he went into dissolved in acrimony. As a result, by the end of the 1980s Asper was a sole proprietor. Lawsuit and countersuit turned into a four-year legal battle with Epstein and Morton that culminated in a four-month trial argued by fourteen lawyers. A court-ordered auction of Global was won by Asper, who bid $131 million for Morton and Epstein's 38 percent. Asper's victory became complete when the judge awarded him more than $1 million in legal costs.[36]

Canadian content

In 1987, in an effort to boost Canadian television content, the CRTC renewed CTV's broadcasting licence for only five years instead of the usual seven. It demanded CTV increase its prime-time Canadian dramatic content by half. The network projected that would create a $100-million loss over the term of the licence, including $48 million in the final year.[37] To meet the CRTC's demand, according to Gittins, CTV spent $403 million on Canadian programming from 1987 to 1991.[38] Its second status to CanWest Global in buying hit US shows saw CTV start to lose the simulcasting game. The reversal of fortune was hastened, noted Gittins, by a series of programming blunders. CTV failed to renew *Cheers* after a price increase and dropped *Seinfeld* after one season, only to watch it take off on CanWest Global. In 1993 CTV was outbid for the *Cheers* spin-off *Frasier* by CanWest, which also got *The X-Files*. The following year CanWest "stole" *NYPD Blue* from CTV by bidding $135,000 an episode for two years compared to CTV's offer of one year at $131,000.

David Asper, who was responsible for much of CanWest's programming strategy at the time, admitted there was more to snatching *NYPD Blue* than met the eye. "We don't like to get into bidding," he told *Report on Business* magazine. "It just enriches the studios. We prefer to work out our deals in advance." The bidding for *NYPD Blue* was less than straight, according to Trevor Cole. "In addition to cash, it gave the distributor a parcel of commercial time, which could be sold for extra profit."[39] The incident prompted CTV to launch a fruitless lawsuit, but that could not keep it from slipping farther and farther behind the crafty Aspers.

CanWest also outdistanced CTV in the early 1990s through marketing and branding strategy, according to Cole. By programming its prime time with series that appealed to the coveted 18–49 audience demographic, it was able to hold viewers from week to week. That meant forgoing the movies and mini-series that CTV and other stations featured. "Its ratings may not spike as high as CTV's occasionally do," noted Cole, "but its promotion — Global, for instance, has perfected the technique of imbedding its name in show titles — has a stronger cumulative effect."[40] Advertising buyers, Gittins pointed out, spent up to 80 percent of their prime-time entertainment budgets on Canada's top twenty programs. Almost all of them were American, and most were playing on stations owned by the Aspers. "Each season, the gap seemed to widen between CTV and its arch-rival with the deep pockets," she noted. "Global's pockets were becoming so deep they appeared bottomless because it already owned the rights to the most popular American shows and had to turn advertisers away."[41]

Another reason for its profitability was CanWest's habit of making spending promises to the CRTC that it never kept. "As part of its commitment to gain the CRTC's blessing to gain control of Global, Asper promised to spend $9 million over five years to establish news bureaus around the country and world," noted the *Financial Post's* Richard Siklos in 1991. "The recent creation of a London bureau — with a single reporter, but no camera crew — is classic Asper economizing and exactly the kind of thing that makes rival executives' blood boil."[42] The CRTC moved to crack down on CanWest Global in 1992, renewing its licence for only four years and demanding it increase Canadian content. By then, however, CanWest had gone global — literally.

A national network

The CRTC's demand for more Canadian programming led CanWest to develop new drama series such as *Jake and the Kid*, based on stories by W.O. Mitchell. *Traders*, a prime-time stockbroking soap opera, was set on Toronto's Bay Street. The initiatives prompted Leonard Asper, who was taking an increasing role in management, to predict a change in CanWest's relationship with the broadcasting regulator. "Our days of being heavily criticized by the CRTC are over," he told *Maclean's*.[43] His boast was premature. The CRTC invited applications in 1994 for new television licences in Alberta, which was one of the few provinces where CanWest lacked a broadcasting presence. CanWest applied, but the CRTC declined to issue the licences when it couldn't find an applicant it liked. In 1996, the CRTC issued another call for licences in Calgary and Edmonton, but it again turned down CanWest's application. Instead it issued them to Asper's old Manitoba rival, Craig Broadcasting. Izzy Asper appealed the ruling to the federal cabinet of his good friend, Liberal prime minister Jean Chrétien. His argument to government ministers, as reported by Peter C. Newman, was supremely ironic.

The commission's decision has destined the Canadian television production industry to longterm domination by foreign broadcasting companies that have absolutely no interest in the pursuit of Canadian cultural goals. High quality programming of export calibre cannot be developed by uncoordinated and frequently haphazard efforts of local independent and unrelated studios scattered across Canada with conflicting corporate agendas.[44]

Meanwhile, according to Matthew Fraser, a CanWest executive attempted to "blackmail" the CRTC by threatening to cancel its new dramatic productions.[45] Chrétien's cabinet turned down Asper's appeal nonetheless. According to Gittins, he had been counting on his political clout in Ottawa to help him finally break into the Alberta market, and was disappointed "his Liberal friends had failed him."[46] If so, Asper must have been heartened when Cabinet issued an Order in Council directing the CRTC to consider licensing "one or more national networks."[47] According to Fraser, however, the CRTC's 1997 hearings on a third network in Canada also meant that "the vise was tightening on Global." CanWest's game of operating as a "system" instead of a network would thus be exposed to public scrutiny, noted Fraser.

CTV and other major industry players stepped forward to argue before the CRTC that Global should be recognized for what it was — a full-fledged television network — and that its Canadian content obligations should thus be adjusted upwards.... Two separate consultants' reports found a major discrepancy — as much as 50 percent — between the levels of spending on Canadian programs by CTV and Global.[48]

'Nothing but toll collectors'

The third network hearings provided an opportunity for cultural nationalists and independent producers to complain about CanWest Global's failure to deliver on its programming promises. Front and centre would be two of the company's loudest critics. The first was Friends of Canadian Broadcasting, a lobby group that had been set up in the 1980s to defend the CBC. It pointed to CanWest's lack of Canadian content and passed out graphic evidence to commissioners in the form of a chart plotting Global's prime-time programming. Then came an even more strident voice in support of Canadian content. Producer Robert Lantos, who had engineered such hit series as *ENG* and *Due South*, urged the CRTC to resist Global's call to deregulate Canadian television. "Our broadcast landscape remains today the most colonized of the major industrial nations, and it is in our cultural, industrial, and national interest to resolve that," said Lantos.[49]

The outspoken filmmaker would make an even harsher assessment of CanWest the following year. On accepting an award at Toronto's Ryerson University, Lantos lambasted the Canadian television industry in general and Asper in particular in a speech that was reprinted in several newspapers. He pointed out that Global's only high-profile Canadian drama, *Traders*, was "cynically" scheduled opposite the US blockbuster *ER*. Lantos called it "part of a careful strategy designed to prove even top quality Canadian programs can't get an audience." This led, he added, to "self-serving rhetoric about why shareholders can't be expected to pay for them."

> But hold the Kleenex! Shed no tears for Mr. Asper and his fellow shareholders: CanWest's net after-tax earnings for fiscal 1998 were $200 million, or about four times the combined profits of all Canadian film and television production and distribution companies.[50]

Referring to "the forces of darkness ... whose greed is surpassed only by their hypocrisy," he made an unmistakable reference to Asper, who in 1995 had received Canada's highest civilian honour. "They walk around, Order of Canada in their lapels, calling their countrymen losers and urging us to adopt the values of a foreign culture. They preach free-market economics for others, but when it comes to their own businesses, they are the first to seek the shelter of government policies and regulations." Lantos was just getting warmed up. "But then, the cynics don't aspire to be broadcasters. They want to be rebroadcasters: It's much cheaper, and requires neither talent nor skill. ... They are nothing but toll collectors, hitching a free ride on the back of the regulator who protects them from true free market competition."[51]

Asper filed a lawsuit for libel that claimed $7 million in damages.[52] He then shot back at Lantos in a speech of his own. "Here is the mollycoddled and government-subsidized producer criticizing Canadian broadcasters," he told Toronto's Canadian Club. "It's the height of hypocrisy."[53] The lawsuit dragged on through the courts for years.[54]

The station 'groups'

As a result of its 1997 hearings, the CRTC ruled that a third national television network was not needed in Canada. Several quasi-network "station groups" such as CanWest, however, were told to increase their Canadian content.[55] The decision not to licence a third network had become moot when CanWest declined to push for that status, surprising some. In its submission to the hearings, it argued that "a broader broadcast policy review" was needed to first create a new strategy for Canadian television. "If there are to be additional national networks," it added, "the market will create them in due time."[56] Some, however, pointed out that CanWest was following a self-serving strategy. "While Asper may want representation in Alberta and a national presence, [CanWest] had no interest in becoming Canada's third national network," noted *Marketing* magazine. "[F]ormal network status would mean that broadcast regulators would require them to kick in a lot more money to fund the production of Canadian programming while gaining few benefits they don't already have."[57]

The CRTC scheduled more hearings for the fall of 1998 to consider the question of regulating the station groups, among others. In the meantime, however, CanWest was making moves so fast it was difficult to keep up. In May of 1998, it paid $40 million for 72 percent of Fireworks Entertainment, a television production company. It was a bid to boost its internal production capabilities onto the lucrative international market.[58] The only problem, according to Matthew Fraser, was that the purchase violated a 1983 CRTC prohibition against broadcasters also owning content producers. According to Fraser, the Fireworks deal "immediately sounded alarm bells in Ottawa."[59]

Broadcasting review

CanWest was also gaining a nationwide reach in a way the CRTC had not intended, through the specialty cable channel Prime TV. Global was granted a licence for Prime in 1996 to serve a demographic under-represented on Canadian television screens — old people. CanWest hired elderly TV personality Mike McManus to make a "passionate plea" to the CRTC, according to Fraser, about the need to serve Canada's retired population. Once the CRTC granted Global a licence for Prime, however, it promptly abandoned the 50-plus demographic group stipulated in its licence. By the time it went on the air, noted Fraser in his *National Post* media column, Prime headed down market by rebranding itself as "Canada's superstation." A few months later, Prime's demographic sights were lowered even further when it was re-christened "Canada's entertainment network" to appeal to younger viewers.[60] According to Fraser, Prime became a "shameless re-run channel for the same flossy American shows" that made Asper broadcast operations so profitable. "Global executives may be arrogant, but they do know how to count."[61] Fraser's assessment of the Prime debacle was even more pointed in his 1999 book on Canadian television, *Free For All*.

> Canadians in this unenviable [50+] age bracket were cynically left out in the cold by the Global bean-counters. One can only admire the testicular fortitude of the Global executives who pulled off this scam. It might be asked, though, why the CRTC stood by and did nothing. . . . The CRTC should have promptly pulled the licence of the rejuvenated Prime.[62]

The 1998 CRTC hearings on the future of Canadian television were its first major review of broadcasting in Canada since the late 1980s. They drew no fewer than 287 briefs, including one on Canadian content spending ordered at CTV's insistence. The accounting was prepared over the objections of CanWest. Compiled by the Canadian Association of Broadcasters, it showed that CTV had spent $147 million, or 33 percent of its airtime revenues, on Cancon in 1996-97. By contrast, Global had invested only 18 percent of its advertising proceeds in domestic productions that season, or $67 million.[63] CanWest asked the CRTC to more than double what it considered "prime time," from three hours (8–11 PM) to seven (5 PM–12 AM).

It argued that would allow it more "flexibility" in providing prime time Canadian content. It also argued that domestically produced "infomercials" — programming paid for by advertisers — should count as Canadian content. Cheaper documentary programs, it added, should be considered the same as more expensive dramatic productions. The Vancouver-based Fraser Institute, a right-wing "think tank," submitted a report by a University of BC business professor. It urged scrapping Canadian content rules entirely because they stifled freedom of expression.[64]

Big profits

When the 1998 hearings began, independent producers pointed out that private television profit margins increased from 12.7 percent in 1993 to 17.3 percent in 1997. Canadian content spending, however, dropped from 30.4 percent of revenue to 26.6 percent.[65] The producers asked the CRTC to require ten hours a week of Canadian prime-time drama from each network and station groups, plus three hours of children's programming. Asper reportedly erupted at the suggestion, labelling it an attempt to "reinstitute a socialistic redistribution of wealth from broadcasters into producers' pockets."[66] *Toronto Star* columnist Antonia Zerbisias reported that Asper's angry outburst appeared spontaneous.

> When the CRTC questioned Global about its portrayal of visible minorities on its airwaves, Asper launched into a speech about "hyphenated Canadians" and how putting them on screen could "foster a divisiveness in our community." Later, he said that "a profession has grown up that makes a living out of promoting, in this country, in the last 30 years, 40 years, promoting difference rather than sameness, and there are some disadvantages to that as a nation."[67]

The CRTC issued rewritten regulations for television broadcasting in mid-1999, lengthening prime time to four hours (7–11 PM). It also re-defined what qualified as Canadian content, although not as liberally as CanWest had sought.[68] By then, however, the industry landscape had changed once more, leaving CanWest Global as not only Canada's most profitable private broadcaster, but also its largest. It was an expansion that was not without setbacks. CanWest had sought in 1998 to remedy its lack of lucrative specialty cable channels with a takeover bid for NetStar Communications. It owned The Sports Network (TSN) and the Discovery Channel, among others. CanWest bid $875 million for NetStar, but it was blocked by the US network ESPN, which owned 32 percent.[69] The Americans took the CanWest offer to CTV, which trumped it by paying $908 million.[70] CanWest claimed it had balked at conditions proposed by ESPN because they would have given the Americans too much control.[71] According to Fraser, however, the subsidiary of the giant Disney Corp. was scared off by Asper's litigious nature. "ESPN was aware of Izzy Asper's reputation for combating his own business partners in court," noted Fraser. It balked at going into business with him, according to Fraser, soon after Asper filed his $7 million lawsuit against Lantos for his "toll collectors" speech.[72]

Stalking WIC

Instead, Asper sought to take his television empire nationwide by finishing a takeover he had been attempting for years. It would give him the stations in Alberta he coveted and would also eliminate a major rival. Asper had made his first takeover bid for Western International Communications (WIC) in 1995, after it had been weakened by the death of its founder. Frank Griffiths had bought New Westminster, BC radio station CKNW in 1956 and added television stations in Vancouver and Victoria. In 1990, WIC paid $157 million for Edmonton-based Allarcom, which owned four television stations in Alberta, along with cable's Superchannel and the Family Channel.

WIC had grown to rival CanWest with its expansion across British Columbia of superstation BCTV. By the mid-1990s it owned eleven radio stations and eight television stations, including CHCH in Hamilton, which it renamed ONTV. After Griffiths died in 1994, however, his sons proved incapable of running the company to the satisfaction of the Allard heirs, who held a sizeable portion of company shares. The Allards attempted a takeover, claiming that the death of Griffiths had triggered "coattail" provisions in the company's bylaws that converted all non-voting shares into voting shares. Their lawsuit was settled out of court, but other minority shareholders also filed suit in an attempt to ride the coattail provisions and convince a judge to declare their shares as voting.[73]

Asper was one of the most persistent WIC suitors, launching a

$636 million takeover bid by offering $24 a share for the company in 1995. The offer was rejected by the WIC board, so Asper went to court in a fruitless attempt to trigger the coattail provisions.[74] CanWest, meanwhile, increased its share holdings in WIC from 9.7 percent to 15 percent early in 1997.[75] By the end of that year it had 30 percent of non-voting WIC shares.[76] In March of 1998, however, Calgary-based Shaw Communications bought most of the company's voting shares from widow Emily Griffiths for $91 million. Asper launched a hostile takeover bid by offering $39 a share for the company, again hoping to get voting status for his non-voting shares.[77]

Then the corporate maneuvering got even more complex. In an attempt to keep Asper out, WIC directors passed a "poison pill" bylaw. It would have made additional shares available for sale to existing shareholders at half price if CanWest increased its holdings in the company.[78] That move was quickly struck down by stock market regulators, but the delay was all WIC directors needed to come up with an alternative to Asper's offer. Instead, they accepted a bid of $43.50 per share from Shaw.[79] That left Shaw and CanWest as almost equal shareholders, and the bitter rivals had little interest in doing business together. To dissolve their unintended partnership, the companies made a deal to divide the assets of WIC. CanWest would get the television stations and Shaw would take the radio stations and cable channels.

The only problem was an unfavourable tax ruling from Revenue Canada. Shaw and CanWest went back to the bargaining table, but they were unable to come up with a new plan to divide WIC for more than a year. The CRTC twice had to cancel hearings into the change in ownership of WIC while its partners battled over who would get what. Finally, early in 2000, CanWest announced it had a deal with Shaw and, more importantly, a plan to ensure its approval by the CRTC. Taking over WIC's television stations meant CanWest suddenly owned more than a national network, with double coverage in Canada's largest markets. In Vancouver, it had market-leading BCTV and CHEK-TV in nearby Victoria to go along with CKVU. CHCH in Hamilton fell within the Toronto metropolitan area, where CanWest also owned a Global affiliate.

Gumshoe intrigue

As his lawsuit against Robert Lantos dragged on through the courts, Asper added television consultant Pauline Couture as a defendant. He claimed she had co-written the Lantos speech that derided CanWest as "rebroadcasters" of American programming. "There may be others involved in this wrongdoing and our investigation is continuing," Asper told the *National Post*.[80] CanWest had indeed been investigating its critics, in particular Friends of Canadian Broadcasting. The group had been founded in the mid-1980s to help defend the CBC from funding cutbacks. Under the leadership of Couture's husband, Ian Morrison, it expanded its mandate to lobby for more Canadian content on private television. Antonia Zerbisias first detected a campaign against Friends in advance of CRTC hearings on regulating Canada's quasi-network station groups. "In 1998, the Canadian Association of Broadcasters, CanWest's lobby group, hired writer Jennefer Curtis to do 'a piece' on Friends," Zerbisias claimed. "When she called me for dirt, I smelled something fishy — and she 'fessed up."[81]

Globe and Mail columnist Hugh Winsor reported in 2000 that CanWest had stepped up the heat on Friends. Leonard Asper, who by then had risen to CEO, sent Morrison a letter demanding answers to a dozen questions about Friends within 24 hours. Asper wanted to know financial details and whether the group "had ever received money or research material from another Canadian broadcaster." The intimidation tactics again came on the eve of CRTC hearings involving CanWest, this time into its WIC takeover. "The polite term is opposition research," noted Winsor. "A less charitable description . . . would be bluster or bullying."[82]

As it was finalizing its purchase of Southam in 2000, CanWest hired former business journalist Brenda Dalglish to investigate Friends. The lobby group posted a message on its website warning that Dalglish had been asking around about it.[83] Friends also posted its financial statements online for all to read. "Many dollars are at stake here," noted Winsor. "So it is not surprising that [Leonard] Asper and his company might wish to rough up their critics."[84] The investigator, who had gone into the "corporate intelligence" business as Dalglish & Company, denied working for Asper. He eventually admitted he was behind the gumshoe intrigue, however, and said he had learned a lot as a result of her investigation.[85] It turned out Couture had also written speeches for CTV president Ivan Fecan and even for former CRTC chair Françoise Bertrand. Couture protested that Friends had "frequently" opposed CTV applications to the CRTC, such as its bid to own both Sportsnet and TSN, which was turned down. "I don't recall the Aspers criticizing Friends or its credibility on that occasion," she wrote in a letter to the *National Post.*[86] By 2001, CanWest's occasional sniping at Friends would turn into an all-out shooting war just as the Aspers faced their most serious regulatory challenge to date.

Public benefits

While it was willing to sell its original Vancouver affiliate, Can-West balked at parting with Hamilton's CHCH, which it proposed to operate as an "independent" station. It pledged $17 million in local programming initiatives as part of a "public benefits" package proposed to gain CRTC approval. The public benefits system had been established in the late 1980s to ensure that takeovers that increased ownership concentration also provided positive outcomes. It required that 10 percent of the value of any broadcast media takeover be devoted to worthwhile initiatives, usually programming. As part of the WIC deal, CanWest pledged $84 million in public benefits, or more than 12 percent of the value of the stations it acquired.[87]

The CRTC scheduled hearings for April, which were held in Vancouver instead of as usual at CRTC headquarters in Hull, Quebec. CTV argued the Aspers should have to sell CHCH and CHEK because it would otherwise enjoy "duopolies" in southern Ontario and southwestern BC. By the time the hearings commenced, however, the Canadian media landscape had been convulsed once more with CTV's takeover by Bell Canada Enterprises (BCE). That reduced considerably the persuasiveness of CTV's argument. "How can King Kong, who is storming the city . . . say this is bad for Canada," Leonard Asper asked on the eve of the Vancouver hearings.[88] When they convened, it was the youngest Asper, now CanWest CEO, who faced regulators for the first time while his father took a back seat, literally. In his pitch for approval of the WIC deal, the 35-year-old argued that the relatively small Canadian market could sustain only a few national broadcasters. "We maintain Canada may have room for five to 10 large media groups," he told commissioners, "but only three major conventional [TV] players.[89]

The CRTC reserved its decision on CanWest's application until early July 2000, and when its verdict was issued Izzy Asper finally had his long-sought national network. The CRTC even allowed Can-West to keep its double coverage in southern Ontario and south-western BC, but it was required to move its Hamlton and Victoria stations into a subsidiary network called CH. When the CanWest patriarch was asked for his reaction, it was reported as being made in "typical bombastic style" when he remarked cryptically: "This will not be CanWest's biggest announcement this year."[90]

CHAPTER 3

Going Global

Izzy Asper took his template for financial success in the television business to another level in the 1990s, launching CanWest onto the international stage. The globalization was due in part to the CRTC's reluctance to grant him a national network for most of the decade. It also resulted from opportunities abroad that were perfect for Asper's proven profit formula. The CanWest patriarch increasingly involved in his expansion plans a second generation of Aspers who would eventually take over his broadcasting empire. By applying the same cost-effective principles that worked so well in Canada, Asper achieved what others had been unable to accomplish. CanWest proved particularly adept at starting up private television networks in competition with entrenched public broadcasters. It also showed a talent for turning around television networks that had started out as money losers. Just as in Canada, however, the financial success inevitably came at the cultural cost of a torrent of cheap American programming. Some countries took eagerly to the lowbrow fare, others did not.

CanWest's international expansion was financed by a share offering on the Toronto Stock Exchange in 1991. Asper had hoped to raise $125 million by selling 35 percent of CanWest to the public, but a lukewarm reception from investors and some untimely market turmoil combined to produce less than half that. Stock analysts and investment fund managers looked askance at the offering, which was for "subordinated" shares that carried little voting weight and ensured 95 percent control for Asper, who kept most of the voting shares. A coup in Moscow also led to a sharp drop in stock markets in the weeks prior to CanWest's offering. As a result, the $17 share price CanWest hoped to realize instead amounted to only $13.[1] The $58.5 million stake, however, would be invested wisely by Asper.

Taking New Zealand

CanWest's first international investment was made late in 1991. Westpac bank had taken over TV3, New Zealand's first commercial television network, after it proved unprofitable. A consortium including NBC had put TV3 on the air late in 1989 in the South Pacific island nation of 3.5 million. Higher than expected startup costs prompted NBC to bail out within six months, however, leaving Westpac in charge. In its first 18 months of operation, the network lost a reported NZ$69 million. The conservative National Party won election in late 1990 and removed foreign ownership restrictions allowing CanWest to "rescue" TV3.[2] The deregulation of broadcasting in New Zealand was part of a larger program of economic reforms enacted to resuscitate the country's ailing economy. The measures included lowering taxes, slashing government spending, privatizing public enterprises, and cutting debt — all longtime Asper favourites. Introducing private competition to government-owned TVNZ, which operated two networks, was part of the economic restructuring. Within five years, New Zealand had perhaps the most deregulated broadcasting system in the world.[3]

Westpac went looking for a partner to help it turn TV3 around, and its officials met with CanWest executives in 1991 at a Los Angeles television gala. By year's end, a deal was made for CanWest to buy 20 percent of TV3 for NZ$14.8 million (Cdn$10 million) and take over its management. The purchase gave CanWest more leverage in buying American programming for international distribution. It also took advantage of a reciprocal agreement that considered New Zealand productions as Canadian content, and vice versa. "CanWest needs more territory and this adds the equivalent of British Columbia," Asper told the *Globe and Mail*.[4] To help turn TV3 around, Asper flew to Auckland to personally deliver its employees a wake-up call that soon passed into CanWest lore.

> Izzy gathered about 200 of them together in the staff cafeteria, pointed to a newsperson and asked, "You. What business do you think you're in?" "I'm in the news department," came the reply, according to Izzy, "and the business we're in is to make sure our audience gets the most carefully researched news and information possible." Izzy asked the same question of others in different departments, and got similarly reasoned responses

— we're in the business of entertainment, fine drama and so on. "Wrong," said Izzy. "You're all wrong and that's why you're bankrupt. You're in the business of selling soap."[5]

Under CanWest management, TV3 went from red ink to black within a year. It cut costs by laying off staff, axing an expensive children's programming division, and airing cheaper US shows. From a loss of NZ$22 million in 1992, the network turned a modest profit of NZ$650,000 the following year. In 1996 it earned NZ$28 million. The formula worked so well that CanWest paid Westpac NZ$78.8 million in 1996 for another 48 percent of the network, bringing its ownership of TV3 to 68 percent.[6] It took further advantage of the deregulated New Zealand market in 1997 by launching a fourth network called TV4 that focused on entertainment. It also paid NZ$30 million that year for More FM, the country's third-largest radio network of nine stations.[7] "New Zealand became the testing ground to see whether our skills were transportable," noted Peter Liba. "We discovered that indeed they were, because it's television and it's run with the same product — programming — and its funding comes from the same source — commercials — and everything else, if you like, is housekeeping."[8]

Crossing the Tasman Sea

When Asper arrived in Australia in 1992, he was carrying a briefcase full of introductions to people there he referred to as "the Canadian and Jewish mafias."[9] He had his eye on Network TEN, the third-place television broadcaster that had gone into receivership. Westpac was trying to unload TEN, which was losing A$2 million a week, for its book value of A$240 million. "When CanWest executives heard they could buy a network, in a country that had only three commercial networks and no pay-TV for about A$240 million, they couldn't get on a plane fast enough," Peter Viner told Australia's *Business Review Weekly*. "To buy one station in North America in a city that might have 30 channels would cost about A$350 million."[10] Asper flew to Sydney, according to CanWest's official historian Allan Levine, and told Westpac officials he would not leave until he could make a bid for TEN. Days went by as Asper waited in a hotel room while the bankers weighed selling Network TEN to a Canadian. Strict limits had been placed on how much of Australia's media could be held by foreign owners, and in television the maximum was 15 percent. Finally, the bank agreed to let Asper make an offer, according to Levine, "as he was about to catch a plane."[11]

That's when Asper's briefcase full of contacts came into play. While CanWest was flush with cash from its 1991 share offering, it needed to attract domestic partners to comply with Australia's foreign ownership limits. "Asper opened and closed his briefcase of introductions dozens of times as he cast around for fellow travelers," reported *The Australian* national newspaper.[12] First he met Jack Cowin, who had come to Australia with a franchise from Burger King, but finding that name taken called his chain Hungry Jack's instead. "He said most people who get involved with TV think it's about starlets, champagne and limos when in reality it's about selling commercials," remembered Cowin. "You have to keep enough people watching and spend the minimum amount of money as you can to do that."[13]

Cowin introduced Asper to ad man John Singleton and each threw in for 10 percent of Network TEN. "He proudly described himself as a bottom feeder," Singleton recalled of Asper's sales pitch.[14] Asper's contact list also included the partners in TEN's regional broadcaster in Brisbane, Telecasters North Queensland. They came in for a share, as did others, but when Asper added up the commitments he had, they amounted to little more than 40 percent.

Then his most important contact paid off, with an introduction to the Sydney law firm Freehills. It devised an arrangement that allowed CanWest to make an end run around the country's foreign ownership laws by taking equity in TEN as debt instead of as shares. As a result, CanWest contributed 57.5 percent of the A$230 million purchase price, which included A$85 million in cash and A$145 million in assumed debt. A re-organized network issued an equal number of voting shares, of which CanWest took 14.99 percent, and non-voting debentures. The debentures, all of which went to CanWest, would pay an interest rate that reflected TEN's financial performance. The long-term debt instrument, similar to a bond, essentially made CanWest a creditor of Network TEN, explained *The Australian* in 2003.

> In effect, CanWest would have 57 percent of the "economic" interest, but stay within the law with just under 15 percent of the voting stock. It was a brilliant design, and many potential foreign buyers of media assets pleaded to be able to "do a

CanWest" to get around pesky foreign ownership limits. After two inquiries, the federal government put a stop to any further "CanWests." It remains a unique structure.[15]

Canadian invasion

In Australia, media ownership was nowhere near as wide-open as in New Zealand, and many there looked askance at TEN's innovative arrangement. With Asper's bold takeover, Australia had seen a second front open up in what must have seemed like a Canadian media invasion. Conrad Black had arrived the year before to lead a consortium that took over the Fairfax newspaper group, despite similar strict limits on foreign ownership of print media. Black caused controversy not only with his actions — axing hundreds of staff in a bid to boost the bottom line — but with his words as well. Izzy Asper showed a similar penchant for stirring the pot with his comments both in the Canadian and Australian media. No fewer than three Australian government inquiries into media ownership would result, two of them involving Asper. Only Black, however, would quit the country as a result.

CanWest's reputation soon reached Australia, where its clashes with the CRTC became grist for the media mill. Asper went into damage control mode. According to the Melbourne *Herald Sun*, he "tried to recover ground by telling audiences the planned $30–$50 million Australian investment would be used to offset the big costs of local production."[16] It was a comment Asper made to the *Globe and Mail* back in Canada, however, that really raised Australian eyebrows. "There is no question we will be involved with management," Asper said of TEN.[17] The comment was bound to get the attention of the newly-formed Australian Broadcasting Authority, noted the *Herald Sun*. Asper seemed to be speaking, the newspaper pointed out, "more like the man who will run the Australian group than a minority 15 percent shareholder, to which his group is restricted."[18] Not only was foreign ownership limited, but non-Australians were prohibited from exercising control over broadcasting media. The ABA had been established earlier in 1992 with a mandate that included enforcing those provisions. If Asper's October comment didn't get the attention of the new regulators, a report in the *Herald Sun* a month later may have.

It is understood that Ten's managing director Gary Rice recently moved to bring the Brisbane manager Mike Lattin to Sydney to take the key post of network programming chief. But the putative buyer — that can only mean Canwest — reportedly said: wait a minute. Under the sale contract, Ten is naturally prohibited from making any major spending decisions without consulting the buyer(s). Canwest is prohibited by the Broadcasting Act from having any direct management influence. At the moment it is the only name on the Westpac contract, and in exercising its right to consider the appointment, raised the question of whether it has stepped over the line.[19]

Rice, a respected veteran Australian television executive, had taken over management of TEN for Westpac. He had turned a loss of A$100 million in 1991 into a profit of A$15 million in 1992 by cutting a quarter of the network's staff and slashing its programming costs. His choice of Lattin as programming head for TEN would be confirmed, but Rice would be out within months as more ardent cost cutters arrived from Canada. It would be Lattin's clashes with his new bosses from Canada that would finally prompt an ABA inquiry.

Peter Viner was a Global Television original, having joined the fledgling Ontario network in 1974 as its vice-president of marketing. Six years later and still only 35, he was named general manger of CKVU in Vancouver and remade the third-place station as youth-oriented U.TV.[20] In late 1992 Viner went to Australia as part of CanWest's pre-purchase "due diligence" team inspecting Network TEN's operations. He soon returned with Izzy Asper and Stephen Gross, CanWest Global's president, to help recruit investors from Asper's list of contacts. His familiarity with TEN made Viner a logical choice for one of CanWest's two seats on its board of directors, with Gross filling the other. In early 1993, he moved to Sydney to take up duties as chairman of the board's executive committee. According to the *Globe and Mail*, however, Viner was "generally regarded as a chief-executive-in-waiting."

Sources say the departure signs [are] attributed mainly to Mr. Asper's desire for a friendly, familiar hand at a one-seat helm, and Mr. Viner's determination to cut spending further. Mr. Viner has been gladhanding advertising agency executives, reassuring them that Ten intends to stick to Mr. Rice's survival script, with no imported format remakes. [21]

Viner's appointment as CEO of Network TEN was made official in June of 1993 and backdated to April 1, after his resignations from U.TV and CanWest Global became effective. His influence on TEN's operations, however, had been apparent since the New Year in his unofficial role of "consultant." Mike Lattin, TEN's new programming director, noticed it particularly. In January, Viner approached Lattin to discuss whether he wanted to continue to head up programming. Rice was still CEO, and while he signed off on Lattin's job description, he had no role in negotiating it. According to Lattin, Viner was obviously in charge of a February 24 meeting of the Budget Review Committee, even though Rice was still CEO and was present. While Viner would approve programming decisions after he was appointed CEO, it was usually only after extensive consultations by phone and fax with Gross back in Canada. "Mr. Gross was ever present at most negotiations," claimed Lattin. "Mr. Gross acted like a senior executive and had to be consulted on programming issues. You had to convince Mr. Gross you were right."[22]

Programming decisions were often based on CanWest's best interests, according to Lattin, not Network TEN's. He made a decision to drop the soap opera *General Hospital* from the network's daytime lineup, only to be over-ruled by Gross. Lattin claimed Gross told him at a meeting in Los Angeles that TEN had to take *General Hospital* so TV3 in New Zealand could get programming from its distributor. Network TEN even broadcast the Toronto Father Christmas Parade to help CanWest in a sponsorship deal in Canada with McDonald's. In late 1994 Viner met Asper in Toronto. He told Lattin on his return that as a result "there will be some reorganisation but I can't tell you anything yet." A fax from David Asper requested programming help for CanWest's bid for a television licence in the UK, which Viner directed Lattin to provide.[23] The inescapable conclusion, to Lattin at least, was that CanWest was in control of Network TEN through Viner and Gross, in violation of Australian law.

Shortly after Lattin's resignation from Network TEN on October 15, 1994, his allegations surfaced in a lawsuit. Charles Curran, who operated the regional network Capital Television in Canberra, alleged misleading and deceptive conduct under the Trade Practices Act. The *Canberra Times* published extracts of Lattin's affidavit filed in support of the lawsuit, but CanWest sued the newspaper and threatened to sue any media outlet that repeated the allegations.[24] A copy of the *Canberra Times* article, however, soon came to the attention of the ABA. An investigation quickly ensued and continued for more than a year. By the following August it had generated 950 pages of testimony and subpoenaed 15,000 pages of documents.[25]

Asper weighs in

Even while the ABA investigation was going on, Izzy Asper fanned the flames with comments in the media. No sooner had the inquiry begun than Asper was quoted in a Canadian magazine in a way that suggested he, not the board of Network TEN, had appointed Viner. "When you buy a company that's 14,000 kilometres away you'd better be sure you send someone you trust to run it," he told the *Financial Post*.[26] The ABA requested an explanation. Lawyers for Asper answered that the quote should be "taken in context of the subject matter of the interview," which was family succession in business.[27] Viner meanwhile set about achieving the efficiencies in Network TEN's operations he had been sent Down Under to accomplish. "Viner did what he had to do; what had to be done," according to Levine's official CanWest history.

> Costs at all levels were reduced or eliminated altogether. He also introduced a North American management style to the station that had a positive impact on the staff. "In Australia at the time," he explains, "the management style was hierarchical and autocratic, not collaborative like in North America."[28]

The positive impact on staff might have come as news to many working at the network. As a current affairs program was dropped, the Adelaide *Advertiser* reported Network TEN was "plagued by staff reshuffles and low morale." Viner claimed he was unaware of any morale problem.[29] If it dampened employee spirits, Viner's cost cutting did wonders for Network TEN's bottom line and hence the fortunes of CanWest. As TEN's profits soared in 1994, CanWest received dividends of A$3.4 million from its share holdings and interest payments of A$22.8 million from its debentures.[30] After earnings doubled to A$103 million the next year, CanWest got another A$6.8 million in dividends and A$45.5 million in interest, making TEN the main driver of CanWest profits. The company had received in less than three years a return of 110 percent on its investment.[31]

One of the secrets to CanWest's success at TEN was in drawing a younger audience. "We reshaped it to look more like a Fox affiliate, more urban and younger," Viner told the *Globe and Mail*.[32] Along with viewers of such popular American shows as *The X-Files, Beverly Hills 90210*, and *Seinfeld* came advertisers after their disposable income. Time-tested Canadian tactics, such as counter-programming, also proved successful according to Levine. "When its opposition, Networks Nine and Seven, were showing current affairs programs at 6:30 PM on week nights, TEN put the Australian soap opera *Neighbours* on the air and people started changing channels."[33]

The financial and programming success of Network TEN seemed to embolden the CanWest leader. As the ABA inquiry dragged on through most of 1995, Asper flew Down Under to raise the stakes in his battle with the broadcasting regulator. He appeared on a Network Nine business program to complain that Australia's media ownership laws made no sense. "I can leave here as a non-Australian and buy a radio station in Sydney as a foreigner," he said. "Why can't I buy a television station?" If the ABA forced it to sell its TEN debentures, Asper promised CanWest would expand into media sectors that were not as tightly regulated. "If we don't win the argument we'll go into cable," he said. Asper also criticized a recent ABA decision allowing pay TV operators to rebroadcast free-to-air network content, calling them "parasites."[34] Then Asper played the technology card, dropping a bombshell. "From 1997 I can put a satellite up from Fiji," he pointed out. "Whatever technology will permit, the laws can't stop. It will be done. If I can reach every home in this country from Fiji, there's no sense passing any laws about foreign ownership."[35]

By month's end, CanWest was off the hook with the ABA, as a 208-page report cleared it of exercising control over Network TEN. "We're confident that the Australian directors on the Ten board and the Canadians are between them in control and that's as one would expect it to be," said ABA chairman Peter Webb.[36] While Lattin's version of events was confirmed, the ABA investigation found it did not prove CanWest was exercising control over TEN. "Mr Viner consulted broadly before making some decisions, including with the directors of [TEN] and the Executive Committee generally," concluded the report. "His consultations with Mr Asper were no different, in effect, to similar consultations with other directors."[37]

Despite the vindication, Asper seemed dissatisfied and a week later even threatened to pull out of the country. Proposed changes to Australia's media ownership laws would allow no increases in foreign ownership, and only minimal cross ownership of newspapers and television. "It may well be if the government of Australia doesn't want, for whatever reason, foreign ownership or foreign investment, or CanWest in particular," Asper told government broadcaster ABC Radio. "Well obviously there are lots of places in the world where one can invest. ... And reluctantly but certainly we would divest our interests in Network Ten and employ our resources where they are welcome."[38] Asper's threat to go elsewhere was hardly a hollow one, as ambitious international expansion plans were already under way at CanWest. Within a year, however, the company moved to increase its ownership in Network TEN, bringing another investigation by the ABA. This one would not conclude as favourably for CanWest and would even involve the Canadian government in what threatened to become an international trade dispute.

Asper had promised the ABA that ownership of Network TEN would go public on the stock exchange or "float," in local parlance. When he presented a float plan to the ABA in late 1996, however, its officials noticed some recent changes in TEN's ownership. Four of the network's six minority shareholders had sold to holding companies based in Australia. When questioned by reporters, Asper denied CanWest had increased its ownership of TEN. "It's been difficult to achieve a consensus among all the shareholders, but I'm optimistic that the events of the last 60 days have brought the advent of a float closer," he said. "CanWest has not bought any shares in Network Ten whatsoever."[39] An ABA investigation, however, showed the holding companies had bought the shares with money borrowed from a subsidiary of CanWest set up in the Netherlands. Alarm bells started going off at ABA headquarters when it became apparent CanWest was in a position to control 76 percent of TEN as a result. Asper assured ABA investigators he was only concerned that TEN stock fell into friendly hands. "Antibodies can come into a company, mischief makers, stupid people," he told them on December 19. "Diabolical, fiendish, cunning fronts for competitors."[40] The holding companies had paid A$241 million for 18.5 percent of TEN, which suggested a value of A$1.3 billion for the network that had been bought for A$230 million three years earlier.

After a four-month investigation unravelled the complex dealings, the ABA ruled CanWest in breach of the law. It gave the Canadians six months to sell the extra shares or face a $2 million fine.[41] A separate investigation by Australia's Foreign Investment Review

Board also demanded divestiture by September 30 "irrespective of price."[42] A change in government to a Liberal coalition led by John Howard had brought proposed changes in media cross ownership laws, but not on foreign ownership. Non-voting shares were also banned, meaning CanWest would have to reduce its ownership of TEN to the 15 percent limit when the company went public. "Asper had been hoping the Government would move the other way and liberalise the foreign ownership restrictions," noted *The Australian*. "He appears to have greatly misread the mood in Canberra."[43]

On a May trip to open TEN's new $40 million headquarters in Sydney, Asper visited the capital to lobby politicians for an exemption from the new rules. "We've certainly requested formal grandfathering," he told reporters. "You don't change the ground rules retrospectively. That is something that we civilised countries do not do. . . . We did not come in through the back door nor did we come in in the middle of the night. We came in proudly and honorably with the blessings of appreciation [sic] of all concerned."[44] His message to Treasurer Peter Costello was more succinct. "Let commerce rule," Asper said he told Costello, "not the law."[45] The dispute threatened to turn into an international incident when Ottawa weighed in on CanWest's behalf. The Liberal government warned Australia "it would consider the demand a breach of international treaty obligations."[46] Asper, as was his habit, took the dispute to court. His lawsuit was dismissed, but a Federal Court judge over-ruled the requirement that CanWest sell at any price.[47] Still Asper pressed the case, appealing the ruling. "The man simply does not give up," marvelled *The Australian* of Asper's "interminable game of snakes and ladders."[48]

Finally in September, Asper announced a deal to include CanWest's excess shareholdings in a public stock offering through TEN's affiliate Telecasters North Queensland. CanWest's majority ownership of TEN was exempted from the changes to media rules outlawing non-voting shares. Speculation sent the estimated value of TEN's shares to $17.90, implying a company value of $1.63 billion, right behind the $1.75 billion value of market leader Network Seven.[49] "It is this formal grandfathering of the CanWest structure," noted *The Australian*, "which makes it so valuable."[50] Broadcasting, foreign investment, and stock market regulators had all "appeared powerless against Asper flouting the Australian law," observed the *Australian Financial Review*.[51] As TEN shares split 10-for-1 and listed for $2.15, CanWest realized a quick profit of A$134 million on its excess stake.[52] TEN's share price soon soared to A$2.70, boosting CanWest's five-year Network TEN investment 27 times over to A$1.4 billion.[53] "With the benefit of hindsight," noted *The Australian*, "this was the bargain of the decade."[54] As a reward for his part in the success, Peter Viner was appointed Izzy Asper's successor as CanWest president and CEO when he stepped aside at age 65 in late 1997. "Our goal is to double the size of the company in the next four or five years," said Viner, whose brother Tony was CEO of Rogers Broadcasting.[55] Viner's most important job, however, would be to train the next generation of Aspers to take over the company's reins.

Seeing La Red

CanWest established an international division in 1993 that was headquartered in Los Angeles. Charles Weber, an American movie industry veteran, was hired as president and CEO of CanWest Global International. Weber told CanWest's annual meeting he was exploring opportunities in Britain, Germany, Switzerland, Scandinavia, Eastern Europe, the US, and South America.[56] The Aspers were particularly keen on tapping into the Hispanic market and its 400 million Spanish-speaking viewers worldwide. CanWest Global International's first move in that direction was into Chile the following year when it bought half of fifth-place La Red television for US$8.4 million. La Red had been launched in 1991 but languished in a market dominated by two stations established by universities as educational broadcasters. Between them they pulled in 60 percent of viewers and 70 percent of advertising revenue. "We'll be creating a different niche," said Victor Rodriguez, hired by CanWest from Toronto's CITY-TV to be vice-president of marketing for La Red. "We're going after women and kids."[57]

La Red already had some popular shows, reported the *Globe and Mail*. "*Desjueves*, or *Thursday*, is political satire considered radical in Chile — 'like *Saturday Night Live* with a Latin twist,' Mr. Rodriguez said. And *Coctel*, or *Cocktail*, is a talk show with a difference. The host, a comedian, interviews guests in a bar."[58] CanWest applied its usual turnaround formula, cutting costs by axing 42 percent of La Red's staff and increasing US programming. The changes seemed to work at first. Audience share increased from 5.7 percent to 7.9 percent after one year and advertising revenue grew by 20 percent to

$356.4 million.[59] But hopes of turning La Red around foundered on unexpected cultural factors. The powerful influence of the Catholic church prevented airing of the Canadian mini-series about sexual abuse by priests, *The Boys of St. Vincent.* The new dance program *Club Latino*, which featured scantily-clad women, was a ratings hit with male viewers. Advertisers would not support the program, however, because their wives objected, so it was cancelled after only six weeks.[60] Hit American shows dubbed into Spanish lost something in translation, according to *Canadian Business.*

> U.S. comedies such as *Frasier* don't get any laughs because sarcasm isn't part of Chilean humor. Instead, Chileans prefer to laugh at the latest antics on the popular nighttime soap operas *Estupido Cupido* (Stupid Cupid) and *El Amor Esta de Moda* (Love is in Style), which run on the two big networks and take almost 70 percent of the weeknight audience.[61]

The hoped-for Chilean beachhead on the Hispanic market quickly evaporated as market research showed irreconcilable cultural differences. "We found out that there is no such thing as the 'Spanish market,'" said David Mintz, the CanWest executive who unravelled the riddle. "The programs, for example, that Telemundo and Univision do in the United States, which is mostly [sic] for Cubans, Puerto Ricans, and Mexicans, don't play in South America. They speak a different dialect of Spanish." National differences also meant programming could not even be shared between countries in South America. "You can't even interchange programs between Argentina and Chile, even though they border one another for hundreds of miles," Mintz told Levine, "because the Argentineans speak much faster."[62] After more than two years of frustration, CanWest sold its share of La Red for US$9.5 million in 1996. The $1 million paper profit was actually a $3 million loss after accounting for development costs.

European expedition

Another continent CanWest was keen on getting into was Europe. The new market economies of Eastern Europe were of particular interest to the Aspers, who visited Poland, Hungary, Slovakia, and the Czech Republic in the mid-1990s. They even had a deal to launch Romania's first private television station until it went off the rails for political reasons. "You'd go to a meeting with a deputy minister in the government one day, and the next day he'd be replaced by someone else," recalled Izzy Asper. "It was untenable."[63] Doing business in the high-rent part of Europe would prove even more frustrating. The rich United Kingdom market was underserved by commercial television and dominated by the BBC. There were only two private television networks in the nation of 60 million, with an annual advertising pie of £4 billion to be split. To boost government revenues, the UK had gone to a "blind auction" process for awarding its 10-year television licences. It required sealed bids to be submitted, along with detailed programming and business plans. The highest bidder, however, would not necessarily be awarded the licence. Factors other than bid size would be pivotal, including program quality and financial planning. The highly political process was rife with intrigue.

CanWest's first try at cracking the British market resulted in a messy false start. The broadcasting regulator Independent Television Commission (ITC) announced it would accept applications for a new UK television licence in 1992. CanWest joined a consortium to bid for the licence with Toronto's CITY-TV, Sony Pictures, and London's Thames TV. The consortium was the only applicant, but a last-minute disagreement over the extent of Asper's participation caused CanWest to drop out. According to the *Globe and Mail*, Asper demanded control of Channel 5 and a "fat fee" for running it. He then announced through the media that his conditions were non-negotiable. Similar to Australian law, ITC regulations prohibited non-Europeans from controlling British broadcasters. Asper's demands were made after CanWest CEO Stephen Gross left London for a holiday in Israel, according to the *Globe and Mail*. "Asper flew into town and drafted a list of conditions," it reported. "When Mr. Gross returned to London later in the week ... he apparently was a little taken aback at what his boss had done."[64] The CanWest founder denied overplaying his hand.

> Asper shrugged off the notion that he may have blown the deal or let his ego get in the way. "That idea is amateur night on the Thames," he said. In fact, he contended that, if anything, abandoning the deal showed his skills were sharper than ever. "It takes determined discipline to know when to walk away, especially when you have invested months of time and research and money."[65]

The ITC declined to grant the licence to the re-arranged consortium, citing quality concerns, and CanWest set about preparing for the next round of applications. In 1994, it took a 24.5 percent stake in Talk Radio UK, the country's first commercial network in that format. It was launched the following year with programming described as "up-front, irreverent and entertaining."[66] Talk Radio UK never did catch on, and unfortunately for CanWest it "attracted a flurry of complaints about vulgarity," noted *The Economist.*[67] David Asper was put in charge of CanWest's renewed Channel 5 application when bids were again invited in 1995. For a year, he travelled back and forth to London from Regina, where he managed CanWest's television station. Due to media leaks during the 1992 bidding, the company's interest in Channel 5 was kept quiet until the May 2 application deadline for fear of increasing other bids. "We stayed out of sight in London, in low-rent hotels, so our presence wouldn't be known until the deadline for filing," Asper said. "We didn't want anyone to know we were there."[68] The *Financial Times*, however, caught wind of CanWest's interest. It noted that a planned British partner, the United newspaper chain, had pulled out of its consortium. Bidding groups had to include a UK company, and European interests had to make up a majority of their ownership. Asper confirmed CanWest would bid, telling the newspaper: "United has already been replaced."[69]

Secret agent Asper

David Asper went to great lengths to ensure CanWest's plans remained a secret. "Details of his business plan are securely stored in a vast green safe which dominates his spartan Covent Garden office," reported the *Independent* newspaper. "Asper opens it with a tiny key he keeps close to his heart."[70] Concerns over corporate espionage were very real as rumours filled the London newspapers. "We had our office swept for bugs several times," Asper told Levine. "We also had our telephone lines swept constantly."[71] So paranoid was the CanWest team in London that multiple versions of the company's Channel 5 bid document were prepared by staff members, according to Asper.

> For security reasons, we decided to have them run 10 business plans with different bid amounts, only one of which was the real one. The actual amount was a closely guarded secret. . . .

> We were up for three straight days and nights and by the third, no one except the four CanWest people knew what was happening.[72]

Many in England were concerned about a media owner from the colonies scooping up what was expected to be the country's last terrestrial television licence. The concern wasn't over Canadians from CanWest, however, but about Rupert Murdoch. The Australian dominated the UK newspaper market. His tabloid *Sun*, the world's best-selling daily, had a circulation of more than three million. On the other end of the quality spectrum, the venerable *Times* Murdoch bought from Thomson was challenging Conrad Black's *Telegraph* in a vicious price war. He had also broken into the UK television market with the digital satellite service BSkyB. Reports circulated that Murdoch was planning to offer £35 million a year for the Channel 5 licence.[73]

CanWest then caught a break in its search for a British partner. It had already lined up the Scandinavian Broadcasting System of Stockholm (SBS), but its ownership was dominated by US investors, including ABC. Network TEN was also on board, but Asper's group needed a strong media presence from the British Isles. Fortunately one fell into its lap at the last minute when a consortium led by the Mirror newspaper group and NBC fell apart. One of its British partners, the production company SelecTV, switched to CanWest's bid, code-named UKTV. SelecTV not only boasted an extensive programming library, but its president was married to comedian Tracey Ullman. That meant CanWest could add her to its list of attractions, which it did with a proposed new series called *Tracey Takes On*. As the deadline approached, the only thing left to be decided was the magnitude of CanWest's proposed annual licence fee. According to the *Sunday Times*, Asper felt it had to be high enough for an outsider like CanWest to be taken seriously. The skullduggery the newspaper described over which bid document would be submitted to the ITC was something out of a spy novel.

> Over pepper steak and lobster at Bibendum, an art deco restaurant in west London, the problem was acute for David Asper and his colleagues from CanWest ... At 10pm, Asper telephoned a partner preparing the papers. In case of spies, he had four different codes. "The steak at the restaurant was delicious tonight," said Asper. His partner knew what that meant: they would bid £36.2m.[74]

When the bids were unsealed, the magnitude of the CanWest offer was not the only surprise. The UKTV bid was £14 million more than the next highest, but second place was a dead heat. Both Richard Branson's Virgin TV and a group fronted by Pearson PLC, which had recently taken over Thames TV, offered the ITC exactly £22,002,000 a year. Even more startling to some was Murdoch's bid — only £2 million a year. Asper issued a press release claiming victory, but the winning application would not be decided until fall. Few in Britain considered CanWest a serious contender. "I've more or less dismissed them from the race," media analyst Louise Barton told the *Globe and Mail*. "I don't see they'd be credible in the eyes of the Independent Television Commission."[75] Barton was more pointed in her comments to the British media. "I don't think they have a hope in hell," she told the *Sunday Times*.[76] Rival bidders pointed to the size of CanWest's offer as evidence the Canadians were out of their depth. "You would never make a return," said Greg Dyke, who led the Pearson bid. "We could not have bid that on the basis of our business plan."[77]

'Tea leaves swilling'

As ITC staff set about analyzing the proposals, Asper discarded his cloak of secrecy. He went on a public relations campaign to sell CanWest's bid to the skeptical British. UKTV's budget proposed spending £150–200 million a year, but analysts predicted it could only bring in £90–120 million a year in revenue. That brought speculation the ITC might disqualify it as not viable, but Asper insisted CanWest had done its homework. "We've been very conservative and cautious," he said. "We wouldn't have submitted our bid unless we were certain the business plan would give a fair return to our investors."[78] Some pointed to the high ratio of proposed co-productions with UKTV's foreign partners. Asper called the approach "the future of independent television."[79] When it was reported that Murdoch's bid was so low because Channel 5 might start off reaching less than half the country, Asper lashed out. The network should eventually reach 70 percent of the population, he said, or 40 million. He blasted as "distracting falsehoods" a whisper campaign by rivals designed to discredit the UKTV application. "There is a tremendous amount of ill-informed comment being made," he said. The UKTV bid was based on an initial audience share of 6.6 percent and 11 percent after 10 years, he explained, and assumed profitabil-ity within three years.[80] Asper even pledged to donate 1 percent of UKTV's pre-tax profits to a charitable trust and offered to have that written into its broadcasting licence.[81]

Soon sentiment swung to see CanWest's bid as not just the biggest, but also the best for Channel 5. Even Louise Barton had to admit that a sea change had taken place. "I was quite skeptical of the bid at first but the industry had come around to feeling they were going to get it."[82] On October 19, ITC staff members cleared all the bid-ders on programming quality and financial grounds and recom-mended the UKTV proposal be accepted.[83] Commission members, who were political appointees, delayed acting on the recommenda-tion, however. "Speculation in an already rumour-mad industry is now reaching fever pitch," reported the *Independent*, "as the ITC's silence is read like so many tea leaves swilling in the bottom of a cup."[84] Finally, on October 28 Pearson was declared the winner. Both UKTV and Virgin TV had been ruled out on quality grounds, the ITC said. A commission insider described as "extremely poor" the pro-gramming proposed by the disqualified bidders. "UKTV relied too strongly on drama serials and entertainment and half of its output would be repeats," reported the *Glasgow Herald*. "The only factual programming it proposed, apart from news, was between 7 PM and 8 PM on a Sunday."[85]

The surprise decision smelled fishy to some who pointed out that prominent Labour party supporters led the Pearson bid.[86] As far as David Asper was concerned, the fix had been in from the start. "I fear that it did not matter what anyone else did because the Pearson group had been anointed." he told Levine.[87] Support for his view grew when it emerged that ITC commissioners had over-ruled staff experts who had passed the programming proposals of all bids.[88] Outrage ensued when a judicial review launched by Virgin TV heard the ITC had allowed the Pearson group to increase its fund-ing commitment by £100 million in September.[89] CanWest quit the country, selling its stake in Talk Radio UK. "When England turned us down we said screw it," Izzy Asper told *Maclean's*. "It's full blast on the Canadian end."[90]

Eyes for the Irish

CanWest finally broke into television in the British Isles in 1998 when it led a consortium that got Ireland's long-sought first private network up and running. TV3 had been in the works for a decade but stalled when a consortium including Northern Ireland's Ulster Television and management of the rock group U2 fell apart. UTV pulled out in 1996 and was replaced by CanWest, which soon bought a stake in the Belfast broadcaster as well. CanWest took 48 percent ownership of TV3, put up most of its US$24 million in start-up costs, and assumed managerial control.[91] "Ireland is key inasmuch as it's on the doorstep to the UK, the largest English-speaking market in the world after the US," said Viner. "We want to be in the UK, to be a player. . . . Once we're in the region and get an understanding of the landscape, we'll be able to develop much more quickly."[92] CanWest also began buying up UTV, paying £3.5 million for a 2.7 percent share in October 1997 and a week later boosting its ownership to 7.4 percent.[93] Six months later, it bought another 18 percent from Scottish Media for £23.9 million.[94] CanWest's goal in moving into both halves of the religiously-divided isle was to eventually create a pan-Irish broadcaster, explained Leonard Asper. "Inevitably, there will be peace and Ireland and Northern Ireland will become economically integrated."[95]

The Irish broadcasting regulator Independent Radio and Television Commission approved the launch of TV3 in November 1997.[96] CanWest moved Rick Hetherington, manager of its original CKND, to Dublin and he created TV3's digital studios from scratch. Hetherington promised 15 percent Irish content to start, rising to 25 percent by 2003, and fourteen hours a week of news, public affairs, and sports. In a bid to boost ratings, TV3 secured the rights to Champions League soccer. Its debut was eagerly anticipated in a market dominated by staid fare from two networks of state broadcaster RTÉ. A reporter for the *Irish Times* toured TV3 studios and found many perky overseas imports among the news staff. "They are young and trendy and some of their accents are Australian," wrote Roisin Ingle. "Then there are the TV3 babes. Attractive bright young things with limited to fair experience, but oodles of potential were plucked from relative obscurity . . . to become the station's main news journalists."

TV3 is committed to providing nothing more sinister than what they call "bright, breezy, watchable news." But the TV3 babes concern Irish Times TV critic Eddie Holt: "What's next, topless news? I mean how far can you go to make the news sexy before it becomes ridiculous?" At least a little further than our State broadcaster anyway, according to TV3 research which found that some news programmes are perceived to be talking down to their audience, or behaving like domineering relations.[97]

Within two years, CanWest had acquired 90 percent of TV3, but in late 2000 it sold half to England's Granada Media for £29.3 million (Cdn$62.2 million). The deal provided CanWest with needed capital to complete its purchase of Southam from Conrad Black. It also linked it with a British partner similarly looking to expand in Europe. It was a programming coup for TV3, as Granada's many popular shows included *Coronation Street*, a staple of RTÉ.[98] Rerun American programs, however, were also prominent on TV3, including *The Love Boat*.[99] Soon Irish enthusiasm for TV3 wore thin, according to Dublin City University professor Farrel Corcoran. Its pledge of local content had caused producers there to be "mesmerised by the promise of a significant increase in worthwhile, indigenous production."[100] What resulted instead was something less than a boost to Irish cultural life, according to Corcoran. CanWest reached its target for home produced material, he noted, "by including repeats of the Champions League, news and weather."[101]

CHAPTER 4

The Black Plague

Conrad Black not only conquered Canada's leading newspaper chain, which he quickly passed on to like-minded owners at CanWest, he changed the nature of its journalism. That was exactly what had long been feared, as for decades Black's political partisanship led Canada's media elite to shut him out of owning major dailies in his native land. When he finally did wheel and deal his way to emerge as Canada's dominant newspaper owner, Black indeed brought partisan journalism back home. By 1996, following his hard-fought takeover of Southam newspapers, Black's Hollinger International was the third-largest newspaper company in the world.[1] In 1998, Black broke the mold of journalism in Canada with his founding of the *National Post*, which reported from a decidedly conservative perspective. Then in 2000, he suddenly left his native land in a huff, claiming he could no longer endure its resistance to his neo-conservative entreaties. Black's newspaper empire would begin unravelling a few years later, but the Southam chain he passed on to politically sympathetic owners would never be the same.

Hollinger International sat atop a complex pyramid of holding companies controlled by Black. By 1994 it counted among its titles the influential *Daily Telegraph* in London, the *Chicago Sun-Times*, the *Jerusalem Post*, and the second-largest chain of newspapers in the US. Black bought the *Telegraph* for a bargain price in 1985, when the venerable daily was underperforming financially. A move to non-union operations at new premises meant almost three quarters of the 3,900 *Telegraph* staff were soon cut from the payroll. The venerable daily's profit picture quickly improved. From an annual loss of £8.9 million in 1986, the *Telegraph* recorded a profit of £41.5 million in 1989. According to Black biographer Richard Siklos, it became a "newspaper cash machine capable of funding its owner's desire to pursue the acquisition of practically any newspaper in the world."[2]

International expansion

Black expanded Hollinger's US subsidiary, American Publishing Co., into a chain of 340 newspapers over the next decade. He and other Hollinger investors paid more than US$300 million in a series of more than 100 purchases in the chain-building exercise. Black and his Vancouver-based lieutenant, David Radler, built their empire in large part through a regular ad in the US magazine *Editor & Publisher*. The ad was aimed at a generation of small-town publishers ready to cash out and retire. Unlike in Canada, Black's neoconservative views were a selling point in Hollinger's US expansion, according to Radler. That was especially true in the American south, where the company was often able to acquire newspapers despite not being the highest bidder. "One of the reasons their conservative owners let us buy them is that they felt more comfortable selling to us than someone else because we're so conservative," Radler told Peter C. Newman. "Our ideological reputation has been a real plus for us."[3] The newspapers they bought became money spinners under the management of Radler, known as the "human chain saw" for his cost-cutting prowess. Radler explained the sight-unseen method he used for reducing labour costs at each new publication Hollinger acquired.

> I visit the office of each prospective property at night and count the desks. . . . That tells me how many people work there. If the place has, say 42 desks, I know I can put that paper out with 30 people, and that means a dozen people will be leaving the payroll even though I haven't seen their faces yet.[4]

While American Publishing would rank as the second-largest newspaper chain in the US by number of titles, it was only ever 12th in total circulation. Most of its newspapers were small but highly profitable because they enjoyed a local monopoly. One major metropolitan daily Hollinger acquired was the faltering *Chicago Sun-Times* in late 1993. Hollinger paid US$180 million for the eighth-largest US daily, then proceeded to follow the same cost-cutting strategy it used at smaller newspapers. Radler took personal charge of the *Sun-Times* as publisher, setting aside $10 million for buyouts of highly-paid senior staff members. He soon doubled the paper's profitability to 15 percent after slashing a fifth of its workforce, but Radler had a higher profit goal — 25 percent. Many at the 500,000-circulation tabloid were unhappy with the cuts, and eight senior editors left within a year. Radler saw a corporate culture clash between *Sun-Times* staff and his executives, most of whom had been owners and publishers of the small-town newspapers Hollinger had acquired.

> Some of my American Publishing executives are a little rough around the edges, okay? In mean, they come from a background of entrepreneurship, okay? They don't have the style or the presence — presence isn't the word, there's probably too much presence — that these kind of people are used to.[5]

One newspaper Black and Radler coveted was the *Jerusalem Post*, Israel's oldest and largest-circulation English-language daily. While Black was raised a Protestant and converted to Catholicism as an adult, Radler took a particularly keen interest in Middle Eastern affairs due to his Jewish heritage. Eyebrows were raised in 1989 when Hollinger offered US$20 million for the faltering daily. The highest bid to that point had been US$8 million. Radler at first promised not to interfere with the traditionally-liberal daily, according to former publisher Erwin Frenkel. Soon, however, he took an active interest in its politics. After the purchase was complete, according to Frenkel, Radler cut costs in the *Post*'s newsroom. A time clock was even installed, on which journalists were required to punch in and out. Radler ousted Frenkel as publisher and appointed a friend of his who had been a colonel in the Israeli army, but who had no newspaper experience. "He wanted a newspaper that served and reflected the prevailing nationalist temper and did not criticize occupation or settlement of the West Bank or Gaza," observed Frenkel of his successor. "Like his bosses, he had little respect for journalists and what he considered their pretensions to know better."[6] The change in editorial direction prompted thirty-two *Post* journalists to threaten to resign. Radler gratefully accepted the cost savings brought by the uncompensated terminations. "It was convenient for me," he said, "because there were 32 too many people, if not more, in the editorial department at that time."[7] Radler eventually took the reins as publisher himself.

Aussie escapade

Not every international adventure Black and Radler embarked on ended in such complete victory. A foray into Australia resulted in a Senate inquiry after Black blew the whistle on his own political back room dealings Down Under. He quit the country as a result, taking a tidy profit with which to turn his attention back to his native Canada. Hollinger bought 15 percent of Australia's Fairfax newspaper chain, the country's second largest, in 1991. Due to the country's foreign ownership limits, however, the Canadian company was prevented from acquiring more. Black set about reducing the chain's workforce, which boosted profits and doubled its stock price. On the eve of a 1992 national election, he met with Prime Minister Paul Keating about raising the country's foreign media ownership limits. Keating, according to Black, urged him to apply to the Foreign Investment Review Board to raise his ownership to 25 percent, promising to "champion" the bid. "If he was re-elected and Fairfax political coverage was 'balanced,'" Black wrote in his 1993 autobiography *A Life in Progress*, "he would entertain an application to go higher." Black added that the leader of Australia's opposition had "already promised that if he was elected he would remove restraints on our ownership."[8] The book caused an uproar in Australia, as Black had indeed been allowed to increase his stake in Fairfax to 25 percent only weeks after Keating was re-elected. After Black's autobiography hit bookstores Down Under, a Senate inquiry was called.

Witnesses painted Black in unflattering terms during a proceeding that foreshadowed his 2007 courtroom comeuppance. Merchant banker Malcolm Turnbull, a legal adviser to Fairfax bondholders, called Black an "extraordinary egoist," which came as no news in Canada or England. "He [has] almost no regard for telling the truth," testified Turnbull. "Black consistently overstates his role in things. . . . I don't believe his word can be trusted on matters where his own involvement is concerned."[9] Former prime minister Bob Hawke concurred. "The simple fact is that Conrad Black does not tell the truth," Hawke said. "He has the habit of distorting events through the prism of his own perceived self-interest."[10] When Black took the witness stand, he testified that the entire affair had been a misunderstanding. "There was no nudge, there was no wink, there was no undertaking," he told the inquiry.

> We are not lapdogs of any regime. Mr Keating was certainly not using the word balance as a euphemism for support or favouritism . . . or as hostile to his enemies. . . . I do not know what else I can do to bury this putrid corpse, short of driving a silver stake through a copy of this committee's terms of reference.[11]

Black hit back at his detractors, calling Turnbull "notoriously unstable" and claiming Hawke had offered to spy on Keating for a fee of US$50,000.[12] Hawke went on television to deny the charge and questioned Black's "mental instability."[13] The Canadian's new wife chimed in that Hawke and her husband's other critics should be whipped. "Although I don't believe in capital punishment I do think that there may be something said for flogging," said *Sunday Times* columnist Barbara Amiel. "I don't think my husband could lie . . . He's relentless in his pursuit of what he wants but he's an honest man."[14] The inquiry issued its report in June of 1994, concluding there had indeed been back room influence peddling between Black and the prime minister. "The committee finds that Mr Keating attempted to exert pressure at Fairfax for favourable election coverage by making a linkage between 'balance' in election coverage and an increased ownership limit for Mr Black."[15] If a political disaster, Black's Australian excursion proved a financial windfall, as he later sold his shares for a profit of A$300 million.[16] It was against this backdrop, however, that Izzy Asper began his own battle with the Australian Broadcasting Authority a few months later.

Failed takeover bid

While he had proven adept on the international stage, Black had been largely shut out of the press in his native Canada. Hollinger's holdings there consisted of the small Sterling chain of newspapers published mostly in British Columbia. This had less to do with Black's abilities as a financier than with the concern of many in Canada about his neo-conservative politics. In 1980, Black had designs on Canada's second-largest newspaper chain, FP Publications, whose founding partners had passed away. It owned the

Globe and Mail, the *Montreal Star*, the *Ottawa Journal*, the *Winnipeg Free Press*, and the *Vancouver Sun*. Also interested in making an offer was Izzy Asper, a former FP columnist. He deferred to Black, however, from whom he had recently bought Crown Trust. "When someone I know and have a regard for, Conrad Black, is negotiating as he apparently is, I don't think it's appropriate to inject oneself between the negotiating parties," Asper said. "I don't want to be the agent whereby anybody's deal is interfered with. That's a matter of business ethics."[17] Black was outbid by Ken Thomson, then Canada's only billionaire, who had inherited the newspaper empire built by his father, including the *Times* of London.

Undaunted, Black set his sights on Canada's oldest and largest newspaper chain, Southam. The former family firm had become a takeover target because its shares were widely held. Four generations of Southams had sold off much of their inherited stock over the years. Family members had made a strategic error in 1945 when they took their newspaper chain "public" on the Toronto Stock Exchange. Instead of issuing two classes of stock and reserving voting shares for family members, they had issued only common stock. Four decades later, Southam family ownership had dropped to 30 percent, making it an attractive target in an era of leveraged buyouts.[18] After anonymous buyers began scooping up large blocks of Southam shares in the summer of 1985, trading became frantic, as did family members.

The death of Southam president Gordon Fisher at the height of the crisis complicated matters. His brother John, an engineer by trade, was pressed into service as his replacement. Elder statesman St. Clair Balfour, who had been company president in the 1960s, also stepped back into service. They scrambled to enact bylaws, known as "shark repellent," that would make the company impossible to take over. That brought a legal challenge from minority shareholders who had bought Southam stock in speculation of a takeover attempt. A more permanent solution was devised, or so they thought. In a "near merger," Southam agreed to trade 20 percent of its shares for 30 percent of the smaller Torstar Corp., which published the *Toronto Star*. The share swap made taking over Southam a practical impossibility for any other potential acquisitor. It included a "standstill period" of ten years, during which time neither side could increase its holdings or sell its shares in the other.

Thwarted again, Black sold his acquired Southam stock for a tidy profit and moved on to London, Jerusalem, Chicago, Australia, and just about everywhere except Canada. Other minority shareholders in Southam, however, were undeterred in their quest for windfall profits and launched a legal challenge to the share swap. In their haste, the Southams had neglected to give the required written notice to shareholders of the arrangement with Torstar. As the lawsuit approached its hearing date in 1988, an out-of-court settlement was reached that reduced the "standstill period" from ten years to five. That meant Southam would be "in play" as a takeover target again in 1990, not 1995.

Southam, take two

Despite his worldwide whirlwind of activity, Black was still keenly interested in acquiring Southam. More importantly, he was infinitely better positioned to do so by the 1990s. The profits from his growing newspaper empire were increasing all the time, as apparently was his business acumen. Black's conquest of Southam in the 1990s was a work of high-finance artistry. Through increasingly ingenious manoeuvring, Black used Southam's own cash reserves to help finance the company's takeover. He then used its credit rating, borrowing in the company's name to buy out other shareholders. First, he had to convince Torstar to sell him its 20-percent Southam stake once the standstill period ended. According to Siklos, Black made repeated offers to Torstar, finally convincing it in late 1992 to sell at $18.10 a share, or a 15-percent premium.[19] Again the Southams panicked. Desperate to keep Black from taking over their company, they again searched for a corporate manoeuvre to keep him at bay. Just as in 1985, however, they again suffered from a lack of leadership, as no family member had stepped forward from its fourth generation to help run the company. The high hopes held for Harvey Southam, a former *Vancouver Province* business reporter, were dashed in 1991 when he committed suicide at age 43. That forced the company to go outside the Southam family for its president for the first time. What the Southams really needed, however, was a white knight to ride in and rescue them from Black.

They found their man, or so they thought, in Montreal businessman Paul Desmarais. His Power Corp. held an estimated $27 billion in assets, including a chain of 41 newspapers in Quebec, the flagship of which was Montreal's *La Presse*. Approaching Desmarais to sound him out on Southam's traditional values of quality news-papering, directors at first found Desmarais sympathetic. Falling Southam share prices had created a problem for the company with its bankers due to an increased debt-to-equity ratio. Raising cash by issuing shares from its treasury to Desmarais would solve that problem, dilute Black's ownership, and create a shareholder with equal power. When Black learned of Southam's plan to sell Desmarais stock at $13.50 a share, he protested that the price was too low and lobbied directors to vote the deal down. According to Siklos, this backroom dealing actually sowed the seed of Southam's eventual demise and allowed Black to finally take the company over.

Black and Desmarais owned neighbouring vacation homes in Palm Beach, Florida, noted Siklos. The two men "shared a fascination with Southam and had discussed their respective ambitions to own it over the years."[20] It was in Palm Beach that Black and Desmarais made a deal. They agreed to give each other the right of first refusal, should either decide to sell his shares. While between them they would own less than a majority of Southam shares, their combined stakes gave them effective control of the company. With just over 40 percent of Southam stock, Black told reporters in March of 1993, "if you can't control a company you should join a monastery or something."[21]

The obdurate rump

Black and Desmarais, however, found that Southam was not easily controlled. While they were entitled by their large share holdings to several seats on its board of directors, they were hardly insiders. In fact, Black and Desmarais found themselves treated as outsiders by Southam. They were denied access to some details of the company's finances on the basis that they were industry competitors. Black and Desmarais grew increasingly frustrated at the slow pace of change at Southam. Money-losing divisions of the company had already been sold off and hundreds of employees dropped from the payroll through layoffs and buyouts.[22] That wasn't enough for Black and Desmarais. One of Black's appointees to the Southam board, Hollinger's Jack Boultbee, saw a fundamental difference in corporate culture. "Southam's philosophy was that they were in the business of delivering news," he said. "We're in the business of selling ads."[23] Finally, out of frustration, Black and Desmarais agreed to break up the historic family firm. Black would take ten of its smaller dailies in exchange for his 20-percent ownership. That move was blocked, however, by five of Southam's independent directors, who had served a combined 81 years on its board.

Never a big fan of corporate governance, which limited his board room manoeuvring, Black excoriated the independent directors as an "obdurate rump." According to Siklos, the ugly episode caused Desmarais to agree in frustration to sell his shares to Black in May of 1996.[24] That gave Black 41 percent of Southam. Black quickly convened a special meeting of Southam shareholders and used his voting power to oust the independent directors and install his own replacements. He then used his control of the board to dismiss Southam's CEO and name himself company head. That was done without even bothering to convene a meeting of the company's reconstituted board of directors. Black's disregard for corporate protocol rankled many. Several of the deposed Southam board members ranked among Canada's leading executives. "Radler and Conrad Black don't believe in a board," said Ronald Cliff of BC Gas. "They don't believe in corporate governance."[25] The assessment would prove prophetic. The *Globe and Mail* headlined the brouhaha: "The pain of the obdurate rump."[26]

Strategic dividends

Black steadily completed his takeover of Southam over the next few years, increasing his ownership to 97 percent. That allowed him to take the company "private" again by having it de-listed from the TSE. Black first bid to gain majority control of Southam, offering $18.75 each for all the shares he could buy. When that offer proved insufficient, he increased it to $20. The resulting acquisition of 8.5 million shares in November 1996 gave Black 50.7 percent of the company, allowing him to make some moves uncontested. He first used his majority control in April 1997 to distribute the firm's accumulated cash reserves in a $2.50 per share "special dividend."[27] This enriched Black most of all, by $47 million. It helped Hollinger make a surprise $923-million bid to buy out Southam's remaining shareholders one week later. To gain control of Southam Black had to go through a man once described by *Fortune* magazine as "The Scariest SOB on Wall Street." The bid Black made was a complex one and consisted of $13.50 a share plus one $10 non-voting share in

Hollinger.[28] But Michael Price, manager of the influential Franklin mutual fund that held 20 percent of outstanding Southam shares, balked at the offer. Price demanded $25 a share in cash, but Black refused to increase his bid. When the offer expired in June, only 15.6 percent of Southam's minority shareholders had accepted it. That gave Hollinger 58.6 percent ownership of Southam, far less than the level required to trigger a forced sale by the remainder of shareholders.[29]

Black's majority control of Southam came on the eve of Hollinger's 1996 annual meeting. The timing allowed Black not only to consolidate his power, but also to gloat about it publicly. On May 29, five days after buying Desmarais out, Black strode to the podium in the art deco conference room in the old Stock Exchange building in Toronto. From there he publicly scolded management of the conquered company. According to Black, Southam directors had "long accepted inadequate returns for the shareholders, published generally undistinguished products for the readers and received exaggerated laudations from the working press for the resulting lack of financial and editorial rigour." Black admonished the former first family of Canadian newspapers for panicking in their 1985 share swap with Torstar, pointing out that the move ultimately backfired. "If Southam's management had been a little more courageous," he crowed to stunned silence, "it might still be a family-controlled company."[30]

Taking Southam private

Suddenly, the worst fears of many Canadians were coming true. Increased concentration of ownership of Canada's influential daily press had worried many since the 1960s. The need for limits on ownership had been debated for much of the 1970s and 1980s. The hands-off management style of Southam and Thomson, however, lessened the urgency of regulation. The Southams, unfortunately, had been replaced by someone who did not hesitate to use his newspaper power and influence. With Black's takeover of Southam, newspaper ownership concentration was again a hot topic across Canada after more than a decade. Instead of faceless corporations, however, this time control of the country's press had a face to it — Conrad Black's.

Black was an open enemy of Canada's federal Liberals, who had seized power from the Conservatives in 1993. That election had reduced the bastion of Canada's political right to ashes, leaving them with only two seats in Parliament. As a result, the Tories had fractured geographically, with a new Reform party founded in western Canada under the leadership of Alberta's Preston Manning. Black wanted nothing more than to see conservatives of all stripes united in opposition to the Liberals. He saw the country's national newspaper, the *Globe and Mail*, as supporting traditional Canadian liberal values. As a result, Black moved to start his own national newspaper that would promote conservatism. He announced in April of 1998 that the new daily would begin publishing that fall, utilizing Southam's nationwide printing and newsgathering resources. In May, Black added two more dailies to Southam's stable, buying the *Victoria Times Colonist* and *Nanaimo Daily News* from Thomson. That gave Black 61 of Canada's 105 daily newspapers, including 14 of the 16 published in British Columbia.[31] He also traded four mid-sized Ontario dailies to the *Toronto Sun* chain that month for $150 million and its 80-percent ownership of the *Financial Post*. Black coveted the business daily as a base on which to assemble his new national newspaper.[32] To some, it seemed like a life-sized board game of Monopoly played with newspapers.

The National Post

As the October launch of his *National Post* loomed, Black moved to clear the last hurdle to complete ownership of Southam. He finally acquired the Franklin mutual fund's 8 million shares for $31.68 each, paying a premium of 22 percent above market price. That raised Black's ownership of Southam to 69.2 percent and set the stage for another bid for the remainder of Southam shares in December.[33] This time it was made with even more creative financing than before. First, Black used his majority control of Southam to declare a special dividend of $7 a share, which was financed by borrowing $532 million. Black then bid $22 a share for the remaining Southam stock. The total $484 million offer cost Hollinger only $106 million after it received $378 million from the special dividend.[34] The offer was rejected by some holdout members of the Southam board, but when it was increased to $25.25 early in 1999, they voted to recommend it.[35] When the offer expired two weeks later, more than 90 percent of the 22 million remaining Southam shares had been tendered. That raised Black's ownership of the company to 97 percent.[36] Under Ontario securities law, it paved the way for him to squeeze out the shareholders who refused to sell, de-list the company, and fold it into Hollinger.

Black's new *National Post* exceeded expectations for circulation, quickly soaring to sales of 272,000 daily and higher on the weekend. Critics pointed to the large number of heavily discounted sales that inflated those figures. More significantly, advertising lagged below projections. *Post* editions often included only 20 percent advertising, compared to the industry standard of 30-40 percent.[37] An all-out "newspaper war" resulted in Toronto, where Black hoped to establish a beachhead in a market dominated by the *Star.* Canada's largest daily had a circulation of 458,000 on weekdays and more than 700,000 on Saturdays. The *Globe and Mail* circulated 330,000 copies nationally from its Toronto base, where it also published a Metro edition with local news.[38] The downscale end of the market was dominated by the tabloid *Sun,* which sold 240,000 copies daily and more than 400,000 on Sundays. The *Star* and *Globe and Mail* both bolstered their news coverage to prepare for the increased competition. The dowdy *Globe* even started colour printing. The *Post*'s operating loss of $44 million in its first year proved a financial drain on Hollinger, whose share price fell almost 20 percent. In a bid to ease the company's $2.4 billion in debt, Black announced he would sell up to half of his accumulated Canadian newspaper empire. He offered the smaller publications for sale while seeking a broadcasting partner with whom to move into the new world of "convergence." In response, Hollinger share prices jumped 26 percent.[39]

The House of Lords

Black's decision to sell resulted in part from a dispute with Liberal prime minister Jean Chrétien that started in 1999. The *National Post* was a harsh critic of Chrétien and championed a right-wing alliance in opposition to the Liberals. When British Prime Minister Tony Blair offered the *Daily Telegraph* owner a seat in the House of Lords, Chrétien blocked Black's appointment. He cited an obscure 80-year-old rule prohibiting Canadians from accepting foreign titles.[40] Black, a dual Canadian and British citizen, sued Chrétien for "abuse of process." He claimed $25,000 in damages for "public embarrassment," but the legal action was thrown out of court in March of 2000.[41] Black could only take his seat in the House of Lords by renouncing his Canadian citizenship. That would make him a foreign owner of the press in his native country. Under Canadian tax laws, advertisers would no longer be allowed to deduct from their income the expense of buying space on his pages. Black wanted desperately to take his place among the wigged lords and ladies of Britain's Upper House. It would mean giving up the newspaper empire in Canada he had spent years pursuing.

One of the first calls Black got from a potential buyer came from Izzy Asper. He found Black one of the few people whose word he trusted on a handshake, while Asper had ended up in a lawsuit with almost everyone else. Asper and Black had done business once before, when Black was piecing off the giant Argus Corp. holding company he had taken over in 1978. Asper had been on a buying spree since forming CanWest. Black had already sold the Dominion Stores supermarkets and Massey-Ferguson farm implements division of Argus. He was shopping Crown Trust and sold it to Asper on a handshake for $17.7 million. According to Gordon Pitts, Asper considered Black as a result the "perfect partner . . . that rare person who could be trusted."[42]

Black and Asper also shared a thread of common heritage, as Black's mother had grown up on the same Winnipeg street where the Aspers lived. They also had a Palm Beach connection, as Asper's vacation home there had once been owned by Black. More importantly, according to Pitts, both "saw themselves as outsiders when it came to the dominant soft-left, soft-nationalist elite."[43] They both loved newspapers, and each had felt the influence that comes with shaping what goes onto their pages. Asper had long coveted his hometown *Winnipeg Free Press*, and was in discussions with Thomson to buy it when he instead moved to dealing with Black. Despite building CanWest Global into a third national network over a quarter-century, Asper was never a broadcaster at heart. "His great passion was newspapers." noted Pitts. "He only got into television because he was continually blocked from owning print."[44]

Black had publicly offered only his smaller newspapers for sale, announcing that he intended to keep the more influential larger dailies. The media world had already been rocked in January of 2000 by the blockbuster AOL-Time Warner deal, however. The largest merger in history sent other media companies scrambling to find a multimedia dance partner. The convergence wave hit Canada the next month when telecom giant Bell Canada Enterprises bid $2.3-billion for CTV. The converged entity would partner with Thomson and his *Globe and Mail* to form Bell Globemedia. In May, Asper and his son Leonard, to whom he had recently passed the torch as CEO of CanWest, met with Black and other Hollinger executives. It soon became apparent to the Aspers that Black was prepared to part with the entire Southam chain.

The Bilderberg Group

Black was an active member of the Bilderberg Group, a secretive trans-Atlantic society thought by some to actually run the world as a kind of private government. Its annual meetings of industrialists and politicians began in 1954 and were held at five-star resorts in Europe and North America. The invitation-only gatherings were conducted under tight security and participants were sworn not to reveal what transpired. The Bilderberg Group would only issue an annual press release listing the three-day meeting's venue, participants, and topics of discussion. That fuelled the paranoia of conspiracy theorists. Some of the wildest Bilderberg fantasies claimed members were actually 12-foot-long shape-shifting lizard aliens, and had been behind the 1963 assassination of US president John F. Kennedy. The truth was more mundane, but intriguing nonetheless.

According to intelligence historians, the first of the annual Bilderberg meetings was secretly funded by the CIA.[45] It was held at the de Bilderberg Hotel in Oosterbeek, Holland, from which the group took its name. Subsequent meetings were bankrolled by private groups, the first of which was the Ford Foundation.[46] The CIA's interest in Europe at the height of the Cold War was in neutralizing the political left, most notably the Labour Party in Britain. It also wished to promote a union of European countries into one economic entity organized along capitalist lines. By the mid-1950s, according to one scholar, the CIA was pouring about $1 million a year into covert operations in Europe.[47] Over the years, the Bilderbergers have been credited with defusing the Suez crisis and righting the world economy following the OPEC oil embargo. They also allegedly had a hand in creating the World Bank and an alphabet soup of other global organizations: the International Atomic

Energy Agency (IAEA), the International Monetary Fund (IMF), the Organization for Economic Cooperation and Development (OECD), and even the North American Treaty Organization (NATO).[48] In the post-Cold War era, many assumed globalization topped the Bilderberg agenda, attracting protesters to its annual meetings.

The Bilderberg mission to foster Euro-American cooperation had no more ardent adherent than Black, who was a military historian and political junkie. He loved nothing more than to debate world affairs with the political elite who attended the Bilderberg meetings. He recruited several of them, including former US Secretary of State Henry Kissinger, to serve as directors of Hollinger. In 1996, just after his takeover of Southam, Black co-hosted the annual Bilderberg meetings at a $60-million resort outside Toronto. As limousines pulled up to the former King City Ranch beauty and fitness spa, protesters were kept well back by security. "Our fear is that they are internationalists and free traders making plans," said Paul Fromm, a teacher from nearby Mississauga.[49] Reporters also attended at the gates, but as usual they were denied entry. "It eliminates the fear of being misquoted," conference spokesman James Hyslop explained to a Canadian Press reporter. Asked how it felt to act as the mouthpiece for a conspiratorial New World Order, Hyslop laughed. "God, I wish it was nearly that glamorous. I've read all of the conspiracy theories. It's truly comical."[50]

A Belgian buyout

As Black and Asper were negotiating the sale of Southam, the annual Bilderberg meetings were set for the luxurious Chateau du Lac Hotel just outside Brussels. Black added Asper to the guest list. Also there were Kissinger and Richard Perle, a former assistant US secretary of defense who headed Hollinger's online arm. So was *National Post* columnist David Frum, who would soon leave to work as a speech writer for US President George W. Bush. Asper, who was vacationing in Israel, flew to Brussels. Late at night, after hours at the Bilderberg meetings, he and Black put the finishing touches on the deal to pass the Southam chain to CanWest. To acquire such a newspaper empire in one move was almost too good to be true. Building a similar television network had taken Asper a quarter of a century. Southam would command a steep price, however — $3.5 billion.[51] The total included $2.2 billion in cash, $700 million in debt, and

$600 million worth of stock, which would give Hollinger 15 percent ownership of CanWest. In return, CanWest Global became the first major television network in the world to own a large national newspaper chain. It included a dozen major dailies, 126 smaller newspapers, 85 other publications (mostly trade magazines), and even half-ownership of Black's *National Post*.

It was a deal that would not have been legal in Canada in the early 1980s, when cross-media ownership was prohibited, as it was still in many countries. The sheer magnitude of CanWest's convergence move stunned many in Canada. They began questioning anew the wisdom of allowing such a monolithic force to dominate the media landscape.

CHAPTER 5
Media Power

The problem of media power troubled scholars for most of the 20th Century, during which conventional wisdom on the subject reversed itself more than once. In the space of a few decades, consensus swung from the media having a very powerful effect on society, to a very limited effect, and almost all the way back again. The demonstrated power of propaganda during World War I convinced early sociologists that mass media were powerful indeed. A propaganda blitz quickly changed American public opinion from staunch isolationism to enthusiastic support for entering the war against Germany. Each new medium that emerged during a century of new media seemed to be more powerful that the last. Film revealed the power of moving pictures to stir strong emotions, as 1915's *Birth of a Nation* inflamed simmering racial tensions across the US. Riots in Boston and Philadelphia on its release there led to the film being banned in some cities. Radio revolutionized politics in the 1930s, allowing demagogues like Hitler and Mussolini to rise to power in Europe by enthralling their masses with a rhetoric of hate. US politics were altered radically when president Franklin D. Roosevelt took his policies of government intervention in the economy directly to the people in radio "fireside chats." His New Deal prescription to counteract the Depression thus gained popular support despite being opposed in the press, which was increasingly controlled by powerful and conservative owners.

When psychologists tried to actually measure media power by using the emerging statistical methods of social science, however, they could detect only very limited effects. By taking surveys and conducting experiments, they had hoped to reach inside the public mind and discover immutable laws of communication that would prove the magic keys to persuasion. The US Army took a particular interest during World War II in mastering the secrets of propaganda. It experimented on recruits by showing them films depicting the Axis nations as evil, but found little effect. A series of studies on voting behaviour in the 1940s similarly showed a very limited influence of mass media. Most people had made up their minds well in advance how they would vote and most did not even get their political information from the media, but instead from other people. The social scientific studies of the 1940s contradicted the assumptions of Powerful Effects theory and reversed the conventional wisdom on media effects. As a result, a Limited Effects paradigm ruled for decades in the study of mass communication. This was particularly convenient for the growing US broadcasting networks, as it enabled them to avoid strict government regulation of the newest — and most powerful — medium of all.

To regulate or not

Television was a German invention that had been experimented with by the Nazis for its propaganda potential in the 1930s. Cable hookups to party members' homes even televised same-day coverage of the 1936 Berlin Olympics. World War II interrupted the development of television, but as the 1940s ended the CBS and NBC radio networks in the US began transmitting moving pictures. In other countries, however, caution was uged in implementing such a pervasive communication tool. Some hoped television's informational and educational potential would be developed in addition to its entertainment and advertising capabilities. In Britain, for example, all broadcasting was done by the government through the BBC for decades to ensure quality programming. In Canada, a mixed system of public and private broadcasting was introduced, with the CBC and CTV networks.

By the 1960s, the Limited Effects paradigm was firmly ensconced in the US as the conventional wisdom on media power, but some scholars were skeptical. In Europe, where the power of propaganda had been felt first hand, researchers remained unconvinced. Rather than trying to measure attitude change using statistical methods, European scholars examined the social, cultural, and political consequences of communication. Their use of qualitative methods, more akin to literary analysis, allowed them to decode media messages and uncover the ideology implicit within them. Another popular approach to studying the media in Europe was political economy, which united a discipline that had been separated in the late 19th Century into the fields of economics and political science. The split had entirely removed the moral aspect from economics and most of the economic aspect from the study of politics. Putting the two back together provided some scholars with a better lens through which to view the exercise of political power by economic interests through the control of communication.

A political economy approach to studying communication was tenuously founded in the US after World War II by Dallas Smythe. A Canadian who had been chief economist for the FCC, Smythe taught the first US course in media economics at the University of Illinois in 1948. Such a critical approach gained scant popularity among Americans during the Cold War, however. The political economy approach stood in sharp contrast to the more popular framework provided by Harold Lasswell, a political scientist at the University of Chicago. Lasswell had devised the template for much of media effects research with his simple verbal model that posed the functional question: "Who says what to whom through what channel, with what effect?" The preoccupation of most social scientists had been with the final part of that equation. Smythe restated Lasswell's formula for political economy purposes as: "Who gets what... when, how and where?"[1]

Returning to the relative safety of Canada following the Cuban Missile Crisis of 1963, Smythe founded the study of communication at Simon Fraser University outside Vancouver. He was critical of most American research on communication, which he dismissed as "administrative" for mostly filling the function of marketing research. Smythe instead preferred "critical" research that examined the power structures implicit in mass media. Under his influence, the study of communication at Simon Fraser became some of the most critical in Canada. Smythe also tugged at the cloak of "objectivity" that gave scholarly research the appearance of being value-free. "All of us have our predispositions, either to criticize and try to change the existing political-economic order, or to defend and strengthen it," he noted. "The frequent pretense of scientific 'neutrality' on this score is a delusion."[2]

McLuhan was the messenger

While Smythe's follow-the-money approach to studying media never caught on in the US, the ideas of another Canadian did revolutionize thinking on communication in the 1960s. As a professor of Elizabethan literature at the University of Toronto, Marshall McLuhan was an unlikely candidate for oracle of the Internet age. McLuhan's interest in media had been sparked by U of T economist Harold Innis, whose research had shown that throughout history, control of a society's dominant communication medium was the key to gaining economic and political power. Innis saw that "monopolies of knowledge" inevitably arose around any medium of communication. The phenomenon repeated itself from the high priests of Egypt, who held even more power than the pharaohs through their ability to decipher complex hieroglyphics, to the growing 20th century newspaper chains. These monopolies, Innis concluded, could only be broken by new media, around which would then inevitably grow a new monopoly. Innis died from cancer at age 56, after which McLuhan dedicated himself to continuing his research.

McLuhan came to see that the dominant medium of communication — whether oral, written, or printed — had also proven central through history to the very form of social organization. Even more important than the information conveyed through communication, however, was its means of communication. "The medium IS the message," McLuhan hyperbolized in his classic 1964 book, *Understanding Media*. Tribes communicated orally around the camp fire, relying on trusted elders whose knowledge had been passed down from their ancestors. The invention of writing changed everything, McLuhan realized. It meant that for the first time the wisdom of the ages could be recorded, disseminated, and preserved for future generations. The mass reproduction of knowledge on the printing press, he saw, led to larger social groups. Printed literatures that arose across Europe in various forms of Latin became entire new languages — French, English, Spanish, German — and led to the rise of nations held together by a common tongue. But the linear logic of the printed line also changed the nature of humankind, according to McLuhan. Separating the eye from the ear, he theorized, separated thought from feeling and alienated the reader from the tribal group, leading to a fragmentation of individual and society.

Broadcasting changed everything back again, according to McLuhan. Radio was the tribal drum that brought nations back to a common camp fire, injecting emotion to communication once again. Television not only reunited the eye with the ear in the most powerful medium of all, noted McLuhan, but it also communicated with a different kind of logic than print. Its powerful combination of the spoken word and vivid imagery evoked emotion more than it appealed to reason. As McLuhan watched the first Sputniks being launched into orbit, he realized that satellite communication would soon unite all of humankind — for better or for worse — in one big "global village." His focus on the medium instead of the message made McLuhan a media messiah in the 1960s. It also started the trend line of conventional wisdom on the perceived power of media effects heading back upwards again.[3] The more that was learned about the subtle and insidious effects of mass media, the higher the curve climbed until it was almost as high as the Powerful Effects theorists had intuitively assumed. From a narrow, microscopic focus on measuring the individual, psychological effects of media messages, McLuhan's way of thinking about media had broadened the study of communication. It began to examine the influence of media at the level of culture.

More philosophy than science, the Medium Theory he inspired relied not on provable scientific results elicited through quantitative methods. Instead it used historical analysis to view media effects. Also known as "technological determinism," it was developed after McLuhan's death in 1980 mostly by Americans, notably Neil Postman and Joshua Meyrowitz. "For the most radical and elaborate American media theory," wrote one US communication scholar in 1982, "one must look to the work of two Canadians." The insights of Innis and McLuhan, according to Daniel Czitrom, revealed the media to be nothing less than "the prime mover behind historical process, social organizations, and changing sensory awareness."[4] The 1981 Royal Commission on Newspapers report observed that the so-called Toronto school "altered mankind's appreciation of the influence of media." Canadians, it noted, were thus well prepared to understand new media through the "solid foundation of theoretical studies" Innis and McLuhan left.[5]

Cultivation theory

Even in the US, some questioned the Limited Effects model of media power as social unrest grew in the 1960s. Violence in the streets shown on the nightly news alarmed many Americans, and some blamed television itself for the burning inner cities and race riots. A National Commission on Causes and Prevention of Violence set up in 1968 inventoried the mayhem on prime-time television. Thus began one of the longest-running studies in mass communication. It quickly moved from government funding to non-profit sponsorship by groups such as the National Institute for Mental Health. An annual inventory of television violence was conducted for decades by George Gerbner, a Hungarian immigrant who fled the Nazis to study journalism in the US. He and his colleagues at the University of Pennsylvania found the level of violence in American media alarmingly high. Breaking down TV violence demographically, Gerbner et al. noted that women and minorities served disproportionately as victims. Television's best customers — white males — were invariably portrayed as holding positions of power. By counting televised acts of violence and tracking viewer habits, they came to the disturbing conclusion that by the age of 12, the average American child had witnessed 8,000 televised murders.[6]

After a decade of gathering data on media violence, Gerbner added another dimension to his research, surveying television viewers for their perceptions of the world. The more people watched TV, he found, the less their worldview conformed to reality. Heavy viewers — defined as those who watched four hours or more daily — tended to give "TV answers" to questions. They tended to over-estimate the crime rate and the percentage of the population employed in law enforcement. Far from making viewers more prone to committing violence, Gerbner found that watching more television instead made people more fearful of violence. Light viewers estimated their odds of being a victim of violence in a week at about 1 percent, while heavy viewers estimated closer to 10 percent. The actual probability was more like .001 percent. Gerbner called this effect the "Mean World Syndrome" and realized it had profound political implications. More fearful heavy viewers of television were more likely to accept repressive political measures to fight crime, even if they meant violating basic civil rights. Violence became seen by many

TV-happy Americans as a solution to all problems, with dangerous global geopolitical consequences. Even more troubling, Gerbner found a convergence of political views that depended more on
the amount of television watched than on differences in age, class,
race, education, etc. This "mainstreaming" of politics cut across
traditional demographic lines and thus placed television foremost
in political influence.[7]

Gerbner then added a third dimension to his research with "insti-
tutional analysis," which applied the principles of political economy to examining the origins of media messages. Most came from
large corporations seeking to market their content not just in the
US but also abroad. Unlike comedy, violence translated well into
other languages and cultures, making it a cheap industrial ingredi-
ent to be packaged and sold to a global audience. More importantly,
cultural and political values were easily transmitted to unguarded
audiences, according to Gerbner. They were not perceived as propaganda but instead as harmless entertainment. Television content was nonetheless highly ideological, according to Gerbner, and
thus was an extremely effective form of indoctrination. Systematic control of culture through media, he concluded, may be even
more effective than the use of overt propaganda because consumers internalized the transmitted values uncritically. Offering an
"ice age" analogy, Gerbner argued that while the individual effects
of media content may be very small, they all tend to move glacier-
like in one direction and thus add up over time. His discovery was
called "cultivation" theory. Unlike most studies of propaganda, it
examined the long-term, indirect effects of media content at the
level of culture and perception instead of short-term direct changes
in individual attitudes and beliefs.

Agenda setting

By combining social science methods like content analysis and survey research, cultivation theory had uncovered significant media
effects. Other researchers also combined research methods, often
with spectacular results. This "triangulation" approach allowed
scholars to get a better fix on media effects than they could by using
just one method of measurement. University of North Carolina
researchers Maxwell McCombs and Donald Shaw, for example,
decided to test an anecdotal observation. Issues displayed promi-

nently in the news media, journalists believed, were perceived by voters as being the most important politically. Issues that were buried on the back pages of the newspaper or as brief items on newscasts — or not covered at all — were perceived as unimportant. Thus, media content could "set the agenda" for public debate, or so the argument went.

Combining content analysis and survey research, McCombs and Shaw tested this hypothesis by comparing media coverage of election issues with voter perceptions of their importance. During the 1968 presidential election, they surveyed undecided voters in Chapel Hill, NC for their opinions on what were the most important election issues. Not surprisingly, McCombs and Shaw found a high level of correlation between public perceptions of issue importance and the prominence of news media coverage. Their research design failed to prove cause and effect, however, as a chicken-or-egg question remained. Did voters perceive issues as important because they were covered prominently in the media, or did the prominent media coverage result from public interest in those issues? Four years later, McCombs and Shaw fine-tuned their study. In an attempt to prove causality, they surveyed undecided voters — this time in Charlotte, NC — at several points during the 1972 election campaign. That way, they were able to determine that public opinion did indeed follow media coverage, at least as far as issue importance went. Thus they were able to articulate one of the simplest yet most profound findings of communication research. "In short, the mass media may not be successful in telling us what to think," McCombs and Shaw concluded, "but they are stunningly successful in telling us what to think about."[8]

This opened up one of the most fruitful areas of research in political communication. Hundreds of agenda-setting studies were conducted into the news media worldwide. A second level of agenda-setting research looked into not only what issues gained prominent coverage, but what aspects of the issues gained media attention. By adding these "attributes" of issue coverage, researchers moved into the more subjective area of frame analysis, which had been explored by cultural theorists for years. The words used to describe news figures, what was focused on, and what was left out of the frame all proved crucial to perceptions. That added another important discovery. Not only can the media influence what issues we think about, they can also influence how we think about them.

A turn toward experimental research then led cognitive psychologists to examine how symbolism in media texts could evoke emotional responses that influenced how issues were perceived. These studies in "priming" showed how media content could help determine, for example, the criteria by which political candidates are judged.[9] As a result of these findings, political campaigning grew ever more sophisticated.

The Fox effect

One development that demonstrated the power of news media to influence public perceptions was the founding by Rupert Murdoch of the Fox News Network in 1996. By both influencing the US news agenda and helping to frame political issues, Fox News proved a potent media force within a few years. The political partisanship it openly exhibited reminded some of the "party press," when journalists openly supported political positions. Despite being available in millions fewer homes, Fox soon surpassed CNN in popularity as its tough-talking commentary resonated with many Americans in the wake of 9/11.[10] Mainstream news coverage thus shifted several degrees to the right politically in the run-up to the US invasions of Afghanistan in 2002 and Iraq in 2003. Other networks attempted to compete with the new-found popularity of conservative television news, and the imitative right-slanting became known as the "Fox effect."[11] The moniker echoed the previously-noted "CNN effect," which was the influence on foreign policy of real-time satellite news coverage.[12]

A 2003 study by researchers at the University of Maryland found that viewers of Fox News had perceptions of the US invasion of Iraq that did not conform to reality. It found that two thirds of those who got their news from Fox believed the US had uncovered links between Saddam Hussein and al Qaeda. Only one in six viewers of PBS believed that. Similar misperceptions were found on whether weapons of mass destruction had been discovered in Iraq, and whether world opinion supported the invasion.[13] The study's political implications were enormous, noted Harold Meyerson in the *Washington Post*, as voters with those misperceptions were sure to cast their ballots to re-elect George W. Bush. "By this standard — moving votes into Bush's column and keeping them there — Fox has to be judged a stunning success," noted Meyerson.[14]

While troubling, the research suffered from a fatal flaw. The Fox audience could be largely self-selecting, composed of more conservative viewers already prone to the types of misconceptions found. Thus there was no smoking gun of cause-and-effect proven, but simply a correlation. A subsequent study, however, remedied this shortcoming through sheer serendipity. Because Fox News was not carried by all cable systems, researchers were able to compare voting results where it was available and where it was not. Using voting data for 9,256 US towns, they investigated whether Republicans gained vote share in places where Fox News was available during the 2000 election. The study found a significant effect of the presence of Fox News, estimating that it convinced 3-8 percent of its viewers to vote Republican.[15] As Bush won the White House by a narrow margin in 2000, the implications for US and world politics were enormous. Fox News became even more widely viewed by Americans following Bush's declaration of war on terrorism following the 9/11 attacks. The political impact of its well-documented distortions of reality can be presumed to have increased.

In early 2001, Fox News journalists reportedly began to receive daily exhortations from network management on how to spin the news.[16] According to former Fox producer Charles Reina, a memo would arrive by e-mail setting out the direction management wanted the day's news coverage to take. "To the newsroom personnel responsible for the network's daytime programming, The Memo is the bible," claimed Reina, who was hired in 1997 as the network's first media critic. "Intentionally or not, [it] has ensured that the administration's point of view consistently comes across on FNC."[17] Subtle hints included in the daily memos often shaped the newscasts and helped Fox News producers decide which stories should make it to air. "Virtually no one of authority in the newsroom makes a move unmeasured against management's politics, actual or perceived," claimed Reina, who resigned from Fox News in 2003.[18]

The Black agenda

The *National Post* that Conrad Black founded in 1998 could be considered analogous to Fox News. It was a media outlet launched in an attempt to influence a nation's politics with its right-wing perspective. The *Post*'s initial front page declared the need to "Unite the Right," by using the phrase no fewer then three times, including in its banner headline. As a result, noted Chris Cobb in *Ego and Ink*, the words "entered the nation's political lexicon the following day."[19] The fracturing of the Conservative party in the early 1990s had left no coherent opposition to the federal Liberals, whose policies Black opposed. The *National Post*, admitted one of its first editors, was conceived by Black for a specific ideological purpose. "He had very clear views on what kind of operations he wanted to run in his commercial and ideological interests," Gordon Fisher told a conference at Montreal's McGill University in 2003.

> The *National Post* from its launch reflected the clear direction of its proprietor. He intended to change the national debate, and he spent many millions of dollars doing just that. A very small group of us worked with Mr. Black for months prior to launch, and we dealt with little ambiguity in what was being undertaken. With his money.[20]

Soon, with the help of Black's *Post*, the Reform movement spread into Ontario, morphing into the Alliance party to become a national political force. The newspaper took unprecedented steps in assisting the efforts of the upstart Alliance. The newspaper's subscription list was passed to Tom Long, a candidate for the new party's leadership in 2000. According to Cobb, this "took newspaper support for a political candidate to a new, unheard of extreme."[21] With its blatant political favouritism, the *Post* became "a party organ in the old way," according to University of Calgary communication professor David Taras. "It incited the whole journalistic community and changed the temper and the climate," he noted. "It became okay to be crusading and passionate and hold strong opinions."[22]

One of the strongest opinions expressed regularly in the *National Post* was that the ruling federal Liberal party simply had to go. Prime minister Jean Chrétien became a regular target of the *Post* for his personal dealings in a golf course in his riding of Shawinigan. Through sheer persistence, the *Post* shone a national spotlight on Chrétien's sale of shares in the Grand-Mère Golf Club. Investigative reporting by the newspaper also raised questions about whether Chrétien exercised undue influence on federal officials. They had approved loans for Chrétien's former partner, to whom he sold his shares, and who also perhaps owed him money. *National Post* reporter Andrew McIntosh won an award in 2000 for his investigative reporting on the story, but according to Cobb even Black admitted the controversy "wasn't much of a scandal."[23] In the tradition of controversies surrounding high-level politicians, at least since Watergate brought down US president Richard Nixon in 1974, the story was dubbed Shawinigate, and in French became known as L'Affaire Grand-Mère.

Post predilections

Fiscally, the *National Post* was as staunchly conservative as its founder, who was a fan of the tax-cutting policies of US president Ronald Reagan in the 1980s. The *Post* regularly ran stories lamenting government waste and misspent tax dollars. One study, however, showed the *Post* often twisted facts and tortured logic "to push forward an ideological agenda . . . that had massive tax cuts as its top priority."[24] Sociologist Larry Patriquin's analysis of the *Post*'s coverage of tax issues found it ran 4½ times more front-page articles on the subject than the *Globe and Mail*. The *Post*, moreover, cast taxes in a negative light through the use of clever rhetorical devices. "*Post* consistently made weak arguments for tax cuts," noted Patriquin, "through the use of misleading data, unwarranted assumptions, and so on."[25] The result was to "invent" — through the use of colourful adjectives — an image of Canadians revolting against over-taxation.

> In its commentaries on taxation, the *Post* informs us that "festering" anger is developing, a version of "tax rage." This is a result of the fact that taxes are now "crippling," "punishing," "punitive," "stifling," and "crushing" — in short, a burden. Canadians are "taxed-to-the-teeth," "taxed-to-the-eyebrows," and "taxed to death," by "greedy governments."[26]

As a result, Patriquin concluded that *National Post* coverage of taxation issues seemed to fit "the classic definition of propaganda." It was, he added, "a discourse in which incomplete information is presented to people with the purpose of distorting their view of whatever is being discussed."[27] While failing in its mission to topple the Liberals in the 2000 election, Patriquin noted that the *Post*'s misleading coverage of fiscal issues did have an effect. It led the Chrétien government to adopt tax cuts as part of its election platform as a defensive measure against the Alliance. "The federal Liberal government did not collapse under the weight of the *National Post*'s immutable logic," claimed Patriquin. "Rather, it was pummelled to the ground by the media equivalent of the schoolyard bully. This is a sad commentary on the low level to which public discourse in Canada has sunk."[28]

Social scientists also objected to what they saw as the *Post*'s blatant misuse of public opinion polls for political purposes. While changes to the Canada Elections Act in May of 2000 were designed to prevent the abuse of polling during election campaigns, a 2005 study found that manipulation continued.[29] A poll commissioned by the *National Post* on the eve of the federal election in November of 2000 in particular raised suspicions. The CBC had reported Alliance leader Stockwell Day's fundamentalist Christian beliefs in creationism, including that the Earth was only 6,000 years old. The *National Post* then commissioned a COMPAS poll that purported to show Canadians were evenly divided between belief in creationism and evolution.[30] The effect of the poll, according to a 2003 study, was to make Day's views seem mainstream and present the Alliance as a reasonable alternative to the Liberals. The poll's methodology was suspect, however, because of the wording of its questions. The number of Canadians who believed the literal word of the Bible, other surveys had shown, was closer to one in six. "It looms as a major problem for both the polling industry and the consuming public when *advocacy* rather than *description* becomes a possible objective of the exercise," warned researchers from the Institute for the Study of Public Opinion and Policy at Wilfrid Laurier University.[31]

A 2007 study also found that public opinion polls were regularly used by the *National Post* for advocacy purposes. "Opinion polls are manipulated to confer popular legitimacy upon the economic conservatism of the *Post*'s editors," concluded Brock University sociologist Jonah Butovsky. "Rather than giving voice to the general population, polls in the *National Post* are routinely used to 'manufacture consent' for the viewpoints of the corporate and political elite, while misrepresenting popular opinion."[32] By framing the issues a certain way, the *National Post* used polls to create a "partial and distorted" view of public opinion, he added. "Some of the *National Post*'s omissions and misrepresentations seem to be deliberate attempts to manufacture public support for its editorial positions," observed Butovsky.[33] He counted 84 polls conducted for the

National Post by COMPAS, whose president Conrad Winn was a pollster for the Fraser Institute. Winn had published research opposing employment equity and financial aid for immigrants, as well as documenting the CBC's left-wing bias. "Winn has sculpted his survey questions to elicit the desired responses," charged Butovsky. "The persistent use of a rightleaning pollster does not, on its own, confirm that the *Post* presents a biased view of popular opinion, but it is consistent with the remainder of the evidence."

> Academic surveys were drawn on very infrequently.... Government polls were also used infrequently. None of the stories concerned poll results produced by a "left-progressive" think-tank such as the Canadian Centre for Policy Alternatives, Canadian Centre for Public Policy, or the Caledon Institute.[34]

The fiscal conservatism of the *National Post* made it completely compatible with the ideology of Israel Asper, which he had often promoted in his column on taxation. The *Post*'s opposition to the Liberal party, however, was something that would have to change under his ownership.

The power of think tanks

The unprecedented success of Reagan's economic policies in enriching the rich led to a worldwide movement among the well off to promote "trickle down" economics. Reagan's policies of deregulation, small government, and reduced social services had led to a stock market boom, albeit at the expense of impoverishing many Americans. Increasing the wealth of some, so the theory went, would benefit all in the long run. Many liberals questioned the logic of trickle-down economics in the absence of taxes designed to redistribute the benefits of runaway capitalism from rich to poor. In order to promote deregulation and globalization, well-funded neoconservative "think tanks" proliferated in the US and Canada by the 1990s. The public policy institutes, as they preferred to be known, popularized their free-market message in large part through the news media. By both producing great quantities of "research" and by monitoring media outlets, think tanks exhibited a push-pull effect on the media. As a result, their work became a major influence on media content, and thus on public perceptions of reality.

The modern think tank was the brainchild of Austrian economist Friedrich Hayek, whose deregulationist theories inspired the neo-conservative movement.[35] Hayek urged the creation of think tanks to win the "war of ideas" with advocates of regulation who favoured interventionist policies. Hayek even helped to personally found the original free-market think tanks — the Mont Pelerin Society in Switzerland in 1947 and the Institute for Economic Affairs in London in 1955.[36] Think tanks spread to North America in the mid-1970s with the establishment in the US of the Heritage Foundation, the Cato Institute, the American Enterprise Institute, and others. A 2000 study showed that right-wing think tanks enjoyed greater media visibility than liberal policy institutes due to their financial resources and access to conservative publications.

> Conservative voices cumulatively speak more loudly than left, liberal, minority, or environmentalist voices. The much louder collective voice of think tanks in the "conservative cluster" seems due at least in part to the greater financial resources available to those institutions. The budget of the entire "liberal cluster" of think tanks is slightly less than that of the Heritage Foundation alone.[37]

A major benefactor of conservative think tanks was Richard Mellon Scaife, the reclusive billionaire owner of the *Pittsburgh Tribune-Review* and heir to the Mellon family banking and oil fortune. He was reported to have contributed more than $200 million to conservative causes over a thirty-year period. About $20 million of that went in 1995 alone to think tanks like the Heritage Foundation ($1.5 million), the American Enterprise Institute ($465,000), and the Cato Institute ($100,000).[38] One of Scaife's favourite projects was the media-monitoring group Accuracy in Media (AIM), to which he gave $2 million between 1977 and 1997. Another Scaife beneficiary was the Media Research Center (MRC), which was established in 1992. Like AIM, the MRC monitored the mainstream media for evidence of bias. With a vastly greater budget, however, it also published research. In 1992, the MRC created the Free Market Project to promote a culture of free enterprise through its *MediaNomics* newsletter. "Inevitably, the constant critiques grind journalists down, and they begin to subtly and not-so-subtly embrace the conservative spin," noted Trudy Lieberman in *Slanting the Story*. "The organization helps create a climate to neutralize honest reporting."[39]

Slanting the news

The rightward spinning of news in the US, Lieberman claimed in 2000, was "one of the most significant political stories of the last two decades." The success of the political right, she noted, was the result of "a variety of aggressive strategies used by well-financed think tanks and policy institutes to influence the media's coverage of political and economic issues." This led, she concluded, to "misleading and one-sided reporting that has given the electorate a distorted view of many important issues."[40] Much of media monitoring quibbled about how news sources were labelled. Together with the rise of right-wing talk radio, this resulted in a shift in the language of news by the mid-1990s. Reese Cleghorn, a former dean of journalism at the University of Maryland, noted the trend in 1994 for that school's *American Journalism Review.*

> Now we routinely see news stories that call right-wing radicals "conservatives." So no longer is there a right wing. Centrists have become "liberals." What to call a real liberal? Nobody knows anymore, except maybe Rush [Limbaugh] and Newt [Gingrich.] They call them Socialists or counterculturists. The extremists on the religious right are seldom called radicals. And their self-definition as Christians is taken at face value.[41]

Lieberman, who was health policy editor for *Consumer Reports* magazine, documented the effect of this rightward shift on issues such as cuts to Social Security and food and drug regulations. The Clinton administration's attempt to implement universal health care failed, she claimed, because of work behind the scenes by think tanks. "Talk show hosts and direct-mail campaigns by conservative groups fuelled the ultimate scare-story that turned the public against reform — the fostered threat that people would not be able to choose their own doctor."[42] This was ironically achieved by using the media-savvy techniques of 1960s consumer crusader Ralph Nader against the left, according to Lieberman. Think tanks emerged partly in response to Nader's success in lobbying for product safety, and they turned the tables on consumer groups by portraying them as special interests. Well-organized conservative groups "turned media framing upside down," according to Lieberman, and "seized the language from the Nader era."[43] Sources journalists once relied on for comment, such as academics and consumer advocates, were "de-legitimized" by accusations of bias. Think tanks replaced them as idea generators and the major influence on public policy. By offering their own solutions and attacking those whose policies they opposed, according to Lieberman, the right-wing groups enjoyed spectacular media success.

> Conservative think tanks have moved beyond framing and have come to use the media as both a friend and a foe to further their objectives. They have become masters at cultivating the press, but are just as quick to charge "liberal bias" when the media they've so carefully pampered do not stick to the conservative line.[44]

Think tanks in Canada

In their 2005 book *Rescuing Canada's Right*, Tasha Kheiriddin and Adam Daifallah prescribed a course similar to that taken in the US since the mid-1960s. In a chapter titled "Rebalancing the Media," they pointed to the spectacular success of US think tanks at influencing news coverage of political issues. "If conservatism is to succeed in Canada, conservatives must inject some semblance of ideological balance into the media," they wrote.[45] Daifallah was a former member of the *National Post* editorial board and Kheiriddin was Ontario director of the Canadian Taxpayers Federation. They urged Canadian conservatives to follow the lead of their US counterparts and create more publications that spoke to the political right. "To rescue the right, conservatives must stop hating the media — and become the media," they wrote.[46] Part of their plan was media monitoring by pressure groups that would complain at any hint of liberal bias in the news and promote a preferred set of labels to frame issues. "American conservatives have built themselves an ideas factory," they wrote. "To build its own conservative culture, the Canadian conservative movement must replicate these models."[47] A *National Post* writer covering the book's launch noted one strategy from the "playbook for balancing the terms of public debate, presently colonized by Liberal locutions."

> For the Liberal terms "medicare" and "public health care," substitute "state health care monopoly"; for the Liberal "social services," substitute "government programs"; for the Liberal

"investing tax dollars," substitute "spending taxpayers' money"; for the Liberal "budget surplus," substitute "amount Canadians were over-taxed."[48]

With established think tanks such as the Fraser Institute and the C.D. Howe Institute, they noted, Canadian conservatives already had a running start toward their right-wing revolution. "The good news is that the seeds of conservative infrastructure have already been planted in Canada," wrote Kheiriddin and Daifallah. "In fact, we are several steps ahead of where American conservatives were in 1964."[49] Little could they have suspected how quickly their mission would be accomplished. Less than two months after *Rescuing Canada's Right* hit bookstores, a rejuvenated Conservative Party came to power with a minority government. The conservative revolution, it seemed, had been under way for some time in Canada. It had been given a boost, however, by the "sponsorship" scandal that saw millions go to Liberal advertising agencies in Quebec that did little or nothing for the money.

The revolution might have begun in earnest when Conrad Black founded the *National Post* in 1998. It may have originated, however, when a group put their heads together in Calgary two years earlier to plot a new course for conservatives. David Taras attended the "Winds of Change" conference of seventy right-wing activists determined to unify conservatives in Canada. It was organized by David Frum, son of the late CBC journalist Barbara Frum, and law student Ezra Levant. Frum, a former *Wall Street Journal* and *Forbes* magazine writer, had established himself as a leading "pundit" — a journalist who crossed over into partisanship — according to Taras.[50] As a journalist, television panelist and author, Frum actively promoted his ardent conservatism in what Taras saw as a new/old kind of journalism. "Far from hiding his political beliefs, he parades them and to some degree has come to symbolize them. His participation in the political process is seen as natural and, one can argue, reinforces rather than detracts from his success as a journalist." The development signaled a return to the journalism of the party press era, according to Taras.

> The conference highlighted a phenomenon that has been taking place for quite some time in American politics, but seems only now to be emerging full-blown in Canada: that an increasing number of journalists have become ardent political activ-

ists. . . . Some journalists have been able to enhance their status by openly championing partisan positions and causes. . . . Politics and journalism are no longer separate estates.[51]

The Fraser Institute

The following January, *Edmonton Journal* columnist Linda Goyette received a plain brown envelope in the mail that contained a leaked document. It was titled "Toward the New Millennium: A Five-Year Plan for the Fraser Institute." In it, executive director Michael Walker outlined his ambitions for the Vancouver-based think tank. "A central focus of our program during the next five years will be the expansion of our penetration of the national media," wrote Walker. The Fraser Institute, according to the 28-page paper, hoped to double its $2.5-million annual budget over the next five years. Walker boasted of the impact the think tank had already made on public opinion. "The Fraser Folio program of faxes to radio talk show hosts has increased dramatically the number of 'hits' for our material on radio talk shows across the country," he wrote. "Some 450 radio stations are now the regular recipients of this service." The think tank planned to develop, he added, "a database of journalists who respond to our material and catalogue the extent to which particular journalists cover our releases." Its *Fraser Forum* newsletter would become a full-fledged magazine, and a new column on health policy would be distributed to newspapers. According to Goyette, Walker described new ideologically-directed research in education, health care, the environment, the legal system, social policy, and regulation.[52]

Goyette was unsure of the report's authenticity until Walker confirmed it, posting a version on the Fraser Institute website to reduce the impact of her scoop. "It was such an ambitious project, I couldn't tell if the thing was real," Goyette told Clive Thompson of *This* magazine. He gleaned a few nuggets of his own from the document, such as its plan to propose its own definition of poverty as an alternative to the minimum income levels set regionally by Statistics Canada. "The [Fraser] institute's poverty lines, in contrast, are roughly one-half as high," noted Thompson. "Whereas Statscan figures $15,479 represents poverty for a single adult living in a midsized city, the institute argues it ought to be only $7,480." A comparison of the leaked document and the version posted on the Fraser Institute's website also revealed an interesting omission. The excised portion, Thompson noted, showed where the Fraser Institute's "invocation to information warfare becomes the most naked." In the leaked version, Walker called for the think tank to "target" studies by the liberal Vanier Institute for the Family, calling it "the Vanier Institute for (well, against, actually) the Family." The paper, Thompson concluded, provided "a compelling look into how deeply the group understands the modern art of media wrangling."[53]

The Fraser Institute was established in 1974, after an NDP government was first elected in British Columbia. One of its first advisors was Friedrich Hayek, then 75. Its initial funding of $100,000 came from executives of the logging company MacMillan Bloedel.[54] A 2004 study found that the Fraser Institute's annual funding had grown to $6.6 million. It came from charitable foundations (52 percent), other organizations (38 percent), and individuals (10 percent). Several large Canadian foundations provided the bulk of the Fraser Institute's funding, according to Simon Fraser University researcher Donald Gutstein. The Donner Canadian Foundation, a cousin of the New York-based Donner Foundation, contributed almost half a million dollars to the Fraser Institute in 2002, he noted. It also gave $100,000 to start CanStats, a Fraser Institute division that monitored media coverage of science and statistics. Other large Canadian foundations that funded the Fraser Institute, according to Gutstein, included the Max Bell Foundation of Calgary and the Toronto-based Weston Foundation.[55]

CanStats was short for the Canadian Statistical Assessment Service. According to Gutstein, it was modelled after the US group Stats, which had been created by the Center for Media and Public Affairs. CanStats was headed by Kenneth Green, who came to the Fraser Institute from the Reason Foundation, a libertarian think tank based in Los Angeles. One of the main activities of CanStats, according to Gutstein, was to discredit any science critical of industry. A favourite CanStats topic was fish farming, which concerned environmentalists due to fears that imported Atlantic salmon might pass disease to indigenous fish.

> In 2004, CanStats published more bulletins on aquaculture than on any other subject, except for its output debunking global warming. Why might the Fraser Institute be so inter-

ested in defending the interests of fish farmers? . . . One of the institute's largest benefactors is the Weston family, which also happens to own a major BC farmed salmon operation, Heritage Salmon.[56]

The Donner Canadian Foundation was founded in 1950 by US Steel magnate William H. Donner, who died in 1953 after moving to Montreal. In 1993, according to the *Globe and Mail*, it convened a meeting of prominent Canadian neoconservatives, including its columnists Robert Fulford, Terence Corcoran and Andrew Coyne.[57] Attendees resolved to promote conservatism through new publications. Over the next two years, the Toronto-based magazines *Gravitas* and *The Next City* were founded with financial support reportedly totalling $3.4 million from Donner.[58] Along with Fulford and Coyne, a founding editor of *The Next City* was Stephen Harper, then a Reform Party MP from Calgary.[59]

According to Gutstein, the Donner Canadian Foundation had assets of $200 million in 2004 and gave $2 million a year to right-wing causes. "In the mid-90s, it established three new libertarian think tanks, the Atlantic Institute for Market Studies in Halifax, the Frontier Centre for Public Policy in Winnipeg and the Montreal Economic Institute."[60] Allan Gotlieb, a former Canadian ambassador to the US and old camp counselor of Izzy Asper, was named chairman of the Donner Canadian Foundation in 1996.[61] The Fraser Institute's growing prosperity was a testament to Walker's coordinated fundraising efforts, according to Gutstein. His six years on the board of the Max Bell Foundation, which had been endowed by a founder of the FP Publications newspaper chain, also came in handy. "By the late 1990s," noted Gutstein, "right-wing Canadian foundations were exhibiting the same coordinated funding behaviour that made the American foundations so formidable."[62]

The media connection

Canadian law required tax-exempt foundations like Donner to list recipients of their grants. Charitable societies like the Fraser Institute were not required to disclose their funding sources, but the names of their trustees were made public. Before CanWest bought Southam, the trustees of the Fraser Institute included Hollinger directors David Radler and Barbara Amiel, Conrad Black's wife. In

1999, when Radler was publisher of the *Vancouver Province*, reporters there launched a protest over what they called "blatant political interference" by editor-in-chief Michael Cooke. Over the objections of other editors, he had ordered a professor's comments removed from a story because they were too "left-wing." The *Province* journalists charged that Cooke was "shameless in his support of the Fraser Institute," and had even ordered it referred to in news stories as "prestigious."[63]

Political bias on *Province* pages under Radler drew another protest from its own journalists following the 2000 federal election campaign. Stories unflattering to Alliance leader Stockwell Day, they claimed, had been removed from the newspaper between editions by new editor-in-chief Vivienne Sosnowski. The advocacy group Campaign for Press and Broadcast Freedom filed a complaint with the BC Press Council on behalf of unnamed *Province* journalists. It asked that *Province* management publicly acknowledge they were "using partisan criteria to edit the news." The complaint was quickly dismissed, however, as dealing with a matter well within management rights.[64]

When CanWest bought the Southam newspapers in 2000, David Asper was a Fraser Institute trustee.[65] A reporter who visited the think tank's Vancouver headquarters a few years later noted that it housed a David Asper Centre for the Study of Law and Markets.[66] Leonard Asper was a trustee of the Toronto-based C.D. Howe Institute, which had been founded in 1973. Its budget, according to the *Globe and Mail*, doubled between 1999 and 2005. The C.D. Howe Institute, the newspaper noted, was among the think tanks "credited with influential arguments in the early 1990s that led to public support for deficit-cutting."[67] According to University of Western Ontario political scientist Donald Abelson, "few think tanks in Canada have attracted more attention in the media."[68]

Fraser Institute alumni soon made their way onto the editorial boards of CanWest newspapers, including Fazil Mihlar at the *Vancouver Sun*, Danielle Smith at the *Calgary Herald*, and John Robson at the *Ottawa Citizen*.[69] Mihlar, who was named editorial pages editor of the *Sun*, was a former policy analyst for the Fraser Institute. Smith, who also hosted the television program *Global Sunday*, was a former intern there. According to Gutstein, one of Harper's first acts after he was elected was to cut the think tanks a tax break.

"Buried in his first budget was a provision to exempt from capital gains tax donations of stock to charity," Gutstein wrote. "Adding this new exemption to the existing tax credit for donations to charities means that the donor pays only 40 percent of the dollars he donates. Taxpayers pick up the rest. The Fraser Institute is a registered charity."[70]

CHAPTER 6

Convergence

The growing power of a shrinking number of large media owners in Canada began to alarm some political observers starting in the 1960s. The first major study of the problem began in 1969. A Senate committee reported the following year that chain ownership of Canada's daily newspapers had grown to 45 percent in 1970 from 25 percent in 1958. When the reforms it proposed to slow the growing concentration of press ownership fell flat, the result was a stake through the heart of Canadian journalism a decade later. The date of August 27, 1980 would live in Canadian newspaper infamy as "Black Wednesday," and the fallout was severe. "This Commission was born out of shock and trauma," began the 1981 report of the Royal Commission on Newspapers, which was called after the simultaneous closures of Thomson's *Ottawa Journal* and Southam's *Winnipeg Tribune*.[1] Each closure gave the other chain a monopoly in that market, and it smacked of collusion. Criminal charges of conspiracy were laid against the two chains, which between them owned 58 percent of Canada's dailies.[2] "Newspaper competition, of the kind that used to be, is virtually dead in Canada," concluded the commission. "This ought not to have been allowed to happen."[3] Almost lost in the indignation was the fact Thomson sold its *Vancouver Sun* to Southam the same day, giving it both dailies there.

The commission reported within a year with proposals for limits on how much of Canada's press one owner could accumulate. One of its strongest recommendations was for a prohibition on cross ownership of media, or "convergence," as it became known. "Common ownership of different media in one community is clearly a restriction on competition," the report concluded, "a lessening of the diversity of voices providing information and expressing opinion."[4] While a Canada Newspaper Act was drawn up to introduce limits on press ownership, a more expedient measure was enacted to prohibit cross ownership. A policy directive was issued to the CRTC by the government of prime minister Pierre Trudeau in 1982. PCO 2294 ordered the CRTC to deny the application for a television broadcasting licence of any owner of a daily newspaper.

More pressing priorities in the dying days of Trudeau's government, however, prevented the Canada Newspaper Act from being passed into law. Trudeau was more concerned about his legacy of repatriating the Constitution and enacting the Charter of Rights and Freedoms, for which media support was needed. Soon the federal government turned Conservative with the election of Brian Mulroney as prime minister in 1984. The cross-media ban was rescinded the following year. According to Allan Bartley, the CRTC directive was a half measure that was just as easily erased as it had been put in place. "It was a political direction rather than a policy direction," noted Bartley, "chosen because it appeared the least troublesome of the available options."[5]

The convergence tsunami

The wave of multimedia deal making that swept the world in 2000 was prompted by the seismic shock of the AOL-Time Warner merger that January. It was the largest corporate merger in history. It grafted America Online onto a monolith that had been created by the largest media merger in history a decade earlier. Time Warner was actually the junior partner in the pairing, and its owners got only 45 percent of the new company. The AOL-Time Warner deal had its greatest impact in neighbouring Canada, where the floodgates had been left wide open for convergence with the rescinding of PCO 2294. The tsunami transformed the country's media landscape with no fewer than three major transactions by year's end.

The frenzied business climate of the late dot-com boom saw media executives scramble for strategic alliances to keep from being left behind by technology. The advent of the Internet and digital broadcasting in the 1990s led many to see media convergence as inevitable. They predicted that the content of newspapers, magazines, radio and television would eventually all be delivered only online. In the immediate aftermath of the deal making, two key questions about convergence emerged. Would it work, and should it be allowed? The answer to the first question soon came clear, at least in the short term. The 2001 bursting of the stock market technology bubble cut the value of converged media companies to a fraction of their peak worth. The dot-com collapse quickly made convergence a dirty word among media executives as the bellwether AOL adventure fell apart.

The answer to the second question became more difficult to divine. The Internet revolution was expected by technology enthusiasts to render national regulatory agencies redundant. Companies worldwide pushed for the lifting of restrictions that prevented them from owning outlets in multiple media. The growth of satellite broadcasting, they argued, rendered obsolete the traditional problem of bandwidth scarcity that had once justified regulation. Consumer advocates, on the other hand, sought to retain limits where they existed and to implement them where they did not. They argued that more concentrated ownership of news media would reduce diversity of opinion and political pluralism. Owners claimed an unregulated Internet ensured diversity and pluralism. The political conflict that ensued turned the debate over cross ownership into a major confrontation in some countries. In Canada, however, the question was already moot.

In the US, the Federal Communications Commission prohibited cross ownership of newspapers and television stations in the same city in 1975. A dozen or so existing arrangements were allowed to remain in place. As a result, newspapers like the *Chicago Tribune* shared newsrooms with television stations like WGN in "grandfathered" operations. In Australia, the Labour government of prime minister Paul Keating placed an absolute ban on common ownership of newspapers and television stations in 1987. That forced many of the country's major media owners to divest their interests in one medium or the other, including Rupert Murdoch. In the UK, major newspaper owners were prohibited by the 1990 Broadcasting Act from owning more than 20 percent of any ITV licence.

The AOL-Time Warner merger was exempt from the FCC's cross ownership ban because it did not include a daily newspaper and a TV station in the same market. It married an Internet pioneer with a multimedia giant that had movie, cable television, music, book and magazine publishing divisions, but no newspapers. The convergence game was suddenly on, and the fever hit nowhere harder than in Canada. A month later, telecom giant Bell Canada Enterprises (BCE) announced a $2.3-billion takeover bid for the CTV network. The Aspers bit in July, taking Southam off Conrad Black's

hands for $3.2 billion. In September, BCE/CTV went them one bet-
ter, partnering with Thomson's *Globe and Mail* to create a $4-billion
multimedia entity called Bell Globemedia. Thomson took only 29.9
percent ownership of the converged firm, which prevented immediate CRTC scrutiny of its involvement.[6] None of those deals, however, was the largest multimedia transaction in Canada in 2000.
That was engineered by Quebecor, which published almost half the
daily newspapers in Quebec. It had recently expanded nationwide
by taking over the *Toronto Sun* chain of fifteen dailies. In a seven-
month takeover battle, Quebecor outbid Rogers Communications
for Quebec cable television giant Groupe Videotron. The $5.4-bil-
lion takeover included the province's largest television network,
TVA.[7] Rogers was the country's largest media company when 2000
began, but it ranked only fourth when it ended.

'Put up or shut up'

While Conrad Black could at first bring himself to part with only
half of the *National Post* he had created, the Aspers were intent on
having their say in its content. Their defence of Chrétien began
when David Asper wrote a column that appeared in the *National
Post* and other Southam newspapers in March of 2001. "The media's
coverage of the accusations against the Prime Minister has crossed
a line that delineates solid investigative reporting from adjective-
driven innuendo," he wrote. Asper even singled out the *National
Post* for criticism. "This newspaper and others across Canada,
including other forms of media, have had a remarkably unfair 'go'
at the Prime Minister."[8]

Newspaper owners had seldom been so strident, publicly at least,
in criticizing their writers. To some Canadians, Asper's column
crossed the line journalistically. To others, it crossed it politically.
Some saw it as blatantly currying political favour on the eve of
CRTC hearings into renewal of CanWest's seven-year broadcasting
licence. Many political observers saw it as influence bought and
sold. Given the political activism of the Aspers, many wondered if
convergence was such a good idea after all. David Asper's unprecedented show of support for the prime minister signalled to some
a new era of political favouritism in the press. MPs from all three
federal opposition parties combined in the House of Commons to
demand an inquiry into media ownership. Heritage Minister Sheila

Copps at first promised a "blue-ribbon panel" of experts to study the matter.[9] Within days, however, Copps reversed herself. She announced that media ownership would instead be studied by a committee that had already been formed to examine broadcasting policy.[10] The standing Heritage Committee chaired by Montreal MP Clifford Lincoln had been tasked the previous month with considering the future of an industry convulsed by convergence. "Ours is not a race against convergence," Lincoln said as the committee began its work. "We're going to lose that race. . . . We've got to find out what the impact is."[11]

David Asper's shot across the bow of the *National Post* obviously didn't sit well with journalists at the newspaper, who fired back. An editorial printed alongside his column disagreed strongly with Asper's argument. "The onus is not on newspapers to 'put up,' but on Mr. Chrétien to convince us of the propriety of his actions," it noted defiantly. "This newspaper will continue to follow the story, and it encourages all other Canadian media and all opposition politicians to do the same."[12] The next day, columnist Mark Steyn also defended the newspaper, pointing out that Asper's column had refuted "not one specific fact or allegation made by the *Post*."[13]

'Ridicule and dishonour'

Correspondence entered into evidence at Black's 2007 fraud trial in Chicago revealed the depth of Izzy Asper's outrage at the *Post*'s editorial insubordination. In a strongly-worded fax to Black several days later, he complained about the "outrageous . . . and savage attack" on his son. "I assume the *Post*'s conduct, both before and after the publication of David's piece, and the firestorm its staff helped unleash across the media, was caused by your personal orchestration, or done with your acquiescence and approval." Asper threatened "unilateral action to address the slurs and abuse that has been heaped upon us from a variety of quarters." He claimed the *Post*'s rebukes violated their partnership agreement that promised CanWest would get advance notice of any editorial positions adverse to its interests.

> Given that we view this as a blatant and defiant breach of the letter of our agreement, and more saddening and provocative, the spirit of our arrangement, I consider the situation both currently and foreseeably, as in crisis. Neither you nor I would

profit from a public battle, which would give great pleasure to those who wish neither of us well, but regrettably, you have chosen to publicly throw a gauntlet, administer a public slap in the face which has both embarrassed, humiliated and held up to ridicule and dishonour both my family and my company.[14]

Black assured Asper he had nothing to do with the reaction of *Post* journalists to his son's criticisms or the political fallout from them. "I did not orchestrate anything," he wrote the following day. "The 'firestorm' orchestrated itself." On the contrary, Black told Asper, he had intervened with the *National Post*'s editor to tone down the reaction. "I was shown Ken Whyte's editorial comment and I asked him to remove one sentence that I thought was inadvertently insulting to you and your family and he did so." Black pointed out that he and other *Post* executives had warned David Asper that criticizing his own journalists would produce "great resentment." They also told him, in Black's words, that his column "would appear to anyone in that country still interested in an independent press to be servile toadying to a rather corrupt regime." Black noted he had told the Aspers "many times" that ownership influence on news coverage had to be accomplished in a more subtle manner. "If he [David Asper] wished to alter the tenor of the coverage this should be done, at least initially, in comprehensive discussions with the individual metropolitan editors."[15]

Black argued that it was not he who had violated their partnership agreement, but instead the Aspers. Another letter entered into evidence at Black's trial showed that he had complained about David Asper attempting to influence news coverage earlier in 2001. "I am aware that considerable pressure has been exerted by David on *National Post* editorial personnel on behalf of Chrétien," he wrote to Izzy Asper on January 5. "This is not reconcilable with our agreement."[16] His letter of March 14 made it clear Black felt the Aspers were shooting themselves in the foot by interfering so obviously in political coverage. "I believe it is, in fact, contrary to the spirit of our arrangement and to CanWest's corporate interests for you people to tinker so recklessly by these interferences with the credibility and therefore the value of these franchises which my associates and I so swiftly built up."[17]

The sabre rattling by the Aspers also didn't daunt columnist Lawrence Martin, who covered Ottawa for the Southam chain. It was his 5,000-word feature in late February that prompted David Asper's defence of the prime minister. His column only seemed to egg Martin on. Before March ended he wrote three columns on Shawinigate in one week. As more revelations emerged in April, Martin trumpeted the disclosures. Then in May, he wrote an unflattering column about another prominent Liberal politician known to be close to Izzy Asper — John Turner.[18] Martin also went on the record as favouring increased funding for the CBC, and he even urged moving its headquarters to the Aspers' hometown of Winnipeg.[19]

More public benefits

Like CanWest's purchase of WIC, the other convergence deals of 2000 were subject to the CRTC's public benefits program. Public benefits had been pioneered by magazine publisher Maclean Hunter. It offered a $36-million package of inducements in 1989 to gain CRTC approval for its takeover of Selkirk Communications. The $600-million purchase was then the largest in Canadian broadcasting history.[20] The deal made Maclean Hunter a multimedia pioneer with 200 magazines, 35 cable systems, Selkirk's television and radio stations, and the *Toronto Sun* newspaper chain. In 1994, it was in turn taken over for $3.1 billion by Rogers, which provided a public benefits package of $101.9 million.[21]

Quebecor had to ante up $48.9 million, which was 10 percent of the estimated value of TVA. BCE had to find worthy causes in the amount of $230 million. Its public benefits package included $45.5 million for movies of the week, $18 million for documentaries, and $25 million for dramatic series.[22] It also included $2.5 million to endow a BCE Chair in Convergence and Creative Use of Advanced Technology at Ryerson University in Toronto.[23] CanWest was not required to provide public benefits from its purchase of Southam because the CRTC did not have jurisdiction over newspapers. Like CTV and TVA, however, the Aspers would soon have to face questions from the broadcasting regulator about how far convergence could go.

Making public benefits payments did not mean the networks would be allowed to combine their newsgathering operations with those of their newspapers. As a condition of licence renewal, the CRTC wanted a "firewall" of separation between television and newspaper journalists. It asked CTV and CanWest to draw up a "code of conduct" to protect consumers from anti-competitive behaviour.[24] A precedent had been set in 1997 when Quebecor volunteered just that kind of a separation after taking over the Quebec network TQS.[25] In applying for CRTC approval of its TVA takeover, Quebecor agreed to abide by the same strict code.[26] The six-page prohibition could only be changed with CRTC approval. It promised that Quebecor's print and broadcast journalists would "at no time transmit, receive, exchange or discuss information by phone, fax, Internet or other technology."[27]

When CTV and CanWest came up for licence renewal, however, they balked at making similar assurances. Like Quebecor, CTV had no plans to integrate its newsgathering operations with those of its print partner. "You can't do a TV and a print job all day, every day," said CTV president Trina McQueen. "You just can't."[28] One Quebecor executive even ridiculed the idea of expecting its journalists to report for newspapers and television. "That would be silly," said Quebecor vice-president Luc Lavoie, a former reporter.[29] CanWest's vision for the future of converged journalism was quite different, however. It had been articulated by Leonard Asper earlier in 2001.

> In the future, journalists will wake up, write a story for the Web, write a column, take their cameras, cover an event and do a report for TV and file a video clip for the Web. What we have really acquired is a quantum leap in the product we offer advertisers and a massive, creative, content-generation machine.[30]

According to Antonia Zerbisias of the *Toronto Star*, the differing concepts of convergence received differing support from scholars. "CTV's plans for editorial separation explain why prominent journalism academics, at the network's request, support its licence-renewal application (but not CanWest's), while denouncing the imposition of a code."[31] When the CRTC convened its hearings in April, a lineup of witnesses from universities appeared to speak against restrictions on converged newsgathering. "The TQS code may be justified in Quebec," testified Donna Logan, director of the Sing Tao School of Journalism at the University of BC "In English Canada such a restrictive code would be excessive and detrimental."[32] Logan went even farther in her praise for the concept of convergence. "One of the things that has always disturbed me about journalism in Canada is that there were too many reporters chasing so few stories," she told the CRTC. "Converged journalism offers an opportunity to break out of that mould by freeing up reporters to do stories that are not being done and are vital to democratic discourse."[33]

Even stronger in his condemnation of the code of separation Quebecor had already agreed to was Peter Desbarats. "There is no way, short of placing secret agents in newsrooms, that any system can effectively monitor all forms of communication between journalists working for the same organization," he wrote in the *Globe and Mail*. The Maclean Hunter Chair of Communications Ethics at Ryerson University called the Quebecor code a "Big Brother" mechanism. According to Desbarats, it created "a degree of cross-media state intervention that may well be unprecedented in modern democracies." While such a code might save the jobs of a few journalists in the short term, he warned, it could not prevent the manipulation of news.

> As every journalist knows, the primary means of influencing the character of the news produced by a media organization is through recruitment and promotion of key personnel. This can be done as effectively within a split as within a unified organization. Trying to build a "firewall" between print and TV newsrooms is an exercise in futility.[34]

'Demonstrable' benefits

Leonard Asper dismissed the CRTC's concerns over increased ownership concentration at the 2001 licence renewal hearings. "I think the concerns about convergence and ownership consolidation are theoretical," he testified, "while the benefits of convergence are demonstrable." Asper pointed to the vigorous defence mounted a month earlier by *National Post* journalists after his brother's column. That episode, Asper claimed, showed the "fierce independence" of their newspaper journalists. "I think that example is the most stark that CanWest is not going to try to stifle or chill the journalist and that freedom of expression is alive and well within the CanWest system."[35] Any imposition by the CRTC of a code of conduct, Asper suggested, could be subject to a legal challenge. "We believe that is bordering on if not unconstitutional and a serious imposition against freedom of speech."[36]

One document filed with CanWest Global's licence renewal application that drew the interest of CRTC commissioners was its

National Post partnership agreement. It required "advance notice of ... any editorial position which could reasonably be viewed as embarrassing, damaging or adverse in the interest of CanWest or affiliates." In the event that anything adverse to CanWest was printed, the clause required the *National Post* to publish a reply. The Asper response would run "in the op-ed or editorial pages of such other prominent location as CanWest shall reasonably request." Leonard Asper denied that the clause was a smoking gun that proved collusion. "The reason that clause is in there is because we believe in three principles: balance, fairness, [and] diversity of opinion," he testified.

> Our concern in that discussion was that the *National Post*, with all due respect to its owner, would not meet those principles and we would be embarrassed by that. It was not meant to address an issue where an article about CanWest was appearing, that CanWest is doing something wrong, or whatever.[37]

While refusing to accept a binding code, CTV and CanWest Global agreed to maintain separate management structures for their print and broadcast news operations. They pledged adherence, however, only to a voluntary one-page "Statement of Principles and Practices." They also refused to allow the CRTC to monitor their operations for compliance, insisting they could "self-police." As the *Toronto Star*'s Antonia Zerbisias pointed out, however, the networks "drafted their own codes for maintaining separate 'news management structures,' although they did that drafting in concert with one another."[38]

Also appearing before the CRTC were cultural groups, including the Friends of Canadian Broadcasting. They argued that the separation of management offered by the networks did not go far enough. "If only one reporter is sent to cover an event for print and television," Ian Morrison told the commissioners, "it will not matter to the public how many clear and distinct management structures there may be." Morrison echoed the concerns of many in western Canada, where CanWest dominated the news media. "Diversity of voices will dissipate in communities all across Canada," he warned. One compromise proposed was for the CRTC to grant only provisional licences for two or three years, by which time the effects of convergence would be more apparent. "They should not renew these two corporate giants for seven years because by then convergence of ownership will be so solidified there will be no breaking it," said Megan Williams, national director of the Canadian Conference of the Arts. "The CRTC will have effectively lost regulatory control."[39]

The convergence conundrum

The CRTC was charged with preserving the right of Canadians to get "differing views on matters of public concern," but convergence created a quandary. Technology was changing so fast that a clear picture of how it would transform media had not yet emerged. "Convergence has overwhelmed Ottawa with new problems," noted the *Globe and Mail*. "Mergers have turned media companies into powerhouses with massive political and corporate sway."[40] The problem was that if the CRTC renewed the CTV and CanWest licences for a full seven years, the media world might look much different by the time they appeared before it next. "That places the CRTC in a tight spot," noted *Maclean's*.

> By 2008, it may be too late to act should the convergence experiment prove a disaster to the country's free and independent media, argues the CRTC's [spokesman Denis] Carmel. "If we allow this convergence to happen without restrictions and it has unforeseen consequences, everyone will blame us."[41]

Ottawa's response was to allow events to continue unfolding without restricting them while studies looked into how the media landscape was shifting. "They prefer to get a good handle on how convergence will develop before trying to regulate it," noted the *Globe and Mail*. "Technology is changing so quickly, they say, that there is no clear indication how cultural diversity will fare."[42] As a result, the CRTC renewed CTV and CanWest Global's licences for a full seven-year term in August of 2001. No restrictions were imposed other than what the networks had already volunteered. CTV president Trina McQueen called it "one of the best things that has ever happened to Canadian journalism."[43] Not everyone was as jubilant. "The CRTC has abdicated its role as a regulator," declared Peter Murdoch of the Communications, Energy and Paperworkers union, which represented 20,000 media workers.[44] The International Federation of Journalists, which had more than 450,000 members, pointed out the perils. "If the private sector can get away with cheapskate news and call it quality, there is no doubt that more pressure will be brought on public broadcasters to follow suit," said Aidan White, secretary general of the Brussels-based organization. "Canadian consumers are going to be the losers in the process."[45]

'Petty corruptions'

After the 2001 licence renewal hearings, a series of revelations by the *National Post* raised serious questions of conflict of interest. Several prominent journalism educators, it turned out, had written letters to the CRTC supporting BCE's takeover of CTV. The *Post* reprinted letters from Desbarats and Bill Wilton, executive director of the Canadian Journalism Foundation (CJF). In his dual capacity as an academic and as chair of the Foundation's research committee, Desbarats supported the takeover "as in the national interest."[46] BCE had promised $3.5 million in public benefits money — if its takeover of CTV was approved — to establish a Canadian Media Research Consortium. The mandate of the group founded by the CJF, Ryerson, York University, UBC, and Université Laval was to "focus on the development of Canadian data for use in media planning."[47] Academic support for the research included enthusiastic letters and testimony to the CRTC. "The Canadian Media Research Consortium would add significantly to the resources available in Canada for media research," wrote Desbarats. "The presence of the Canadian Journalism Foundation in the Consortium would help to ensure that research sponsored through the Consortium would be used to encourage public dialogue on Canadian media issues."

> As a journalism educator and researcher, I have long been aware of the skimpy resources available for media research in Canada, particularly compared with the United States (even when population differences are taken into account) ... The Canadian Media Research Consortium would add significantly to the resources available in Canada for media research.[48]

The CJF was best known for handing out journalism awards at an annual gala dinner. It had been founded in 1990 by a group of corporate executives. According to Terence Corcoran, they felt Canadian business was "getting a raw deal from a wildly pinko media whose primary objective was to trash the corporate sector."[49] Not all of Canada's major news media outlets joined the CJF, however.

While active in the group's planning stages, the *Globe and Mail* opted out before it was formed, noted Anthony Westell. The *Globe* bailed "apparently because it fears that the foundation's money will be tainted by big business interests," wrote the director of Carleton's journalism school.[50] According to Chris Cobb of the *Ottawa Citizen*, the CJF grew out of a study of media by the Niagara Institute in the mid-1980s that was funded by the Jackman Foundation, the Southam Foundation, Molson, and others.

> Since then [the CJF] has successfully courted media and non-media corporations which have, in some form or another, paid for CJF-sponsored get-togethers in various parts of the country. The invitation-only occasions have been for high-profile journalists, their upper managers and other movers and shakers from outside journalism. Most of the get-togethers, be they dinners, lunches or weekend retreats, are off-the-record.[51]

The problem with the CJF, according to Eric Reguly of the *Globe and Mail*, was that it had "more to do with lobbyists and PR" than journalism. "The trick is to find the real working journalists among the clutter of names. In the past, the Foundation gala has been a sea of corporate PRs and lobbyists boozing and having fun all in the name — of course — of promoting 'excellence in journalism.'"[52] Reguly's revelations resulted in the next CJF board meeting being moved from the Royal Bank's offices to the less corporate University Club in Toronto.[53] The *National Post* disclosures were also troubling to some. The CMRC funding, Corcoran pointed out, potentially recycled the public benefits money paid by acquiring corporations back into marketing research done to their own benefit. "If the major corporations ... want research into the media, then surely they can spend their own money up front rather than cash extorted ... via a regulator." According to Corcoran, the CMRC was "founded for the sole purpose of skimming a graft off the CTV takeover." The price BCE would have to pay for approval included "official bribes to assorted courtiers, favour-seekers and hangers-on who circle the CRTC," he charged. "Lining up for part of the payoff are some of Canada's leading journalism academics."[55] Corcoran urged the CJF to "leave the academics to wallow in their own petty corruptions" and outlined the conflicts of interest if it did not.

> That leaves the foundation, set up by major corporations to raise ethical standards in the media, in the position of having

participated in the extortion of money from BCE in return for providing a fawning defense of its takeover of CTV. . . . All of this should make good fodder for the next foundation educational session to help raise the standards of journalistic ethics and reporting.[55]

The next day, the CJF did exactly what Corcoran had urged and pulled out of the CMRC.[56] Its exodus left the research consortium solely in the hands of university journalism educators.

The school of converged journalism

Another letter to the CRTC by a consortium principal supporting BCE's takeover of CTV wasn't printed in the *National Post*. It came from Donna Logan, director of the Sing Tao School of Journalism at UBC. "I am particularly concerned by questions that have been raised by the Commission with respect to a potential reduction in diversity of editorial voices arising from media cross-ownership," Logan wrote. "The claim that media mergers result in fewer voices is largely a myth perpetrated by the critics of joint ownership."[57] Since being named founding director of UBC's journalism school in 1997, Logan had downplayed concerns over concentration of news media ownership. "If the dangers of media ownership concentration were as dire as some critics would have us believe, the people of Vancouver would be rioting in the streets," Logan wrote after Can-West took control of the city's largest media outlets in 2000. "There is no rolling back the clock on who owns the media in Canada," she insisted. "And it is a waste of time to call for increased government regulation."[58] Earlier that year, Logan made her position perfectly clear in an interview with the *Vancouver Sun*. "What gets me upset," she said, "is when people automatically say concentration of ownership is bad and divestiture is good."[59]

Some, however, found CanWest's relationship with Logan's journalism school uncomfortably close.[60] The two most senior *Vancouver Sun* editors served as faculty members. As part of the company's public benefits package offered after its purchase of WIC, CanWest had promised $500,000 to UBC. "We believe very strongly in the principle of journalism and enhancing it," Leonard Asper said on a 2001 visit to Logan's journalism school. "We're better off if there are better journalists." CanWest would be making more than thirty similar gifts to post-secondary institutions over the next five years "to assist media studies in Canada," Asper added.[61] "We're going to become the premier news organization in the country," he boasted. "And we're going to invest in the nuts and bolts of that by starting with journalism."[62]

That fall, Logan's school hosted a conference on convergence. It was described as an "invitation-only Summit meeting of journalists, and media and news executives from across the country."[63] It was, however, notably lacking in scholars critical of convergence. The conference was billed as providing "opportunities to get beyond the polarized rhetoric that has dominated the debate about convergence."[64] Putting talk about convergence into action, the proceedings were later broadcast on Global Television. The following year, nine CanWest Global print and television journalists were similarly televised from UBC's journalism school discussing the invasion of Afghanistan. The broadcast, noted *Maclean's*, was "part of the new world of convergence, at least as Leonard Asper sees it."[65]

With friends like these

While the CRTC licence renewal hearings were under way, Friends of Canadian Broadcasting again came under attack. Not only was the timing again convenient for CanWest, but the journalists who undermined the cultural lobby group's credibility just happened to be closely aligned with the Aspers. First, author Peter C. Newman resigned from the Friends steering committee. A founding director of the group, Newman complained that it had become "tragically diverted" from its original purpose. "I have watched with mounting horror the evolution of this once pure and modest organization into a special-interest lobby group," wrote Newman in his resignation letter to Friends president Ian Morrison. He also pointed to ties between Morrison's wife Pauline Couture and CTV as a blatant conflict of interest.[66] Some critics pointed out that Newman had a few conflicts of his own. He had started writing a regular column in the *National Post* and his book *Titans*, in which Izzy Asper was favourably featured, was being serialized on Global Television. Newman claimed the timing was coincidental, but some critics were skeptical. "It would be easy to believe it was merely a 'coincidence' if it was the first time that Friends had come under fire while CanWest was facing the CRTC," noted Antonia Zerbisias of the *Toronto Star*.[67]

Newman urged other high-profile directors of Friends to follow his lead and resign, but instead they rallied to its defence. "I have no intention of resigning," said Daryl Duke. "I'm curious as to his timing. He has had nothing to do with our directions and deliberations for over ten years." Pierre Berton, who like Newman was a founder of Friends, also defended the group. "It still supports public broadcasting very well," said Berton. "It's broadened its mandate to include concerns over what is happening in the private sector, and I think that's appropriate because public broadcasting is affected by private broadcasting." Duke added that cultural watchdogs like Friends were more essential than ever under convergence. "I believe it's needed more than ever as the consolidation of newspapers and media organizations becomes ever stronger, and gets a greater percentage of our broadcasting assets."[68]

A few days later, *National Post* media columnist Matthew Fraser began a weeks-long campaign against Friends. Fraser had previously portrayed Morrison favourably in his column and in his 1999 book on television.[69] Then CanWest bought half of the *National Post*. In the wake of Newman's resignation, Fraser derided Morrison as the "unofficial court jester" of the CRTC. "In recent years, Mr. Morrison's Friends charade has become emblematic of the absurd circus of specious posturing and backroom deal-making into which Canadian cultural lobbying has degenerated."[70] In his next column, Fraser attacked Morrison's online accounting of how the $1.7 million in donations Friends received the previous year had been spent. "We learn, for example, that Friends spent $555,179 last year to 'brief supporters,' $416,169 communicating to the general public and $274,333 on media relations," noted Fraser. "Many of these are duties performed by Mr. Morrison himself." He pointed out that Couture was also vice-chair of the Canadian Journalism Foundation, of which Morrison's son Patrick was a manager. "The family ties between Mr. Morrison, his son Patrick and Ms. Couture place operational control of both Friends and the CJF squarely in the Morrison-Couture household."[71]

By month's end Couture was forced to resign from the CJF board after Southam newspapers threatened to boycott the group's annual awards dinner.[72] Fraser was "disinvited" from co-hosting the gala, only to receive an apology from the CJF.[73] Soon, however, Fraser would come under criticism for conflicting interests.

Asper family values

In July of 2001, Southam News columnist Lawrence Martin was suddenly sacked. An announcement from CanWest said Martin's contract would not be renewed in a cost-cutting move. Southam News editor-in-chief Murdoch Davis admitted, however, that there were some non-monetary reasons for dropping the columnist. "We're going to try other approaches at commentary, and [Martin's column] just didn't fit," he told the *Globe and Mail*.[74] Three weeks later, it was announced that CanWest had bought Black's remaining half-ownership of the *National Post*. Izzy Asper then met with its editor Ken Whyte and managing editor Hugo Gurdon. According to Chris Cobb, they came to an understanding about the *Post*'s editorial direction.

> "He had a list of twelve things that he believed in," Whyte says. "Hugo and I made notes, and afterward we went out and checked and balanced to see if we'd editorialized on any of them. On all of them we were in agreement. We thought they were too close to the Liberals and they thought we were too close to the Alliance, but on the issues, we agreed."[75]

Perhaps not coincidentally, a list of Asper family values had been outlined in a speech that April by David Asper. In responding to critics of his column, he assured the Calgary Chamber of Commerce that his family would not be deterred from its policies of editorial intervention. "If necessary, we will continue to contribute to the papers, or recruit others to do so, in our determined and unwavering commitment to balance and diversity," he said. Asper also responded to criticism of his column in the House of Commons by Conservative leader Joe Clark. "To Mr. Clark and the other wailing Chicken Littles," he said, "Get used to it."[76] The Asper agenda he recited contained fifteen points ranging from immigration policy to the Supreme Court of Canada. The former should favour western Canada, he said, while the latter should include one member from each province.[77] (See Appendix)

The problem with convergence was that no one was quite sure what it was or how it was supposed to work. Cross ownership was aimed at achieving "economies of scope" across media. That was on top of the cost-saving "economies of scale" already achieved through concentrated ownership within each medium. A reduction of costs was expected through the "synergy" of sharing staff and content between outlets in different media. Revenues were expected to increase through the sale of multimedia advertising packages. The cross-promotion of media outlets owned by the same company was also seen as another major advantage of convergence. One thing journalists balked at, however, was Leonard Asper's vision of journalists filing stories for three different media. To them, his dream was a journalistic nightmare. When pressed in a meeting with *National Post* journalists for his vision of convergence in September of 2001, Asper responded with a questionable comparison. "It's kind of like pornography," he said. "You know it when you see it." The remark, according to one report, was "met with a collective groan."[78]

After the first frenzied round of convergence deal-making, the reality began to sink in that what the new media moguls had in mind might not actually be workable. Some had been skeptical from the start, noting that the idea had been tried before. In the early days of radio, newspapers were among the first owners of the new medium that was predicted to spell their demise. Business writer David Olive, then with the *National Post*, was one of the first to point out that history might be repeating itself. "Convergence is not a potential boondoggle," he wrote as the deal making climaxed. "Convergence is a proven failure, tried and found to be largely unworkable by everyone from William Randolph Hearst in the pioneer days of mass media to Ted Turner, Rupert Murdoch, Michael Eisner and Edgar Bronfman Jr. in the modern era." The new technology of the Internet, he warned, was simply an excuse for increased ownership concentration. "Despite the sheen of newness that architects of these new media models try to give them," he wrote, "these models of hyper-interactivity are merely the same old vertical integration (and mass production and mass marketing) by a different name." The economies of scale that had made increased ownership concentration more profitable had a limit, however. Organizations that got too large became unwieldy and hard to manage. Adding an extra dimension with multimedia operations would magnify the challenges for media executives. "The managerial distraction of running a multimedia colossus," Olive predicted, "is beyond the talents even of supercharged moguls."[79]

Skepticism was also widespread among media managers, according to research done in the UK. The question of cross ownership had been brought to the fore there by the 1997 election of prime minister Tony Blair. Legislation passed the previous year eased slightly the limits on how much of the country's broadcasting sector could be owned by newspaper groups. Blair came to office determined to remove the restrictions entirely. That prompted concerns about the growth of political power concentrated in the hands of a few large media owners. Media economist Gillian Doyle interviewed newspaper and television executives and found little enthusiasm for convergence at their level.

> Few senior UK media managers seem convinced that there are any real operational synergies between television broadcasting and newspaper publishing. ... Most agree that the skills and techniques involved in newspaper production and distribution are quite *different* from those required in the television industry, and vice versa.[80]

A worldwide push

The impetus for convergence instead came from the level of ownership, Doyle concluded. A host of motivations included increased company size, greater prestige and political power, increased employee morale, empire building, and defence against takeover. The technological possibilities of convergence, however, were instead advanced as a politically-expedient reason for allowing increased concentration, according to Doyle. "Whereas deregulation of cross-ownership could otherwise have been seen as a response to special pleadings from influential media owners, 'convergence' provided ministers and policy-makers with a convenient and much more respectable argument for change."[81] A 2006 study by Canadian researchers found support for Doyle's conclusion that, in the absence of economic advantages, the major effect of convergence was to increase corporate size. "More than technological convergence," they noted, "the mergers and acquisitions that occurred in 2000 seem to have resulted in increased conglomeration in the Canadian news industry."[82] The concentration was justified by a "rhetoric of technological innovation and new-economy demands," they observed. There had, however, been "little public debate of the potential social and political consequences."[83]

A similar push for deregulation took place in Australia. Liberal

Prime Minister John Howard was elected with a coalition government in 1996 after promising a review of the country's cross ownership ban. His first reform bid failed in 1998 due to opposition from coalition partners after Howard resisted lifting foreign ownership limits. A second attempt in 2001 foundered in the Senate. Australia was the media market considered most similar to Canada's due to its small population and vast geographic size. Foreign ownership restrictions in both countries resulted in their media being dominated by a small number of large owners.[84] Australia had even surpassed Canada in the 1990s with the free world's highest concentration of press ownership. By then Rupert Murdoch alone owned two thirds of its newspapers.[85] Allowing cross ownership of media would permit even more power in his hands, and some Australians saw foreign ownership as a more palatable alternative.

Even as political leaders were working for an end to restrictions on cross ownership, media managers were proving unable to make convergence work. Some were finding it a catastrophic experiment. At AOL-Time Warner, differences in corporate culture made integrating old and new media businesses impossible. The problem was worsened by accounting irregularities uncovered at AOL and the bursting of the dot-com bubble. The result was a loss for AOL-Time Warner of US$98.7 billion in 2002. That almost doubled the previous record for corporate red ink spilled in a year. From a post-merger high of $55 a share, AOL-Time Warner stock hit a low of $8.70 in July of 2002. That made its stock market value less than what Time Warner alone was worth before the merger. The pioneering convergence venture thus went down as one of the most disastrous corporate marriages of all time.[86] The company's board voted to remove AOL from the corporate name, which reverted to Time Warner, and put its online division up for sale.[87]

Convergence woes

The financial fortunes of CanWest Global Communications would follow a similar downward trend. Before 2001 ended, CanWest posted a quarterly loss of $37 million. Advertising sales slowed with a deepening recession and the company struggled with the cost of servicing a debt that approached $4 billion.[88] Its biggest problem was the *National Post*, which lost an estimated $10–15 million in the three months ended August 31. Firing 120 of its staff the next month and axing several of its sections would save the *Post* about $45 million a year, according to Leonard Asper.[89] Included among the section casualties were Sports, Arts and Life, Review, the Weekend *Post*, and its *Saturday Night* weekly magazine. The move turned out to be a disaster. Complaints from advertisers resulted in several of the severed limbs being quickly re-attached the following week.[90]

Another way CanWest cut costs was by centralizing much of the work done at its media outlets across the country. A "call centre" at company headquarters in Winnipeg combined the circulation and telemarketing functions of its newspapers. It was expected to eventually employ 400, and the company also planned to move another 800 administrative and managerial jobs there. The centralization was planned in part due to provincial and local inducements that could reach as high as $6.6 million. The Manitoba government promised CanWest $2,900 for each job it moved to the province. Winnipeg provided property and business tax relief of $93,000 for the first 400 jobs, and up to $3.1 million if CanWest brought all 1,200 jobs to the city.[91]

The company's attempts at "multi-platform" advertising sales in print, television and online media, however, stalled with the recession. One of the few customers CanWest was able to sell on its concept of advertising convergence was the Royal Bank. It bought a multimedia package that included four 12-page quarterly supplements in the *National Post* and other Southam dailies. It also included 60-second spots on Global newscasts, a 30-minute special on Global's specialty channel Prime, online coverage, and several magazine articles. The "State of the Nation" project was a quarterly report on the economy. The first installment focused on the extent to which Canada had supposedly fallen behind the US. Analysis articles by *National Post* writers based on opinion polls claimed that 35 percent of Canadians were prepared to accept the US dollar as Canada's national currency.[92] Topping the list of impediments to productivity in Canada, according to Terence Corcoran, were unions and provincial health care "monopolies."[93]

The Royal Bank's senior advertising manager saw the package as a convergence not just of media, but between advertising and information. "It's not just a brand message," said Peter Tutlys, "but a brand story."[94] A *Regina Leader-Post* reader was less impressed. He protested in a letter to the editor that the insert was nothing more than a "lengthy right-wing editorial." Its inclusion in the *Leader-*

Post, he added, meant the daily "must discontinue all claims that it is an objective newspaper."

> An objective newspaper does not release 12 pages of one-sided, biased propaganda. ... The press is a source of information for Canadians. It is not a political pulpit. The perversion of the press into a vehicle for political propaganda must be stopped. It is time for the federal government to take action.[95]

CHAPTER 7

The Gazette Intifada

The first sign that anything was amiss at the *Montreal Gazette* in 2001 was the sudden resignation of publisher Michael Goldbloom at the end of August. He had headed the city's only English-language daily for seven years, and the Aspers had owned it for a year. Goldbloom made it clear in announcing his departure that he was quitting because of differences with them. "When a company changes hands, there is inevitably a period during which the new owners and the existing executives have to determine whether they see eye to eye and if they are a good fit," he wrote in a memo to staff. "CanWest has a more centralized approach to its management, and there are some aspects of the operations where we have had different perspectives."[1] The *Globe and Mail* had its reporters ask around about possible reasons for Goldbloom's departure. "Sources at *The Gazette* confirmed yesterday that senior editors at the paper were told earlier that month to run a strongly worded, pro-Israel editorial on a Saturday op-ed page," it reported.[2]

The contentious editorial argued that Canada should back Israel's response to recent Palestinian attacks, no matter how harsh. The New York-based *Columbia Journalism Review* reported that the editorial was ordered to run in newspapers across the Southam chain.[3] It was written in the wake of the suicide bombing of a Jerusalem pizzeria that killed fifteen unarmed civilians. Israeli troops surrounded Palestinian headquarters in East Jerusalem. Some cabinet ministers called for the assassination of trapped leader Yasser Arafat in retaliation. The editorial appeared to agree.

> Canada must recognize the incredible restraint shown by the Israeli government under the circumstances. . . . Howsoever the Israeli government chooses to respond to this barbaric atrocity should have the unequivocal support of the Canadian government without the usual hand-wringing criticism about "exces-

sive force." Nothing is excessive in the face of an enemy sworn to your annihilation.[4]

CJR reported that the editorial was accompanied by a no-rebuttal order from CanWest head office. "Papers in the Southam chain were told to carry neither columns nor letters to the editor taking issue with that editorial, according to journalists at two Southam papers, who said the order came via a conference call."[5] A letter to the editor did appear in the *Gazette* a few days later, however. It described the editorial as an "overwrought response . . . all but calling for Israel to rain vengeance upon the Palestinian people."[6] While he would not discuss his resignation at the time, Goldbloom gave an interview to the *New York Times* in 2002. "There is no question in my mind the Aspers feel they own the newspapers and the newspapers should reflect their views," he said. "It's not just what you see in the paper but what you don't see."[7] According to the British magazine *The Economist*, editors of CanWest newspapers were given instructions in March of 2001. They were told "to avoid talk of an imminent recession (in case that might hasten one), and to provide pro-Israeli coverage of the Middle East. Criticism of the broadcasting regulator was also said to be off-limits. CanWest's television licences . . . are up for renewal."[8] The *Gazette*'s editorial page editor, Peter Hadekel, asked for and was granted reassignment. Goldbloom's successor as publisher was former Canadian Football League commissioner Larry Smith, who had no experience in the newspaper business. Smith, a former fullback with the hometown Alouettes, pledged his full support for CanWest's editorial policies.[9]

National editorials

The controversy that would become news in Canada and around the world erupted on December 6, 2001. Henceforth, CanWest decreed, regular "national" editorials written at company headquarters in Winnipeg would run in all of the Southam dailies. The only exception was the tabloid *Province* in Vancouver, where CanWest also owned the *Vancouver Sun*. The duplication of editorial views in "competing" local newspapers, after all, would have appeared odd. At first scheduled to appear once a week, the head office editorials were planned to eventually run thrice weekly. The first dealt with tax treatment of donations to private charitable foundations. While the Aspers were indisputably generous, their philanthropy was also often strategic, designed to gain influence. Being a tax expert, Izzy Asper well understood that because it was usually tax deductible, charitable giving was also financed in large part — or even mostly — by other Canadians. That was a subject, some pointed out, in which the Aspers had a direct interest, as the family had its own private charitable foundation. CanWest not only ordered the editorials carried without dissent, it did so under the Southam logo. As one study of the national editorials pointed out, the use of such corporate logos, known as "branding," was a common practice in television. Importing the concept to newspaper journalism, it concluded, proved less than successful.

> The inclusion of the Southam torch logo, which accompanied all national editorials, appears to be a case of direct borrowing from television of the concept of network branding. The question becomes: how appropriate is the transfer of this branding concept from one medium to another? We suggest that the concept did not travel well at all.[10]

The Southam "brand," which had been bought by CanWest from Conrad Black, stood for quality journalism. Just as importantly, it stood for local independence for publishers. Southam head office had historically taken pains to allow its newspapers to reflect the temper of their communities, even if that meant disagreeing with ownership.[11] While the Aspers owned the assets of Southam, their attempt to capitalize on its brand — while flouting the principles it stood for — rankled many journalists. None were more miffed than those at the *Montreal Gazette*. Their community stood not only on the French-English front line, but also a bastion of multiculturalism that included sizeable Jewish and Muslim populations. The *Gazette* had been founded at the height of the American revolution with the help of Philadelphia's Benjamin Franklin. It boasted, one *Maclean's* writer observed, "attitude weighed by the ton, panache to match, weird idiosyncrasies, crazy writers and wicked editors." The newspaper was a throwback, noted Benoit Aubin, to the days before strong local editors were replaced by "faceless corporate managers from out of town."[12] Unbeknownst to most, the newspaper's journalists had already been tested twice in 2001 by head office interference. The ordered national editorials proved the proverbial straw that would catapult the CanWest controversy onto the national stage.

Byline strike

The protest by journalists at the *Gazette* was immediate. Many went on "byline strike" by withdrawing their names from atop stories. "We're skittish about anything that smacks of corporate centralization," reporter Doug Sweet told the *Globe and Mail*.[13] *Gazette* reporter William Marsden went on CBC radio's *As It Happens* to outline the trouble that had been brewing behind the scenes. He told of television critic Peggy Curran, whose review the previous week of a CBC documentary was held back by editors. It didn't run until she filed a union grievance, and then only with one major change.[14] The documentary *In the Line of Fire* by Canadian film-maker Patricia Naylor had aired November 28 on CBC television. Naylor had interviewed journalists wounded on the West Bank while covering the Palestinian-Israeli conflict. The review that appeared in print, which was not flattering to Israeli troops, contained one important diference from Curran's original.

> The documentary, which takes a point-of-view approach in arguing its case, is a chilling portrait of war correspondents and camera crews under siege. Palestinian journalists, many of whom work for Western news agencies, say they feel threatened by young Israeli soldiers who seem unwilling or unable to make a distinction between them and stone-throwing protesters.[15]

The sticking point with *Gazette* editors, according to the *Globe and Mail*, was Curran's characterization of the film. "She was told that the piece had to be rewritten, sources said, adding that editors wanted to label the documentary 'one-sided' until Ms. Curran agreed to call it a 'point-of-view documentary.'"[16] *Gazette* editor-in-chief Peter Stockland told the *Toronto Star* that Curran's column received only "routine editing to ensure balance." Reporter Bill Schiller pressed him on the issue. "'It is a factual error to say the Curran column was altered,' he [Stockland] said, adding it just needed something 'inserted,'" wrote Schiller. "Presumably, Curran pressed the keys that made the changes to indicate the program wasn't necessarily truth, but 'a point of view' documentary.'"[17] The problem, reporter Marsden told *As It Happens*, had been going on for some time. "They do not want to see any criticism of Israel. We do not run in our newspaper op-ed pieces that express criticism of

Israel and what it is doing in the Middle East et cetera. We do not have that free-wheeling debate that there should be about all these issues."

> We even had an incident where a fellow, a professor at . . . the University of Waterloo, wrote an op-ed piece for us in which he was criticizing the anti-terrorism law and criticizing elements of civil rights etcetera. Now that professor happens to be a Muslim and happens to have an Arab name. We got a call from headquarters demanding to know why we had printed this.[19]

Marsden then went bilingual in an interview with the Paris daily *Liberation* in a message even a unilingual Anglophone could understand. "C'est la Pravda!"[20] The *Gazette* journalists put their complaints against CanWest's national editorials into an open letter on December 10. It was signed by 55 of them and published the next day in the *Globe and Mail*. According to Canadian Journalists for Free Expression, the number of signatories to the "Gazette Intifada" would rise to 77 by January.[21] Centralizing opinion writing at CanWest headquarters, the letter pointed out, would "undermine the independence and diversity of each newspaper's editorial board." The *Gazette* journalists argued it would thus provide Canadians with a "greatly reduced variety of opinion, debate and editorial discussion."

> We believe this is an attempt to centralize opinion to serve the corporate interests of CanWest. Far from offering additional content to Canadians, this will practically vacate the power of the editorial boards of Southam newspapers and thereby reduce the diversity of opinions and the breadth of debate that to date has been offered readers across Canada.[23]

The response from CanWest and the Aspers took two forms. The first was a notice posted on the *Gazette* bulletin board threatening employees with discipline for speaking out against the company's editorial policy. "Crucial as free expression and a free press are to journalists, they do not automatically trump every other right nor does the designation 'journalist' negate the right of the owner of a newspaper company to run that newspaper as he or she wishes, consistent with the law," read the memo. It was signed by Stockland and by managing editor Raymond Brassard. "Case law supports sanctions, including suspension or termination, against those who persist in disregarding their obligations to their employer after clear warning."[24] Just how serious CanWest was in suppressing their revolt was made clear to *Gazette* journalists when sports columnist Jack Todd was recalled from an assignment in Boston and suspended without pay for a week. According to the *Globe and Mail*, Todd was disciplined for merely "sending a critical reply to a company-wide email announcing new appointments at CanWest."[25]

The harsh measures made some *Gazette* journalists even more defiant, including cartoonist Terry Mosher. His drawings under the pen name Aislin had long been some of the most trenchant in Canadian newspapers. Twice Mosher attempted to lampoon the national editorial policy, only to have the cartoons pulled by unamused editors. One, according to the *Globe and Mail*, carried a caption reading: "Imagine, a newspaper that looks just like, ummm, Global Television."[26] The other, which found its way onto the Internet, showed a Montrealer eagerly awaiting delivery of his newspaper in the pre-dawn darkness. "Where's *The Gazette*?" he asked. "I can't wait to read their enticing editorial view from Winnipeg."[27]

Gazette columnists who tried to make sense of the furor for readers found their copy altered by editors if it was too critical. Don MacPherson, the newspaper's Quebec affairs columnist, applied his cultural expertise to the problem. He wrote a column arguing that Quebec anglophones were unique among Canadians and required a distinct voice rather than one stamped from a template in Winnipeg. "A policy that forbids a newspaper from deciding for itself where the interests of its readers lie is not only bad journalism, it's also bad business," it concluded, according to the *Globe and Mail*.[28] In print, however, his column came to an entirely opposite conclusion. "A uniquely Canadian policy that allows for editorials written from both local and national viewpoints, and occasional lively disagreement between the two, could be good for business."[29]

Journalistic 'riff raff'

The other way CanWest reacted to the uprising in Montreal was in a war of words that escalated quickly. After the byline strike by *Gazette* journalists became news across the country, David Asper responded in a *Winnipeg Free Press* column. "Our newspapers have and will continue to publish a wide range of views, many of which are anathema to the ownership," he wrote.[30] After the *Globe and Mail* printed the open letter from *Gazette* staff members a few days later, however, Asper made a much less temperate response. "It's the end of the world as they know it," he said in a speech in Oakville, Ontario, paraphrasing a song by the rock band REM, "and I feel fine." If his rock-and-roll reference was lost on some, Asper's defence of CanWest's editorial policy could not possibly have been. He lambasted what he called the "bleeding hearts of the journalist community" in a diatribe that was reprinted in CanWest newspapers. "I seriously wonder whether all the fuss is really about the growth of a western-based source of opinion that is now being fed to the East," Asper said. "I mean, God forbid that a national point of view should come from anywhere but Toronto, Ottawa or Montreal." Asper saved his harshest words for CanWest's protesting journalists.

> Since when do reporters at the *Montreal Gazette* have a right of free speech that is greater than that of anyone else? They have launched a childish protest, with all of the usual self-righteousness. ... If those people in Montreal are so committed, why don't they just quit and have the courage of their convictions? Maybe they should go out and, for the first time in their lives, take a risk, put their money where their mouth is, and start their own newspaper?[31]

Asper also lashed out at the company's competitors, which had been covering the Montreal protest with great interest. "According to them, by distributing these editorials, we have brought the entire world of freedom of expression to a crashing halt. They would have you believe that owners should either never contribute material, or that if they do, it should be done under a cloak of secrecy with a nudge nudge to our editors. This of course is ridiculous." Chief among Asper's targets was the *Toronto Star*. He pointed out that its editorial policy was bound by the principles of social justice enunciated more than a century earlier by founding publisher Joseph Atkinson. "In the case of the *Star*, a dead owner is controlling the show," he told his audience. "At least in our case I'm actually here and accountable." Asper rattled off a list of publishers whose ethics he preferred to those of Atkinson, including Hearst, Black and Murdoch. "All these media titans have established a tradition of owner-opinion." He concluded by making another reference to popular culture that likely escaped his audience as easily as had the first. This one was to animated spaceman Buzz Lightyear. "We fully intend to ignore the rest of the riff-raff who, I suggest, are motivated by selfishness and perhaps a fear that they are not able to keep up with our drive for excellence," said Asper. "For the rest of us, we are headed to infinity, and beyond."[32]

'Fiercely independent'

Asper's insistence that CanWest had the right to impose editorials on its newspapers contradicted what the company had told the Lincoln committee just three months earlier. In its brief to the committee in September, CanWest had stressed the local independence it allowed its newspapers.

> Each of our metropolitan and local newspapers is a strong player in its own community. Each is relentlessly local in its coverage and fiercely independent in its editorial policy. Under CanWest's ownership, that will not change. On the contrary, CanWest understands that the success of our newspapers is due largely to their ability, in their editorial policies, to mirror the interests and values of their local readers.[33]

Soon a chorus of voices rose in opposition to the editorial interference. One of the first reproaches to the Asper edicts came from three former editors of the *Gazette* and was published in the *Globe and Mail*. "Over time, as the corporate editorials cover more and more public issues," they pointed out, "the range of topics on which local papers may express their own views will diminish dramatically." Mark Harrison, Norman Webster and Joan Fraser had served consecutively as editor-in-chief of the *Gazette* from 1977 until 1996. They called the national editorial policy "a potentially dangerous distortion of freedom of the press." The ex-editors argued that there were good reasons for allowing local editorial boards the freedom to gauge the temper of their own communities. "That is why Southam Inc., when it owned these papers, had a strict policy of guaranteeing the editorial autonomy of each newspaper. If CanWest Global wants to express its views, let it do so on op-ed pages, where competing opinions are also free to appear."[34]

Columnists had a field day with the controversy. Graham Fraser of the *Toronto Star* pointed out why Montreal had seen the first protest over CanWest's editorial policy. "Any reporter in Quebec understands that issues considered above politics elsewhere — such as the nature of the country — are highly divisive. Respect for the opinions of those who profoundly disagree is essential, particularly when many are convinced that the English press is engaged in a propaganda war against French-speaking Quebec."[35] *Globe and Mail* columnist Lysiane Gagnon concurred. "If the Aspers decided to call for the abolition of the Official Languages Act, the *Gazette* couldn't defend the interests of its readers, who, as Canada's only English-speaking minority, want federal protection."[36] Charles Gordon of CanWest's *Ottawa Citizen* pointed out that freedom of the press did not just mean freedom from government regulation. "If the unsigned editorials, the voices of the individual newspapers, cannot deviate from a line established elsewhere, then something is lost," wrote Gordon. "Whatever you call it — and it would be nice to think that the freedom from government is not the only freedom journalists have — it matters."[37]

A 'perversion' of journalism

Quebec's minister of communications, Diane Lemieux, announced she would introduce legislation to ensure "a plurality of opinion and ... sources of information."[38] A motion passed the next day by the Quebec National Assembly asked Southam to publish a "statement of principle and of commitment to the quality and diversity of news." It also asked it to "maintain and preserve the original character and autonomy" of the *Gazette*. Public hearings on newspaper ownership concentration had been held in Quebec the previous year. Two companies — Quebecor and Paul Desmarais' Power Corporation — owned all but one of its French-language dailies, controlling 97 percent of circulation. The inquiry issued a report in November calling on newspaper companies to publish statements of principle and to ensure diversity.[39] Lemieux said the Aspers' national editorial policy posed "troubling questions" about the control of news. "The actions of CanWest these last few days are without precedent and lead us toward a certain uniformity of information."[40]

Media critics also attacked the national editorial policy and no-rebuttal order. Ryerson journalism professor John Miller called them "one of the worst abuses of corporate ownership in recent history."[41] Peter Desbarats admitted he was "in a state of shock," as were most Canadian journalists. "What was a personal crusade and amusement for Mr. Black has become, in the hands of the Aspers, a crude and potentially dangerous weapon of political influence," he wrote.

> In less than a year, they have succeeded in antagonizing many of the journalists whom they employ, alarming the intellectual community and triggering a debate about the extent of their media empire and its enormous influence on our political life. Surely this debate can only have negative consequences for them politically and commercially.[42]

Concordia University journalism professor Mike Gasher pointed out that CanWest had "inadvertently given new life to criticisms of corporate concentration."[43] Worse by far for them, according to the former Southam journalist, CanWest had degraded its own product by disdaining basic ethics of newspapering. "CanWest is threatening Southam's journalism with a loss of credibility, something that CanWest managers don't seem to fathom."[44] Even marketing experts agreed. "The Aspers' journalists and critics are doing them a big favour by warning them of the dangerous ground they're tramping on," observed *Marketing* magazine. "In the long run, the commercial viability of these newspapers will diminish if they are not trusted as credible sources of unbiased news and sources of diverse opinion by readers."[45]

The Washington-based National Conference of Editorial Writers also weighed in on the issue. "It is an unwise business decision," wrote NCEW President Fred Fiske in a December 28 letter to CanWest, "likely to backfire with readers who are accustomed to editorials on national and international subjects that take account of the diversity of views in their communities and the ways those subjects are linked to local and regional concerns."[46] The group also addressed the controversy at length in its magazine, *Masthead*. "Major dailies that are unabashed mouthpieces for their owners are anathema to anybody remotely acquainted with the profession's modern ideals," it observed. "That editorials should be delivered from on high is a perversion of legitimate journalism. Readers expect their editorials to interpret issues in accordance with their local and regional concerns."[47] The US equivalent, it argued, would be the Knight Ridder chain forcing its newspapers to run an editorial calling for normalized relations with Cuba, which would alienate many readers of its *Miami Herald*.

Some American media conglomerates provide unsigned editorials to be used at the editors' discretion, but none requires that they be run. Readers will turn to newspapers that better represent them if they feel Southam dailies deliver opinions from a corporate-length remove. After all, why read the local newspaper's editorials if they take the same position as every other newspaper in the country?[48]

The battle spreads

Soon the controversy over CanWest's national editorial policy was attracting major media attention. "You can fit everyone who controls significant Canadian media in my office," Vince Carlin told the *Washington Post*. "This is not a healthy situation." Carlin, an American who chaired Ryerson's journalism school, explained the differences in media for his countrymen. "There is competition in the United States," he pointed out. "There is no competition here."[49] Izzy Asper defended the national editorial policy at CanWest's annual meeting, which convened in Vancouver early in 2002. "As publisher-in-chief, we are responsible for every single word which appears in the papers we own," he said, "and therefore on national and international key issues, we should have one, not 14 official editorial positions." He claimed CanWest's critics had "irresponsibly and deliberately misstated and mis-characterized" the national editorial policy. He promised not to back down. "No owner is going to back off from doing the right thing," Asper said.[50]

The debate had gone on without comment from the disgruntled *Gazette* journalists since CanWest's threat of firing. They had discontinued their byline strike after only two days, filing a grievance with their union for the right to continue it. They had even taken down a website erected to publicize their complaints. David Asper's "riff raff" rant, however, was more than they could endure without responding. Fearing reprisal, they did so anonymously in *Media*, the quarterly magazine of the Canadian Association of Journalists. Calling David Asper an "ideological pitbull" who "went off like a Roman candle," the Intifada lashed out. "What kind of idiot tries to tell reporters what they are allowed to think?" scoffed the *Gazetteers*. "None of this has reduced our determination one iota."[51] They called for a Royal Commission on Media Concentration to supplement the Lincoln committee inquiry already under way. "We've seen the ugly face of censorship at the *Gazette* and, in case you're wondering, the face looks an awful lot like Izzy Asper's."[52]

Soon the battle for editorial independence would be joined across the country. In January, Stephen Kimber quit CanWest's *Halifax Daily News*, for which he had written since 1985, after a column he wrote on the national editorial policy was killed. For Kimber, who also taught journalism at the University of King's College, the issue brought a crisis of conscience. As a teacher charged with imparting the higher ideals of journalism to his students, he said he felt he had a duty to speak out.

> I should not only know better than to acquiesce in such blatant censorship and manipulation but also ... I had a tenured job at a university. Unlike fulltime journalists who would be risking jobs, houses, even families, if they spoke out, I had the luxury of not only being able to quit my column at the newspaper but also to do so publicly.[53]

Kimber said he had an earlier column killed for "playfully mocking the *National Post*'s fawning front-page coverage" of Global's new national newscast. Another on the Middle East conflict had been radically altered. "I cited the failure of Israel's policy of escalating revenge in response to acts of terror as an example of why George W. Bush's single-minded war on terror was also doomed to failure," Kimber recalled. "In the published version, that argument vaporized."[54] As a result of the editing, Kimber said he found himself consciously steering away from certain subjects. "I began to censor myself," he told the *Washington Post*. "I would remember, 'No, I'm not supposed to write about that.'"[55]

While he had avoided topics that might be sensitive to CanWest in the weeks since, Kimber admitted misgivings. "The courage of the *Gazette* reporters — and the Aspers' oppressive retaliation — forced me to rethink my too-comfortable position."[56] He decided to force the issue by writing a column he knew CanWest would never publish. When editor Bill Turpin called to tell him the column had been killed, Kimber resigned. "Secretly, I knew, he was hoping my resignation might finally force the nasty story of what was happening inside CanWest papers from coast to coast onto the public consciousness," recalled Kimber. "It didn't take long for the floodgates to burst."[57]

Kimber's resignation was reported in the *Globe and Mail*. Southam

News editor-in-chief Murdoch Davis denied his column had been killed by head office in Winnipeg. Davis said local editors "decided on their own to spike the column based on the CanWest policy that management decisions are not to be debated in its newspapers."[58] Two days later, Kimber's rejected column appeared in the *Globe and Mail*. "I've had more than one recent column sliced and diced," he revealed. "I can only assume it was done to remove opinions that did not correspond with those of the new owners."[59] Kimber went on to deconstruct the CanWest agenda.

> Winnipeg's Asper family, which made its fortune in the television business, appear to consider their newspapers not only as profit centres and promotional vehicles for their television network but also as private, personal pulpits from which to express their views. The Aspers support the federal Liberal Party. They're pro-Israel. They think rich people like themselves deserve tax breaks. They support privatizing health-care delivery. And they believe their newspapers, from Victoria to St. John's, should agree with them.[60]

'A hill to die on'

Suddenly the controversy over Asper family interference in the news had renewed life. Perhaps more importantly, now the stifled journalist could actually speak publicly about the issue, unlike CanWest's muzzled employees. The issue of convergence was being examined in other countries, and Kimber found himself sought after as a witness to relate his experience with censorship. He was invited to Washington, DC in February to brief staff of the Federal Communications Commission. He appeared at Senate committee hearings in Australia, where restrictions on cross ownership were also being reconsidered. "Many Canadians are beginning to ask whether we need regulations in our country to prevent companies from owning competing media in the same community," Kimber testified. "It would be unfortunate if Australia chose this moment to abandon a rule that has helped protect the publication and broadcast of a variety of diverse and competing views on local issues."[61] His sudden popularity had unexpected benefits, Kimber told *Maclean's*. "If I'd known it would turn into a media circus and would give me all these frequent-flyer points," he said, "I would have quit a long time ago."[62]

Back in Halifax, Kimber's resignation brought a reaction from other journalists at the *Daily News*. Columnist Stephanie Domet wrote a defence of Kimber that was killed, prompting her resignation. Other columnists at the *Daily News* had already quit or been dropped from the paper after running afoul of CanWest's no-contradiction rule. They included Saint Mary's University philosophy professor Peter March, who had written a weekly column for the paper for ten years. Parker Barrs Donham, an award-winning investigative journalist and columnist, went into public relations in part because of the way his columns had been edited.[63] Staff columnist David Swick later confessed he had been instructed on what topics were off-limits and had been practicing self-censorship. After the *Daily News* was sold in 2002, he filled readers in on what had been going on. "Now that CanWest no longer owns the paper, the tale can be told," Swick wrote.

> Following the Sept. 11 terrorist attacks, I wrote a few columns about that event. I was soon informed I was no longer allowed to write anything to do with the Middle East. The reason: I was not perceived to be adamantly pro-Israel. The Aspers are adamantly pro-Israel, and their papers must reflect this sentiment.[64]

Daily News editor Bill Turpin objected to Murdoch Davis blaming him for spiking Kimber's column. He put his job on the line by placing the blame where he thought it really belonged. Turpin wrote a letter to the editor in response to comments by Davis in Southam's *Regina Leader-Post*. It was not published, but it was posted on the newsroom bulletin board and soon made it into the *Toronto Star*. "I and other editors had been urged repeatedly by Mr. Davis to get his advice on any prospective commentary that might run contrary to Southam Publications' rapidly changing editorial policies," read Turpin's letter. "To my profound regret, I did so in Mr. Kimber's case. Mr. Davis told me in colourful terms that publishing the piece would be a career disaster, at least as far as Southam was concerned."[65] Davis told *Maclean's* that the colourful language had been, "unless you're looking for a hill to die on." He also had some advice for CanWest employees. "To those journalists who say they've been chilled, I say: grow a backbone."[66] Turpin resigned a few weeks later.[67]

Columnists dropped

Columnists across the country who criticized CanWest's edicts found themselves out of work if they were on the Asper payroll. Peter Worthington, a *Toronto Sun* columnist, was syndicated not only across that chain but also in dailies as far-flung as the *Whitehorse Star*. The founding *Toronto Sun* editor was one of CanWest's harshest critics. "The Aspers just don't seem to get it," he wrote in one column. "Rather than having all Southam papers straitjacketed into one unanimous, identical view on national affairs, wouldn't it be more effective if the papers could espouse a similar view differently?"[68] The next day, Worthington learned that his twice-weekly column in a Southam newspaper had been dropped. "I got a rather embarrassed call from the *Windsor Star*," he told a reporter, "saying they had been ordered to drop my column and not run the columns under any circumstances on anything." The irony, according to Worthington, was that he had been practicing self-censorship in the *Windsor Star*, which was the only Southam newspaper for which he wrote. He had been careful in that newspaper not to criticize CanWest. "I don't think the Aspers know anything about newspapers," Worthington told the *Toronto Star*. "I think they just see them as something they can use to support their bids before the CRTC."[69]

The toll on columnists even reached Newfoundland. Michael Johansen quit the CanWest-owned *St. John's Telegram* after the paper refused to run his column criticizing the company's editorials. The newspaper's publisher, Miller Ayre, told a radio reporter that columnists for the chain were not free to criticize management decisions.[70] From the Maritimes, the CanWest controversy spread to the Prairies. Doug Cuthand, a native columnist for the *Regina Leader-Post* and the *Saskatoon StarPhoenix*, also had a column killed. It compared the plight of Canada's first nations with that of the Palestinians. "Over the years I have maintained a sympathetic point of view toward Palestinians," Cuthand wrote in the spiked column, which he posted on his website. "I see them as the Indians of the Middle East." It was the first time in ten years of writing a column for the Saskatchewan dailies that Cuthand had a column killed, but he couldn't afford to quit. "I'm going to carry on and continue writing," he told the *Toronto Star*. "But it will never be the same . . . I'll always be looking over my shoulder."[71] His squashed column, which lived on in cyberspace, drew some uncomfortable parallels between the Palestinans and indigenous Canadians.

> The Israelis also built their nation on other people's land but they regard any sign of dissent as terrorism. This is common practice. The demonization of a people and their leadership is a blunt instrument used to get the public on side. As First Nations people we have witnessed attacks on our leadership by groups such as the Taxpayers Federation and the Canadian Alliance.[72]

By the end of January, Davis felt compelled to respond to the growing protest. Those critical of CanWest, he claimed, were going on misinformation, false assumptions, and "goofy rumour" from the *Gazette* newsroom. "They are drinking their own bathwater, feeding their own rumours, repeating each other's errors," he wrote in a column. "They link random, normal editorial decisions at individual newspapers into a conspiracy. But they never bother to check it out." According to Davis, Kimber got key facts wrong, Cuthand's spiked column was "flimsy, badly researched, historically inaccurate and trite," and Worthington was worse. "In his world, you can lazily defame someone without checking facts, and still demand space in his newspaper and a regular cheque from him. And if he says no, it's censorship."[73]

The toll on columnists came full circle to the *Montreal Gazette* in 2002. Lyle Steward, who had been hired to provide a progressive perspective to the newspaper before CanWest bought Southam, quit after three of his columns were spiked. In Montreal's *Hour* magazine, Steward blamed his resignation on "the two local thought police in the CanWest ministry of truth." Editor-in-chief Stockland and newly-appointed editorial page editor Brian Kappler, he pointed out, killed the columns even though they did not touch on Asper-sensitive topics. "I issued no pleas to recognize Palestinians and other adherents to Islam as something more than subhuman," protested Steward, "I made no calls for Jean Chrétien's resignation, and above all, I failed to utter a single criticism of the Southam editorial-cum-untouchable party line." The last straw was the spiking of his interview with Vandana Shiva, an internationally recognized agricultural activist from India. She told him how genetic diversity of traditional crops was being lost there due to the introduction of genetically modified seeds. "This was the kind of story I was hired to bring to the *Gazette*'s overwhelmingly rightwing op-ed page," reasoned Steward. "Kappler, however, had never heard of Shiva, nor did he think her credible. Considering his political biases, that wasn't a great surprise, but he summarily killed the column before we even had a chance to discuss it. That was that."[74]

Prairie protest

In February, CanWest announced cancellation of its plan to increase the frequency of chain-wide editorials, which would continue to appear once a week.[75] That did nothing to stop the controversy over censorship from spreading, however. Haroon Siddiqui, retired editorial pages editor for the *Toronto Star*, gave the annual Minifie Lecture at the University of Regina's journalism school in early March. The recent clampdown on dissenting opinion at CanWest newspapers, said Siddiqui, a Sunni Muslim from India, had been "chilling."

> CanWest media are often critical, rightly so, of undemocratic Arabs who practise censorship against democratic Israel. Yet here we are in Canada witnessing creeping censorship against the Arabs. The Aspers have argued they have a right to their views. But that was never the real issue. Rather, it was their censorship of other views.[76]

Covering the speech for CanWest's local *Leader-Post* was reporter Michelle Lang. She filed a story that began by correctly summing up Siddiqui's main point. "CanWest Global performed 'chilling' acts of censorship," she wrote, "when it refused to publish several columns containing viewpoints other than those held by the media empire, a *Toronto Star* columnist said Monday." Following a consultation with the paper's night editor, according to the *Star*, editor-in-chief Janice Dockham was phoned at home. After a 15-minute conversation, alterations were ordered. The version that ran in the *Leader-Post* the next day, without Lang's byline, began differently. "A *Toronto Star* columnist says it's OK for CanWest Global to publish its owners' views," it read, "as long as the company is prepared to give equal play to opposing opinions."[77] Journalists at the *Leader-Post* followed their counterparts in Montreal and withdrew their bylines in protest. "It's censorship," *Leader-Post* reporter Colleen Silverthorn told the *Star*. "For Pete's sake, they wanted the word censorship out of the article."[78]

Again CanWest reacted harshly. Four reporters who commented publicly were given five-day suspensions. Six others who withdrew their bylines were given letters of reprimand and warned of worse punishment if they did it again.[79] As far as Izzy Asper was concerned, they got off lightly. "I will not tolerate an employee who is not loyal to his employer," he told *Maclean's*. "I happen to think the sanction should be much more strong." Leonard Asper was unsympathetic when told some Southam journalists feared for their jobs. "There's only so much time and effort we can spend trying to comfort people," he said.[80] The byline strike was described as "a tempest in a teapot" by Davis. "Some reporters don't understand the difference between editing and censorship," he told the *Globe and Mail*. "They should go back to school. The point is they ran the story."[81] He also attacked Siddiqui, one of the few working journalists ever to be awarded the Order of Canada.[82] "Who is Haroon Siddiqui? We're not talking about one of the great icons of Canadian journalism," said Davis. "He's just a guy."[83]

Davis gave CanWest's side of the story to *Leader-Post* readers in a column that claimed inaccuracies. "Siddiqui made his speech without a single effort to speak to anyone within Southam Publications or CanWest Global Communications to determine our actual policy or practices, or to review the range of views published in our newspapers," Davis wrote. "Instead, he relied on false reports in competing media (including his newspaper) and a few loopy rumours circulating in journalistic circles."[84] In a column in the *Toronto Star*, he attacked "the brave Mr. Siddiqui" for sloppy research. "That he equated our journalism with the anti-Semitic ugliness of media in totalitarian Arab states is, frankly, just ignorant," wrote Davis. "Siddiqui seems to have issues there. He lists 'the Israeli-Palestinian conflict' as one of our national editorial topics, but we haven't written on it."[85] Davis and Southam president Donald Babick sent a memo to their publishers and editors outlining how to deal with protests such as had erupted in Montreal and Regina. "Southam will not in future allow byline withdrawal in such situations," read the memo. It claimed that the byline strike at the *Leader-Post* was an "abuse of the right or privilege of withdrawing bylines." That form of protest, the memo said, "is intended to apply when the writer has concerns over handling of his or her work. It is not permitted as a way to show disagreement with general editing or management decisions."[86]

By April, calls were growing for an inquiry into CanWest's management of the news. A petition signed by 53 prominent Canadians was tabled at the annual conference of the Canadian Association of Journalists. They included authors Margaret Atwood and Pierre Berton, former federal NDP leader Ed Broadbent, and former Conservative cabinet minister Flora MacDonald. "We call on the Canadian government to commission a public inquiry into the effects of concentrated media ownership," read the petition, which was reprinted in the *Globe and Mail*. "The purpose is not to encourage any government intrusion into the nation's newsrooms, but to allow a full airing of concerns about the media system's responsibility to democracy."[87] On hand to defend CanWest was consultant Raymond Heard, who dismissed the petition. "All these names are great names [but] I would like to see this list fleshed out," he said, adding it would be more credible if some names of bankers were included. "I'm not signing it," said Heard.[88] *National Post* columnist Robert Fulford similarly dismissed the call for a government inquiry.

> Should people who put that much blind faith in authority be taken seriously when they use the word "freedom"? Freedom remains the question here, no doubt about it, but perhaps not in the sense that the signatories believe. What they need is freedom from their own distorted perceptions — freedom from groupthink, freedom from reflexive opinions, freedom from over-the-top responses to minor provocation, freedom from a neurotic reliance on government. My slogan: Free the Globe 53![89]

Also tabled at the CAJ conference was a report by Canadian Journalists for Free Expression. It chronicled the CanWest dispute and made recommendations for resolving it. In releasing the 32-page report, CJFE executive director Joel Ruimy called CanWest "clumsy" and "oafish" in its attempts to stifle dissent among its journalists.[90] The report urged CanWest to take back its former columnists, rescind its ban on byline strikes, drop its threats of discipline, and apologize for the personal attacks on its critics. "Suspensions, threats of dismissal and reprimands to journalists exercising the right to dissent are unacceptable and unnecessary curbs on freedom of expression," it said.

> Media organizations have a special responsibility to ensure that their activities do not weaken support for free-expression

rights in general. Journalists must retain the freedom to speak publicly on professional issues, including those involving their own employers. CanWest Global's confrontational style and dismissive tone will only make it more difficult to unite all media professionals in the task of defending free expression.[91]

Senate inquiry forms

Despite the ongoing Lincoln committee hearings, several senators who were former journalists began to plan their own investigation. "We're not talking about holding a Senate inquiry on national editorials in the Asper papers," said Joan Fraser. "What that did was remind us all of the enormous changes that have been going on in the structure of the industry."[92] Fraser had resigned as editor of the *Montreal Gazette* after Southam had been taken over in 1996 by Conrad Black, who also weighed in on the CanWest controversy. "It's just regrettable [it is] such a public relations and journalistic problem," Black told the *Globe and Mail*. "Personally, I think they could probably have accomplished just as much by having a signed op-ed piece when such things came up. But I know that his [Izzy Asper's] intentions aren't sinister."[93]

CanWest executives launched a campaign aimed at correcting what they considered a misperception about the magnitude of the company's influence. Ken Goldstein, CanWest's executive vice-president and chief strategy officer, wrote a response to the "Globe 53." He noted that when Hollinger sold twenty-seven of its dailies to CanWest, "it actually represented a decrease in concentration of newspaper ownership." The call for a government inquiry into press concentration was therefore based on a false premise, he pointed out. "What the critics really want to do is to stifle an approach to journalism and a set of opinions that differ from theirs. So they have invented a 'concentration' argument to try to persuade the government to create an inquiry into the media." The Canadian Journalists for Free Expression report was dismissed by Goldstein for its argument that CanWest should not judge whether criticism was fair or even factual before printing it. "Here we have a group of journalists arguing that fairness and truth are not important, and that publishers of newspapers should be obligated to publish dissenting views, even if those views are completely false."[94]

A similar defence written by Raymond Heard appeared in thirteen CanWest dailies. It noted that the figure of 60 percent mentioned by Siddiqui was a vast over-statement of CanWest's share of the Canadian newspaper industry. While CanWest owned 27 of the country's 102 dailies, its share of circulation was only 34 percent, Heard pointed out. "Those are the facts, and any journalist who takes five minutes to check can verify them. So how did the '60 percent' myth begin?" The overstatement of CanWest's size and influence, Heard wrote, had been "swallowed" by the senators planning hearings.

> I am one of those Canadians who believe unelected senators have no more right to poke their noses into the newsrooms of the nation than our bedrooms. . . . Since CanWest is much smaller than claimed, and ownership is wider that claimed, what, exactly, would such an inquiry look into? Editing? Whether petitioners should check facts? Whether reporters should challenge speakers to provide accurate sources?[95]

If the calls for government action to protect freedom of speech in Canada seemed loud that spring, however, they would grow deafening by summer.

CHAPTER 8

L'Affaire Russell Mills

In early 2002, a series of political scandals rocked the Chrétien government and CanWest's ability to keep its journalists in line faltered. Public works minister Alfonso Gagliano was banished to Denmark as ambassador in January after his department improperly awarded millions of dollars in advertising contracts. Defence minister Art Eggleton was fired in May for giving a $36,500 research contract to a former girlfriend. Even with the suspensions in Regina, not all Asper journalists were willing to go along with CanWest's unwavering support for Chrétien. As political pressure on the prime minister increased, some Southam dailies began to break ranks and run editorials dissenting from those distributed by CanWest headquarters. The first to go after Chrétien was the *Ottawa Citizen* on June 1. Not only did the newspaper run an editorial calling for the prime minister to resign, it backed it up with four full pages of evidence against Chrétien. The *Citizen*'s deputy editorial pages editor, Graham Green, examined the litany of ethical lapses by government ministers. He concluded that the culture of corruption plaguing Cabinet had been inspired by Chrétien's own actions. His analysis of statements made by the prime minister as events unfolded showed that Chrétien had been less than truthful.

> A detailed review of the prime minister's own words . . . over the past three years shows that he has been the one ignoring the facts. It is hard to reach any other conclusion but that Mr. Chrétien knew what he was saying was untrue and chose to deliberately offer explanations he knew to be false. Only one word — lying — describes this behaviour, and that's what Mr. Chrétien has done in these situations.[1]

The *Citizen* editorial called for Chrétien's "swift, unceremonious departure" due to the "disgraceful behaviour" detailed in Green's indictment. "Mr. Chrétien, we ask you to step down — for the good of your party and of the country. If you will not, we urge the Liberal party to throw you out."[2] Two days later, a Southam editorial in the *Citizen* and other CanWest newspapers contradicted the call for Chrétien to step down. "On any calm and reasoned analysis there is no evidence of government 'rot' or 'corruption,'" it insisted. "For media and opposition critics to attempt to elevate ministerial mistakes, or behavioural lapses, or civil service mismanagement, to the level of 'corruption,' is irresponsible."[3] Another national editorial the next day addressed the dropping from Cabinet of finance minister Paul Martin, Chrétien's rival and heir apparent. That move brought renewed calls for the prime minister to step down. The second national editorial insisted there was "no government crisis, nor, except as goaded by the chattering classes, is the Liberal party in disarray." It added that Chrétien, who had earlier announced his intention to step down in 2003, should not give in to calls from within the Liberal party to hasten his departure. The prime minister, it concluded, had "earned the right to decide for himself when to go."[4]

Some newspapers in the Southam chain cast their lot with those at the *Citizen* who had defied the company line from Winnipeg. The *Vancouver Sun* commented in a local opinion printed alongside the mandated national editorial. Chrétien, it said, had "failed over the past nine years to articulate much in the way of a vision for the country [and] may be now forced to come up with one himself."[5] The *National Post* was even more stinging in its rebuke of the prime minister, describing him in an editorial that day as a "political shark, living to devour and survive."[6] In Ottawa, *Citizen* readers got an explanation from columnist Clark Davey the next day of the opinion warfare that seemed to have broken out on the paper's pages. "The Southam papers were told, in an article signed by David Asper ... that if they had proof of malfeasance on Mr. Chrétien's part, they should put it up. Otherwise shut up," Davey wrote. "The *Citizen* obviously took up the challenge." The subsequent "Holy Writ from Winnipeg" no doubt left *Citizen* readers confused, wrote Davey.

> So given the two wildly contradictory views expressed on the Citizen's editorial pages you, the reader, have to make up your own mind who you want to believe: your "fiercely independent" local editorial leadership (as promised by CanWest) or the

absentee ownership who made promises about independence that they haven't kept and who have a strong vested interest in the continued health of the Liberal Party.[7]

Davey wrote the column knowing he had another card up his sleeve, and it would be played on the pages of newspapers across the country the next day. The retired *Citizen* publisher was one of forty former Southam executives who had decided to make their views public. Calling their group DOVE — standing for Diversity of Voices Everywhere — they put their names to a full-page ad submitted to newspapers across the country. The Southam chain, however, refused to publish it as written. CanWest spokesman Geoffrey Elliot said CanWest wanted only "a minor adjustment" to the copy.[8] Instead the ad, which was paid for by members of the Southam family, appeared in BCE's *Globe and Mail*, and the independent *Winnipeg Free Press* and *Halifax Chronicle-Herald*. It was headlined "Is Freedom of the Press Being Lost, One Newsroom at a Time?" The ad claimed CanWest's national editorial policy and no-rebuttal order limited press freedom and had "already skewed editorials, commentary and news coverage about domestic and foreign issues." It argued that cross ownership of newspapers and television had altered the balance between the right of owners and the public interest.

> CanWest has reawakened us to the daunting potential of concentration of newsroom ownership — particularly when a single firm owns other media. We can picture a time when one or two multi-channel media giants, each with a single national voice, will be telling us all what to think. One danger, of course, is a loss of diversity of voices — a diversity that is the essence of free speech, democracy and pluralism.[9]

Many of the former executives had fought attempts to limit concentration of press ownership throughout their careers. Now, they declared the situation to have changed so drastically with cross ownership that it warranted government intervention. Tax incentives for media companies with independent newsrooms might encourage more responsible editorial policies, the Doves suggested. "Tax policies protect our culture from foreign ownership of Canadian media," they noted. "Might they protect our diversity of voices from converging media companies?" The hawks at CanWest, however, found the notion of government intervention to be a "much more Draconian threat to freedom of the press."[10] Besides, according to Elliot, there was no threat to press freedom from the company's national editorial policy. "We don't take very kindly to the accusation that we impose a single point of view and have skewed editorial and news coverage," he told the *Toronto Star*. "There is no relationship between CanWest national editorials and news coverage . . . it is sacrosanct."[11]

The 'Gordie Howe' of news

That first week of June was a tumultuous one for CanWest and the Aspers, but it was only a hint of the uproar they would provoke next. Ironically, the person who would be its focal point was one of CanWest's staunchest defenders from the call by DOVE for government action. *Ottawa Citizen* publisher Russell Mills had been a fixture at the newspaper since 1971, when he started there as a copy editor. He was named editor-in-chief in 1977 and became publisher in 1986. He was replaced by Davey from 1989–93 while he served as president of Southam Newspapers. He had been dubbed "the Gordie Howe of Canadian newspapering" for his longevity in surviving successive ownership regimes, including Conrad Black's. "He could also use his elbows when he had to," quipped Norman Webster in making the comparison to Mr. Hockey.[12] Mills was also a pillar of the community, supporting numerous charitable and artistic groups, and he had been honoured for promoting literacy and the city's public library.

On June 15, Mills addressed the annual graduation ceremonies at Ottawa's Carleton University. He was presented with an honorary Doctor of Laws degree for his service to journalism and the community. His speech, which outlined the dangers of government interference in the press, was reprinted in newspapers across the Southam chain. The ad placed by former Southam executives "may have been well-intended," Mills told his audience, but their proposed tax measures were a "misguided" remedy. "Government, by its nature, cannot be the guarantor of press independence," Mills said. "Good information about government can best be provided by privately owned media that are not under government control or influence."[13]

His defence of CanWest from calls for government regulation was made with the knowledge that he was almost certain to be fired the next day. Mills had received a letter from David Asper that week. The chairman of CanWest's newspaper group said he was "angry and disappointed" that headquarters in Winnipeg had not been informed in advance of the *Citizen*'s call for Chrétien to resign. Asper had announced his intention to visit Ottawa on June 16 to discuss the matter and he summoned Mills to a meeting at the Westin Hotel that night. There, Mills was informed he was out as *Citizen* publisher, he told one of his former reporters.

> They offered me an opportunity to retire with a monetary settlement, but I would have had to present this as a retirement and sign a confidentiality agreement which would be very binding and would prevent me from discussing this in any way. I simply wasn't prepared to portray this in an inaccurate way, and to say this was a retirement when in fact I was fired.[14]

A brief article in the *Citizen* the next morning announced that Mills had "left the newspaper" and that "few details were available."[15] News of his firing spread rapidly across the city, the country, and the world. A rally outside the *Citizen* building was hastily organized by former city councilor Karin Howard. Prominent community figures denounced Mills' firing and demanded his reinstatement. "Russ Mills was fired in order to set an example to every other employee of the Southam-CanWest organization," said Christopher Dornan, director of Carleton's journalism school.[16] Davey told the crowd estimated at 300 that the firing was "a terrifying example of the ultimate in the political corruption of freedom of the press in this country." He claimed Izzy Asper had met with Chrétien at the annual Press Gallery dinner on June 1, the day the controversial editorial was printed. "What they talked about can only be conjectured," Davey added.[17] *Citizen* journalists decided to withdraw their bylines for the rest of the week in protest and by the end of the day at least 500 readers had cancelled their subscriptions. By the end of the week, that number would reportedly reach 3,000.[18]

Corporate reorganization

The fallout at CanWest head office also came swiftly. A major corporate reorganization signalled that, despite his retirement, Izzy Asper was taking a more active role in the company's management. The *Globe and Mail* reported that the appointment of a new chief operating officer suggested the elder Asper was unhappy with the leadership of his youngest son. Richard Camilleri, a former music industry executive, had no news media experience but enjoyed a reputation as a cost cutter, the *Globe and Mail* noted. Leonard Asper admitted that CanWest's rapid growth had left him "unable to focus on everything I want to. . . . I just simply had too many people reporting to me."[19] The company's financial fortunes had taken a downturn even during the traditionally lucrative holiday advertising season. A $21.7-million loss for the three months ended February 28, 2002 was a sharp reversal from a $10.5-million profit during the same period a year earlier. By then, increased losses from full ownership of the *National Post* were starting to weigh down CanWest heavily.[20]

The mass cancellation of subscriptions brought by the Mills firing worsened the company's outlook even further. From a high of $22 in 2000, its share price fell to $8.50 in June 2002, its lowest level since 1996.[21] CanWest stock soon dipped below $7 as a labour dispute shut down its *Vancouver Sun* and *Province* dailies, costing it an estimated $1 million a day.[22] CanWest suspended its quarterly dividend and sought to sell assets, including its overseas broadcasting operations and small-market newspapers. Analysts noted that much of the debt CanWest had taken on in buying the Southam chain had been financed by issuing high-interest (12.125 percent) notes. The interest on them could be paid with additional debt. As a result, the $767 million in promissory notes were compounding at more than $110 million a year and had by mid-2002 reached $920 million.[23]

CanWest first sold its daily newspapers in Halifax, Charlottetown, and St. John's for $255 million in July of 2002 to magazine publisher GTC Transcontinental. The sale suggested to some that CanWest was abandoning its convergence strategy, as it cut ties between those dailies and Global's television operations in the Maritimes. Leonard Asper claimed the properties "were not central to the company's over-all media integration strategy."[24] The CanWest CEO had grown defensive after analysts questioned the wisdom of convergence in the wake of the dot-com collapse. The company's attempts at "multi-platform" advertising sales in print, television and online media had stalled with the economic downturn. One expert predicted such multimedia packages would never amount to more than a few percent of the total advertising pie. "I don't think convergence selling is a business plan," Mark Sherman, president of The Media Experts in Montreal, told *Marketing* magazine.[25] Asper defended convergence in a letter to the editor. "Convergence is on a roll but it is the new normal," he said of the lack of visible results.

> The process each company is undertaking to some extent is proprietary information and, to be quite frank, none of us care to delve into too much detail about our particular business models.... There is no more need for pregame hype. The game is on, so the players are concentrating on playing, not talking. At the first period intermission we'll be happy to provide the colour commentary.[26]

'Dangerous to democracy'

The political backlash from the Mills firing was intense. Opposition MPs in unison demanded an inquiry into press ownership. "Russell Mills was fired because the Prime Minister's buddy happened to be his boss," NDP leader Alexa McDonough told the House of Commons. "That is downright dangerous to democracy. We need a full public inquiry into media concentration, ownership and convergence."[27] Chrétien shrugged off that suggestion and denied any involvement in the firing of Mills, but some Liberals found it troubling. MP John Bryden told Parliament the incident showed "alarming immaturity on the part of the Aspers ... who appear not to have the foggiest notion of the concept of press freedom in a democracy."[28] The head of the federal Competition Bureau insisted the matter was outside his jurisdiction, which extended only to economic issues. He suggested the CRTC be empowered to ensure a diversity of voices in the press. "Right now, they are talking only about broadcasting," said Konrad von Finckenstein. "But there's no reason you couldn't extend that to print media if you wanted to. It would require legislative change but it is certainly possible."[29]

The senators who had been considering a news media inquiry grew more convinced that one was needed. "I think the events of the weekend have sharpened awareness of the issue," said Senator Joan Fraser.[30] The firing of Mills drew international attention in the *New York Times* and the *Financial Times* of London. The press freedom watchdog International Freedom of Expression Exchange added it to its weekly report of journalistic abuses in such countries as Albania, Haiti, China, Jordan, and East Timor.[31] The Vienna-based International Press Institute called it "an attack on press freedom by an unholy coalition between politics and big business." IPI director Johann Fritz pointed out the perils Mills's firing could have for journalism in Canada.

> Many believe that it is only in autocratic countries of the Third World or in countries in transition that democracy and a free press are in danger. But the Mills affair will have a chilling effect on critical reporting in Canada and will bring an increase in self-censorship.[32]

As usual, columnists in competing newspapers fell over each other to opine on the latest CanWest controversy. Jeffrey Simpson of the *Globe and Mail* longed for the good old days. "Little did any of us expect when the Aspers bought Southam from Conrad Black how creepy would be their journalistic standards," he wrote. "It almost makes you pine for Lord Black."[33] *Toronto Star* columnist Richard Gwyn thought Izzy Asper acted in a "vulgar and bullying manner" in the firing of Mills. "He's brilliant at making money; in this instance he acted stupidly. Specifically, he seems to hold his readers in contempt. Asper's abiding concern is with himself, with his own ideas, opinions, values, biases, prejudices."[34] *Globe and Mail* columnist Margaret Wente wrote that Asper made a "classic mistake" in turning the family business over to his sons. They had "no idea," she concluded, how to run a sophisticated media business. "The boys have botched it," wrote Wente. "David, the son who fired Mr. Mills on Sunday, figures you can run newspapers the same way he runs the Winnipeg Blue Bombers. You push people around, and if you don't get enough respect, you can them."[35] Andrew Coyne of CanWest's *National Post*, however, deemed the issue a "red herring" and declared the dismissal of Mills well within management rights. "A newspaper owner, in my view, has an absolute right to hire or fire whomever he likes," wrote Coyne. "Likewise, he is entitled to decide the editorial line of a given newspaper, or to insist that every newspaper in his possession take the same line. The notion that journalists are or should be free to write whatever they please, without regard for the views of the people who pay their salaries, is an arrogant fiction."[36]

The stage was set for the protagonists to square off on the issues. Mills went first with a guest column in the *Globe and Mail*. He warned that his dismissal could send "a serious chill across the newsrooms of CanWest papers." Mills said it could cause editors and other journalists to be "excessively cautious in their political coverage and commentary." That kind of self-censorship would be politically dangerous, he added. "In a democracy, the people are ultimately sovereign and they need thorough and accurate information and commentary if they are to judge the conduct of the public officials they have put in power." The most important function of the news media in a democracy, he wrote, was to provide independent information and analysis of government.

> Such information and commentary cannot be provided through government apparatus, nor can it be provided properly in the climate of fear that must exist at the CanWest newspapers today. Journalists should not have to fear risking their livelihoods and the security of their families in order to do the essential job of keeping Canadians informed.[37]

Despite the seriousness of the situation as he saw it, Mills said government intervention could make things worse. As a longtime journalist, he was loath to allow any appearance of government influence over the news. "News media that relied on government for protection from proprietors would lack independence from government." Instead, he urged Canadians to "take action to protect the quality and independence of the information and commentary they require in order to be good citizens of a democracy."[38]

Corporate damage control

In the days since Mills was fired, CanWest had been tight-lipped about his departure, saying only that it was a private matter and declining comment. The column by Mills, however, prompted a corporate response to the flood of negative publicity that was causing cancelled subscriptions across the country. Leonard Asper sent a memo to *Citizen* staff, who had taken out a full-page ad in their own newspaper to express their concerns over its management. Asper promised to address "any ambiguity or uncertainty" about the company's core editorial values. "We want our readers to have full confidence in the editorial integrity and independence of our newspapers," Asper wrote. "Readers must be reassured that our newspapers are reliable sources of both the news, and a full spectrum of informed opinion on the stories of the day."[39]

Asper also did a round of media interviews, including one with Kevin Newman on Global's national newscast. He denied that Mills had been fired as a result of a phone call from Chrétien or anybody else. "Even if the prime minister had called to complain," he told Newman, "nobody runs our newspapers and decides the content of our newspapers but us."[40] Asper told CBC television he was trying to set the record straight "and not let the torrent of abuse continue."[41] In an interview with the *Toronto Star*, he said Mills was "not fired for any one act." Instead it was the result of "an accumulation of situations in which he . . . did not comply with some basic journalistic principles." Appropriating the mantra of CanWest's critics, Asper claimed Mills had been guilty of failing to provide a range of perspectives in the newspaper he published. "In our view, the *Ottawa Citizen* was not providing a diversity of voices or a diversity of viewpoints over a prolonged period of time." He said the *Citizen* had "started to become a paper whose news coverage was being affected by the opinions of its local editorial board. We had asked Mr. Mills to address that on several occasions and he hadn't." Asper added that Mills had failed to exercise the "common courtesy" of notifying CanWest that his newspaper would be running such a strong editorial. "What he did, in our name, was say the prime minister should resign," Asper pointed out. "All we're saying is, the unsigned editorial space belongs to the proprietor and we have the right to decide what goes in that space in our name."[42]

In a telephone interview with the *Citizen*, Asper insisted the company had no "special relationship" with the prime minister. "No politician has any ability to influence how we run our business," he said. "We simply have certain principles which we believe in and stand up for. And even if we have to take some public flak for them, we're willing to do so." As an example of Mills resisting CanWest's efforts to make the *Citizen* more diverse, Asper cited his reluctance to run a story by *Montreal Gazette* reporter Hubert Bauch. It had quoted York University professor Ian Greene to the effect that the Chrétien government was the cleanest in history. "It was only at the insistence of CanWest that that appeared in the *Ottawa Citizen*," said Asper. The subscriptions cancelled in protest, he added, would be more than made up by new readers the newspaper gained through its diversity efforts. "For every irate subscriber in Ottawa, I promise you, that when these pages of the *Citizen* are opened to a more broad variety of viewpoints, there will be more subscribers joining the paper who welcome its openness to this greater variety and balance of coverage."[43]

Inside the *Citizen*, a 3,000-word transcript of the interview was spread across two pages. In it, Asper claimed that the promise of editorial independence CanWest made to the CRTC, to which critics often pointed, had been misinterpreted. Rather than referring to local independence, Asper said, the company's assurances instead meant "independence from or freedom from government interference, or the interference of any other interests outside the company." It did not mean, Asper added, that CanWest had no right to interfere with a local editorial board. "That is the job of the publisher and the proprietor to set the editorial direction of the company, and then adhere in our view to these other principles of diversity of voices, notwithstanding the editorial policy." Asper also elaborated on his expectations of editors and publishers for prior consultation with Winnipeg on editorial content. "On serious matters of national and international importance, there should be some consultation. Which I think is what every proprietor and publisher expects," Asper added. "I think it's accepted that the proprietor does have the final say over the unsigned editorial space." CanWest newspapers would be free, he said, to publish other views on their opinion page opposite the editorial page.

> The resignation of a prime minister, or the endorsation of a national political candidate, I think it's fair to say, would require some consultation. The reason being because the proprietor has his or her name on that statement. . . . It is an appropriation of my name and my voice to not consult. People are not allowed to put signs on someone's lawn.[44]

Mills scoffed at Asper's assertion that he had been guilty of failing to follow basic principles of journalism. "He must have read different journalism textbooks from those that I learned from," the deposed publisher told a *Citizen* reporter.[45] He demanded an apology from Asper and threatened a lawsuit if he didn't get one. "Mr. Asper's comments . . . were not true and unfairly damaged a reputation that I pride," he told the *Globe and Mail*.[46] Then Mills was handed an even loftier pulpit from which to denounce Asper in their escalating war of words. In a column in *Time* magazine's Canadian edition, he welcomed a Senate inquiry into the news media. The upper chamber's ensconced appointees, he added, were a better choice to study the issues than elected MPs. "The Senate may be flawed as an institution," Mills argued, "but its committees are somewhat independent of the stifling power of the Prime Minister's office and often do good work."[47] He also called for a ban on cross ownership of newspapers and television stations in the same city. "This is a change of opinion for me," Mills admitted. "I had previously believed that government should not be involved in regulating ownership of media. The current system only works, however, if owners do not abuse their power."[48] Now that the Aspers had demonstrated they would use their media power politically, Mills argued it was time to change the rules.

> Self-regulation would be best, but there is a public interest in preserving different sources of information and opinion when self-regulation fails. Our media-ownership rules should be set up to cope with bad owners. It's often claimed that a ban on cross-media ownership is unnecessary because of the wealth of information on the Internet, but, in fact, there is little community-based commentary on the Web.[49]

Crossing the country

The Mills firing not only worried press freedom advocates around the world, but its repercussions even reached Canada's west coast. The news media there were controlled by the Aspers as nowhere else in the country. According to one report, following CanWest's purchase of Southam, Vancouver had the most highly-concentrated media ownership of any major city in a G7 country.[50] So complete was CanWest's dominance of Vancouver media that it could cut back on its journalism without losing readers to the competition, because there was almost none. It did so relentlessly. The lack of competition also meant the CanWest dailies could promote a political worldview that few had the opportunity to challenge. David Beers, a former *Vancouver Sun* features editor who was purged in a 2001 round of cost-cutting, went public in *Vancouver* magazine with his disgust for the daily the following year. His description of working at the *Sun* was enough to send the average journalism student screaming into another field of study.

> For weeks I hunkered in my cubicle in the middle of the newsroom, trying to ignore the nearby business editor who'd shout "[NDP Premier] Glen Clark's a commie!" . . . When I look at the

Sun today — its bias naked in banner headlines, its shrinking news pages home to celebrities and pooches, its swelling advertorials issuing the Orwellian command that we all "Believe BC" — I feel relief at not having to be connected to all that.[51]

Through its Southam-owned Pacific Press monopoly — the *Vancouver Sun* and the *Province* — CanWest owned both local daily newspapers. Its BCTV evening newscast claimed more than 70 percent of the local audience.[52] Even that did not accurately measure CanWest's media dominance in Vancouver. It also broadcast into the market from its CH network station from nearby Victoria, where it also owned the only daily. It published one of the country's two national newspapers in the *National Post*. It even published most of the free-distribution "community" newspapers in BC's Lower Mainland. Included among its dozen such publications was the *Vancouver Courier*, which appeared twice a week within city limits in three zoned editions. With a weekly circulation of 240,000, it was Canada's largest non-daily newspaper. CanWest's thrice-weekly *North Shore News* was close behind, with almost 200,000. The *Now* newspaper chain, which published in three eastern suburbs, was even larger if its editions were added together, with more than 400,000.[53]

Southam had acquired the community newspapers in the early 1990s to "bomb proof" its Pacific Press d0.ailies from competition for local advertising.[54] The acquisitions had been challenged by the Competition Bureau, which was formed in 1986 to replace the

MARKET SHARE AND CROSS OWNERSHIP, 2002

Market	Owner	Newscast share %	Daily paper %
Quebec	Quebecor	47.1	56.2
Toronto	Bell Globemedia	43.8	18.3
Toronto	CanWest Global	33.0	11.5
Montreal (English)	CanWest Global	5.0	100.0
Montreal (French)	Quebecor	37.1	60.4
Regina	CanWest Global	28.3	100.0
Saskatoon	CanWest Global	15.2	100.0
Calgary	CanWest Global	32.2	57.8
Edmonton	CanWest Global	39.7	60.0
Vancouver	CanWest Global	70.6	100.0

Source: Senate Interim Report on News Media (2004)

ineffective Restrictive Trade Practices Commission. Owning those newspapers, the Competition Bureau calculated, gave Southam 86.5 percent of the advertising market in Vancouver and 57 percent on the North Shore. A tribunal hearing the case ordered Southam to sell either its *North Shore News*, or the *Real Estate Weekly* chain of fourteen newspapers. Southam appealed the ruling to Federal Court and the Supreme Court of Canada before agreeing in 1998 to sell only the North Shore edition of its *Real Estate Weekly*.[55]

Political 'free ride'

Throughout the Asper Disaster, westerners who didn't read the *Globe and Mail* or follow the story online were in the dark, noted *Winnipeg Free Press* columnist Terry Glavin. "Few newspaper readers in Western Canada even know it's going on or what it's about," wrote Glavin. "It makes CanWest Global look bad, so CanWest isn't going to allow its newspapers to draw much attention to what they've been up to." The ousting of a scandal-plagued NDP government in favour of the Liberals in 2001, some noted, had been followed by kinder, gentler coverage of provincial politics. The election of Gordon Campbell as premier had been aided by a reported $35,000 campaign contribution from CanWest. Some saw editorial support as well as financial support. "In BC, it's an open secret that it was the news media that elected Campbell," noted Glavin, a former *Vancouver Sun* reporter.[56] The problem brought by media domination, according to Glavin, became obvious after a May 2002 demonstration in Vancouver protesting Liberal cuts to social services. Local news media were "practically unanimous in dismissing" the protest, he noted. It fell to a journalist from not just outside the province to report the magnitude of the protest, but one from another country.

> It took a reporter for the *Seattle Post-Intelligencer* to tell a different story. The *P-I*'s Joel Connelly noted that the event was the largest demonstration in North America west of the Rocky Mountains since the 1999 "Battle of Seattle," but that Campbell enjoys an "opinion-shaping ally" in the conglomerate that owns Vancouver's dailies and its main television news shows.[57]

Political bias was openly admitted by some Pacific Press journalists in unguarded comments. A University of BC education professor noted just such a candid admission in the *Canadian Journal of Education*. A *Vancouver Province* political columnist appeared prior to the 2001 provincial election as a guest speaker in his teacher education class, according to Paul Orlowski. "During the 90-minute presentation, the journalist accepted numerous questions from the group," recalled Orlowski. "One yielded this answer, 'Hey, I am writing my columns to get the NDP out of government. I think they tax and spend way too much.'"[58] Michael Smyth, the columnist in question, did not deny in 2007 making that statement six years earlier, but claimed it "doesn't sound like something I'd say." He pointed out it was his job as a columnist to offer his opinions on provincial politics. "It was certainly my opinion at the time that the New Democrats did not deserve to be re-elected in 2001," he admitted in an e-mail, "and I would not have shied away from saying that."[59]

The perception of media complicity in Campbell's rule was reinforced by the fact the premier's brother Michael was a business columnist for the *Sun*. He incessantly advocated tax cuts in print and on his CKNW radio show. "There's certainly a perception from a large segment of the public that the Liberals are getting a pretty easy ride, and none easier than in the *Vancouver Sun*, from an editorial position," noted political analyst Bill Tieleman.[60] It wasn't just the newspaper's coverage of politics that went soft after the Aspers took over and the Liberals were elected, according to some *Vancouver Sun* journalists. Business coverage also deteriorated to a similar sorry state. "It is very fluffy and full of consumerism, sex and success," admitted *Sun* business editor Harvey Enchin. "That kind of crud."[61] *Sun* business reporter David Baines, described by *BC Business* magazine as "the most hated man in business" for his fearless exposés, regularly won awards for investigative reporting.[62] He told the magazine that business reporting in the *Sun* had become little more than free publicity for corporations. "It's really just 'producer' reportage, where we're willing to re-write press releases, to reiterate the company's point of view."[63]

The mouth that roared

One of the few independent voices in western Canada speaking out against CanWest's control of its media was talk show host Rafe Mair of CKNW radio. He also wrote a regular column for CanWest's *Vancouver Province*. In the controversy over the Mills firing, CanWest cancelled the *Vancouver Sun* and *National Post* columns of Gordon Gibson, a former leader of the BC Liberal Party. Mair quit his *Province* column in protest and urged his legion of listeners to cancel their subscriptions to both newspapers. He also called for an advertiser boycott of the duopoly dailies. Gibson had written a pair of contentious columns that tumultuous first week of June. One in the *National Post* referred to Chrétien as a "dead man walking" and predicted his imminent departure from Canadian politics. "He will be gone within a year," wrote Gibson. "This will be a very good thing for Canada."[64] Gibson's *Vancouver Sun* column on June 5 labelled the back-to-back Southam editorials defending Chrétien "dangerous to democracy" because they attempted to "muzzle" the press.[65] That column was reprinted by Southam's *Montreal Gazette* and *Kingston Whig-Standard*. Gibson was first informed by the *National Post* that his column had been cancelled, then reportedly telephoned by both Leonard and David Asper, who offered to reinstate it.[66] His column did not re-appear until September, however.

By then, Mair was involved in a full-scale brawl with the Aspers, who had lost a reported 1,000 subscribers in Vancouver within a week of his boycott call.[67] In response, according to *Globe and Mail* columnist Allan Fotheringham, the Vancouver dailies threatened to pull out of CKNW charities if Mair didn't "pipe down."[68] That included donating advertising for the radio station's annual holiday season Orphan's Fund drive. Far from piping down, Mair went on the air with an editorial on June 27 that denounced the Aspers as "hopelessly incompetent" and renewed his call for a boycott of Can-West. A former provincial cabinet minister for the defunct Social Credit party, Mair pledged that CKNW would continue to cover the Asper story despite pressure from CanWest. "What do they think we're made of over here — cotton candy? This station didn't get to the position they're in in this community by giving in to bullies or trying to influence their broadcasters."[69] Far from moderating his stance against the Aspers, Mair hardened it.

> I say that the Aspers slant the news and bully their employees and freelancers and that they are in thrall to the Liberal Party of Canada and its leader. Is their answer, "No, we assure you that we have never asked that the news from the Middle East be skewed?" Of course not, for that would be a lie so big that even they couldn't get away with it.[70]

Pacific Press did more than just pull out of the Orphan's Fund campaign. It cancelled its advertising on CKNW as well. It claimed it was not doing so in an attempt to stifle criticism, however. "We're not trying to silence Rafe Mair," Pacific Press spokesman Don MacLachlan told the *Georgia Straight* alternative weekly. "We're saying it makes no sense to run an ad to try to sell a newspaper on a show which is calling for subscribers to cancel the paper."[71] Mair then told his listeners about a letter he had received from *Vancouver Sun* editor Neil Reynolds, urging him to renew his subscription. Reynolds, a former head of the Libertarian Party of Canada, began his letter by saying it was his job to maintain the newspaper's editorial integrity. Mair composed a reply that he read on the air and posted on the radio station's website, along with his June 27 editorial. "Bad start, Mr. Reynolds, because it is not the maintenance of your editorial integrity that caused so many people to cancel their subscriptions — it was the lack of it in the first place."[72]

Mair hosted a discussion of press freedom on his June 28 phone-in show. Columnist Margaret Wente of the *Globe and Mail* and Arnold Amber from Canadian Journalists for Free Expression joined in by telephone. Donna Logan, director of the School of Journalism at Vancouver's University of BC, appeared in studio. While his eastern guests agreed that the Mills firing raised serious questions about press freedom in Canada, Logan downplayed the problem. "I think it might be going a bit too far to say freedom of the press is in jeopardy," said the former *Montreal Star* journalist and CBC executive. "We should really avoid overblown rhetoric."[73] When a caller phoned in to lament CanWest's ownership of most Vancouver-area media, Logan contradicted the woman's contention that there was a lack of local competition.

> I think the situation in Vancouver is one of the things that gets overblown, because we actually are in a very competitive situation here. Yes, the Aspers control both of the newspapers, but we've got two new television stations that have just come into the market. . . . For national news there's the *Globe and Mail*, for local news you've got the CBC, you've got CTV.[74]

When Mair protested that the newspaper market in Vancouver was almost completely controlled by CanWest, Logan was undeterred. She named two giveaway weeklies as competition. "There is the *Georgia Straight* . . . . and there are [sic] the *Westender* . . .

so there are alternatives. I mean, I don't think the situation is as dire as that." After more than one listener phoned in off the air to point out that Logan's journalism school had recently received a $500,000 endowment from CanWest, Logan was unrepentant. "It's not going to influence us," she said when Mair raised the subject. "The Aspers have been supremely correct about the gift. It is fully paid up. They've said they want nothing to do with the selection."[75] Logan's school, which was the first at UBC to be named after a corporate sponsor, had experienced a recent funding shortfall. The Sing Tao newspaper chain of Hong Kong had encountered financial problems that left it unable to keep up its financial commitments.[76]

University administrators managed to keep the matter quiet until just before the first seventeen graduates of the Sing Tao School of Journalism were set to receive their Master's degrees.[77] UBC diverted money from its special purposes funds to make up the shortfall and stripped the Sing Tao name from the school's title. Journalism students learned of the name change from the *Ubyssey*, the independent campus newspaper, and formed a student union to gain greater transparency from administrators.[78]

A guest appeared on Mair's show the next day to read Logan's testimony at the CRTC licence renewal hearings the previous year. Logan's praise for convergence and opposition to newsroom separation as "excessive" raised questions in hindsight. Some critics pointed to the announcement six months later of CanWest's $500,000 public benefits donation to her journalism school.[79] Logan's contention that convergence would allow media outlets to free up reporters "to do stories that are not being done" drew scant media attention at the time. One of the few journalists to even notice was Claire Hoy. "Is she serious?" wrote the veteran political commentator in the *Hill Times*. "They're not interested in freeing up reporters to chase stories they're not doing now. They're only interested in freeing up their bottom lines by doing the same work with fewer reporters."[80]

CHAPTER 9

Dishonest Reporting

Of all the qualities Izzy Asper possessed in abundance, he may have had ferocity and compassion in the greatest supply. His ferocity was legend in the business world, as almost every partnership he entered into broke down in acrimony, often ending with a lengthy and expensive legal battle. Perhaps even greater than Asper's wrath, however, was his compassion. His talent for making money was almost matched by his penchant for giving it away. He considered his Asper Foundation charity a fourth child, to which a quarter of his fortune should be dedicated. The fact Winnipeg was dotted with the Asper and CanWest names was a testament to his largesse. His charitable donations in Israel supported the International School for Holocaust Studies in Jerusalem and the Winnipeg Community Action Centre for disadvantaged youth in Beersheva. A $1-million donation to help establish a school of communication at McMaster University in Hamilton was followed by an honourary doctorate for Asper in 2002. One student journalist found that ironic, given the controversies CanWest had fuelled.

> In courses that examine the role of the media within society, Mr. Asper's actions have become an integral part of the subject matter. In fact, the great challenge is finding an individual whose actions have been more inimical to the idea of communication since the fall of 2001. ... That Mr. Asper's follies have become part of the curriculum in the communication studies programme is perhaps his greatest contribution.[1]

After handing over the reins of CanWest Global to his heirs in 1999, Asper's philanthropy accelerated as he focused on his legacy. A $10-million donation to his alma mater in 2000 resulted in both an I.H. Asper School of Business and an Asper Centre for Entrepreneurship at the University of Manitoba. That same month, he gave $10 million each to the Winnipeg Foundation and the Jewish

Foundation of Manitoba, the largest gifts in the history of both. A $5-million donation to the Hebrew University of Jerusalem in 2001, on whose international board he had sat for two decades, established a chair in entrepreneurship. In a four-year period, Asper was reported to have given $100 million to charity. "His philanthropy was on an amazing scale," said Rabbi Alan Green of Shaarey Zedek Synagogue. "It's just mind-boggling how much he gave away."[2]

Asper's defining monument, however, was to be the Canadian Museum of Human Rights in Winnipeg. He had dreamed of establishing it since the Charter of Rights and Freedoms was enacted in 1981. Pledging $50 million of its estimated $200-million cost, Asper envisaged it as an architectural landmark on a par with the Sydney Opera House, Big Ben and the Eiffel Tower. It was planned for construction at the confluence of the Red and Assiniboine Rivers, known locally as The Forks. Predictions saw it drawing 350,000 visitors annually, including 100,000 students. Asper planned a 70-metre Tower of Hope, a Holocaust Gallery, a Hall of Fame and Shame, and a daily "barometer of hate" measuring intolerance levels around the world. The rest of the money was to come from other private donors and governments, with $20 million each being pledged by the province and city, and $70 million by Ottawa.[3]

A 'huge bureaucracy'

In counterpoise to his compassion, however, journalists across the country increasingly began to feel the CanWest patriarch's legendary wrath. Asper's attacks on the CBC appeared to increase in both frequency and intensity with his acquisition of the Southam newspapers. A spate of coverage in early 2002 called for CanWest's publicly-funded competition to be dismantled for elitism and irrelevance. CBC president Robert Rabinovich protested CanWest was "using or abusing its newspapers' editorial pages to push the business objectives of its television stations."[4] Rabinovitch claimed that, far from being irrelevant, the CBC's share of English-language prime-time viewing was 40 percent. He called CanWest's newspaper campaign against it "the new face of media convergence."[5]

Asper responded in a column accusing Rabinvitch of playing fast and loose with statistics to justify the CBC's annual $700-million government subsidy. Removing sporting events like *Hockey Night in Canada* from the equation, Asper claimed, reduced the CBC's prime-time viewership to only 15 percent. "What Mr. Rabinovitch is trying to protect is a huge bureaucracy," wrote Asper, "of which he is the leader, of overpaid, underworked CBC head office executives who are living well off the taxpayers of Canada." Far from stepping up his campaign against the CBC, Asper noted that he had "written and argued publicly for decades, in fact, well before my involvement in the television broadcasting industry, that the CBC, as currently structured, is an anachronism and a waste of public funds." The recent anti-CBC commentary in his newspapers, Asper claimed, was "based on news reports generated independently of CanWest or Southam, and by independent judgments of the editors of the newspapers that commented." Worst of all, as far as Asper was concerned, was Rabinovitch repeating the "myth" of media concentration.

> A thorough analysis of the facts reveals that there is, in fact, less media concentration, and more diversity, than ever before in Canadian history. Perhaps Mr. Rabinovitch would like to turn the clock back to the 1950s, when, for most Canadians, there was only one Canadian television choice — the CBC. Now, that was media concentration![6]

Asper had indeed criticized the CBC early and often in his political career, in 1973 blaming it for perpetuating stereotypes of Western Canada, among other things. In 1999, he called for the CBC to be transformed from a government-subsidized network into a public broadcaster funded by private donations. "The CBC should not be state-owned, but most Canadians might be prepared to pay, on a privately funded basis, for a real public broadcaster, much as PBS is supported, but not owned, by the US government." The speech, excerpted in the *Ottawa Citizen* more than a year before he became its owner, set out many of the opinions Asper subsequently expressed about the CBC. "It should be broadcasting programming about Canada, to Canadians — programming which would otherwise not be broadcast — and that means no cartoons, no sports, no local news and with the abundance of national news broadcasts offered by the private sector, CBC should ultimately get out of the national news programming."[7]

Asper's criticism of the CBC, according to its chief correspondent Peter Mansbridge, increased in the late 1980s as a result of the failed Meech Lake accord. "He was a strong opponent of the constitutional initiative, and was convinced the CBC was being used by Meech supporters (meaning Brian Mulroney) to get that side's message across to the detriment of the opposition," recalled Mansbridge, who knew Asper well from covering his political career in Manitoba. "After Meech, his attacks on the CBC became more frequent."[8] Asper expressed his feelings about the CBC's coverage of Meech in what was described as "a low, nicotine-infused growl" during a 1996 interview with *Maclean's*. "The CBC, instead of expressing a view that was reasoned, objective, considered and responsible, became a promoter of hysteria and presented an inaccurate and dishonest view of Canada as espoused by the hysterical Mulroney crowd, that what was at stake was Canada," he said. "At the very least, it had a bias." Asked if his planned third television network in Canada would have its own point of view, Asper replied candidly. "You're damn right it will," he said.[9]

The aspect of the CBC that annoyed Asper the most, however, was its news coverage of events in the Middle East. An Islamic "intifada" or holy war there had wracked Israel with a campaign of suicide bombings. He felt the public broadcaster's coverage was biased, and he did not hesitate to say so. While not a follower of Judaism, Asper admitted an affinity with the Jewish homeland established in Palestine after World War II, which displaced indigenous Arabs. "I'm what you would call a secular Jew," he told the *Toronto Star* in 2000. "I do consider myself quite Jewish in cultural terms. Very early on, I became a Zionist. It's been a lifelong pursuit of mine."[10] After the 1973 Yom Kippur War, Asper had been instrumental in raising money and political support for Israel. He helped found an informal organization that eventually evolved into the Winnipeg Jewish community's lobbying arm, the Canada-Israel Committee.[11] Over the years, he had been a sharp critic of Canada's foreign policy toward Israel. After CanWest acquired the Southam newspapers he often made his views known in print.

In a June 2001 speech in Jerusalem, Asper described Canada's UN record of voting to condemn Israel's actions against the Palestinians as "shameful."[12] The speech was given on accepting an honourary doctorate after he contributed $5 million to help establish a business school at the Hebrew University. It was reprinted in the *National Post* and other CanWest newspapers. In it, Asper blamed most of the Western world for allowing the Holocaust that killed millions of Jews. "Britain welching on its word, duplicitously shut down Jewish immigration, and countries like Canada refused to accept fleeing European Jews as immigrants, all combining to trap Europe's Jewish community and leave it intact for Hitler's inferno."[13]

In September of 2002, the Asper Foundation co-hosted a four-city speaking tour in Canada by former Israeli prime minister Benjamin Netanyahu. After 200 window-smashing protesters disrupted Netanyahu's speech at Montreal's Concordia University, forcing its cancellation, Asper accused them of Nazi tactics. "The minority of a rabble, a rioting group of essentially thugs [and] lawbreakers employed the techniques introduced 70 years ago by Adolf Hitler and his Brown Shirts," he said after five arrests were made.[14] The incident was made into a CanWest Global film the following spring, *Confrontation at Concordia*, by Middle East correspondent Martin Himel. It compared the window smashing at Concordia to the 1938 *Kristallnacht* that saw Jewish shop windows smashed across Germany and presaged the Holocaust. Numerous groups, including those representing Muslims and Palestinians, complained to the CRTC and the Canadian Broadcast Standards Council that it defamed them. *Globe and Mail* television critic John Doyle described it as "absurdly pointed" and an "outrageously aggressive" point-of-view documentary.[15] The film did not mention the Asper Foundation's role in sponsoring Netanyahu's tour.[16]

'Lazy, biased journalism'

A month after the Concordia incdent, Asper made his strongest criticism yet of the news media and those who did not support Israel. In a Montreal speech, he attacked journalists in Canada and around the world for their coverage of the Middle East conflict. The speech was excerpted in the *National Post* and other Southam newspapers. "Both Israel and the honour of the news media are under grievous assault," he told a dinner hosted by the fundraising group Israel Bonds. "Dishonest reporting is destroying the trust in and credibility of the media and the journalists, and the same dishonest reporting is biased against Israel, thus destroying the world's favourable disposition toward it."

> The first and worst lie is what this war is all about. Dishonest reporting tells you that it's about territory, and Jerusalem, and Palestinian statehood, and alleged refugees. Honest reporting would tell you that it is a war to destroy Israel and kill or expel or

subjugate all the Jews. That is proved by the words and deeds of all the key Arab Palestinian leaders. But the media has bought and reported dishonestly and relentlessly the big lie. That big lie is that this war could be ended by Israeli land concessions.[17]

Asper named many international media outlets, including CNN, the BBC and the *New York Times*, in providing examples of alleged bias. He promised he would not mention Canadian media outlets by name because they were competitors to CanWest, with one exception. "That exception is the CBC," he said, "because all Canadians own it and the governments we elect are responsible to us and it for its quality, and integrity." He singled out the CBC's Middle East reporter Neil Macdonald for not correctly identifying Palestinian suicide bombers. "The CBC, along with ... other left-wing media, will still not label the Palestinian murderers as terrorists. By any world recognized definition of terrorism, they are terrorists, but the CBC, particularly in the person of Neil Macdonald, simply refers to them as 'militants.'" Reasons for the biased media coverage, according to Asper, ranged from negligence to malice. "Firstly, too many of the journalists are lazy, or sloppy, or stupid. They are ignorant of the history of the subject on which they are writing. Others are, plain and simple, biased, or anti-Semitic, or are taken captive by a simplistic ideology." Asper announced he had a remedy for the problem, however. "The solution starts on the campus, and in the journalism schools, then it goes to the boardrooms of the media owners, and finally, and most importantly, with the public."[18]

He urged his audience to take action to influence not only media coverage, but also the education of journalists. "You, the public, must take action against the media wrongdoers," he said, suggesting the cancellation of subscriptions and the withholding of advertising from media guilty of "dishonest" reporting. He urged formation of "honest reporting response groups" to call offending media to account. He urged political activism to influence government policy in favour of Israel, which he called "the only beacon of democracy in a swamp of hate, violence and terrorism." One way of helping to change media coverage of the Middle East, he told his audience, was for them to join the boards of universities. Once in a position of influence, he added, they should "demand that the administrators of higher education retake control of the teaching process."

> We must demand that the journalism schools do a better job of teaching integrity more forcibly. Then, we must demand that our media owners invest more money in educating their journalists, and media operators. . . . And we should withhold our financial support from those institutions that fail this obligation of educational integrity.[19]

Asper even issued a warning to the journalists he employed. "If any CanWest media outlets happen to fall within this indictment, then they, too, should take notice that I will always do all in my power to stamp out dishonest reporting, and biased reporting on any subject."[20] Honest reporting, he said, included fulfilling certain responsibilities. "The responsibility to report everything that the public needs to know about a given matter and not just selectively, so that the public may be fully informed; to report everything honestly and not slant the news, biased toward their own point of view. That is, news is news, and should appear as such, and opinion is opinion, and must be clearly designated as such." Dishonest reporting, on the other hand, came in several forms, including the use of misleading terminology. "The term 'terror' has been well defined by major recognized laws," Asper said. "But many biased media describe the Palestinian perpetrators of clear acts of terror against Israel, merely as 'militants,' 'resistance fighters,' 'gunmen,' 'extremists.'" A blatant example of this type of dishonest reporting, he added, was in a report on National Public Radio in the US. According to Asper, it described "a group of Arab murderers who crept into an Israeli home, at night, and murdered a mother and her children, as 'commandos!'" Similar terms, such as "cycle of violence," "moderate Arab states," "peace process," "occupied territories," and "illegal settlements," he said, had become tools. They were used by "journalistic propagandists in their desire to create undeserved sympathy for the Palestinians and opprobrium for Israel."[21]

Reaction to Asper's sweeping criticism of the news media ranged from outrage to stunned amazement. The *Globe and Mail* reported that "bewildered journalists . . . struggled to respond" to his "long, angry speech." Tony Burman, the head of CBC television news, called Asper's accusations "bizarre" and said he would seek space to respond. "To suggest that most of the world's media are involved in a conspiracy against Israel, it's just a totally extreme conception on Asper's part," he said. Burman added that the charges were especially interesting given the ongoing allegations of editorial interference against CanWest Global. "There is something profoundly ironic about being told off about media bias by someone like Izzy Asper."[22] Burman's request to reply was refused by the *National Post*, he said, so it appeared in the *Globe and Mail* instead. He called Asper's charges "unsubstantiated and completely false." Burman added that the CBC's ombudsman had reviewed its Middle East coverage and "his findings. . . . do not support Mr. Asper's view of our journalism." He pointed out that CanWest Global did not have a correspondent in the Middle East, despite being the largest news media organization in Canada. Burman defended the CBC's use of neutral language to report the emotionally-charged conflict. "Rather than use the words adopted or advocated by either side," he said, "we try to use precise and objective terms, such as 'gunman,' 'militant' or 'bomber' and allow our audiences to make their own judgments about the nature of the act."[23]

Most agreed that Asper's accusations went too far and that he had unfairly attacked the CBC. "Mr. Asper, of course, doesn't bother with the kind of meticulous study his accusations demand, but simply handpicks a few examples that he believes shine the best light on his assertion and rests his case," noted Concordia University journalism professor Mike Gasher. "He only rarely speaks of specific stories on specific dates, inhibiting anyone who seeks to take his thesis seriously from independently testing its validity." Gasher described Asper's urging of sanctions against critical media outlets and educational institutions as a call for "vigilante action." He labelled as "threatening" Asper's warnings to his own journalists. The threat was hardly a hollow one, Gasher pointed out, as CanWest had already proven "quite willing to intimidate its own journalists with firings, suspensions and gag orders."[24]

Canada news desk

In December, CanWest announced an out-of-court settlement of the lawsuit brought against it by fired *Ottawa Citizen* publisher Russell Mills. Terms of the settlement prevented Mills from discussing it. A CanWest press release, however, quoted him as urging the 6,000 Ottawans who had by then cancelled their subscriptions in protest to resume buying the newspaper. Mills was quoted as saying he was "confident that CanWest intends to be a good owner of the paper." *Globe and Mail* columnist Hugh Winsor was unconvinced.

"We would like to hear it from the former publisher himself and not through a self-serving company press release," he wrote.[25] Whatever terms of silence Mills had agreed to would not stop him from soon speaking out about an even greater threat to press freedom. The company announced a resumption of the editorials, which had disappeared in the wake of the Mills firing, but only once a week rather than three times as originally planned. "Don't expect the Aspers' views to disappear from their editorials and, maybe even subtly, their news coverage," warned University of Western Ontario journalism professor David Estok. "Expect only that we may hear and see less of it so openly."[26]

The *Globe and Mail* revealed an internal CanWest memo setting out plans for a centralized news desk in Winnipeg. It called it "this country's most aggressive attempt to centralize editorial operations across a newspaper chain."[27] The plan to co-ordinate news coverage at company headquarters was "an appalling idea," complained Carleton University journalism professor Roger Bird. "The only possible reason to do this would be to have a closer control on the news flow by head office," noted Bird, who had been an editor at Southam News for fifteen years.[28] According to Stephen Kimber, the control features of the Canada News Desk included an internet-based datebook system. It would allow editors in Winnipeg "to track every reporter on every assignment in every CanWest paper, while 'bat' phones provide Winnipeg with direct entry into every newsroom in the chain."[29] Mills broke his silence in a *Globe and Mail* column, warning that the move was potentially dangerous. "With the national news editing being done down the hall, there will be an irresistible temptation to meddle," he wrote.[30] Mills called for the matter to be investigated by the Senate inquiry into the news media set to start later that year.

> With the editing being done in Winnipeg, corporate executives could have an opportunity to control the news before it is sent to the newspapers. And the very fact that editors will be working next door to the company's head office may affect their news judgments. . . . The more content that is produced and edited centrally, the less distinctive the individual voices of the newspapers will be.[31]

CanWest ridiculed the concerns as "the journalistic equivalent of a child insisting there might be a bogeyman under the bed." Southam's news service had been headquartered in Ottawa since the

1920s. The move to Winnipeg would free its staff in the nation's capital to concentrate on reporting federal affairs, said Gordon Fisher, CanWest's president of news. "Taking that desk outside the Ottawa bureau will, if anything, ensure that an Ottawa-centric view doesn't pervade the content," wrote Fisher in the *Globe and Mail*. "Only a Central Canadian snob could say that something is lesser by virtue of being in Winnipeg than in Toronto or Ottawa." The call by Mills for the Senate inquiry to investigate the move was a sign that he was turning into a conspiracy theorist opposed to journalistic progress, Fisher added. "It is sad to see Mr. Mills becoming one of those sour, curmudgeonly former employees calling on the Senate, of all organizations, to assure him there is no bogeyman."[32]

Then CanWest announced Southam Newspapers would be no more. The venerable chain got a new name — CanWest Publications — and the long-respected Southam News Service became CanWest News. "We couldn't find any reason to keep the name," explained Leonard Asper, "and we wanted to put our stamp on the company."[33]

Financial woes increase

With the recession deepening, it became increasingly obvious the Aspers had picked a bad time to undertake media empire building. In October of 2002, CanWest shares fell to a low of $3.32.[34] The company scrambled to cut costs and lower its debt. In early 2003, it sold four Ontario dailies, along with twenty-one weeklies and other small publications, to Osprey Media for $193.5 million. The dailies included the St. Catharines *Standard*, Brantford *Expositor*, Niagara Falls *Review* and Welland *Tribune*.[35] Next to go were journalists. The jobs of film, television, and music reviewers were eliminated at ten CanWest dailies, with only two remaining critics providing reviews for the others.[36] The axe fell again at the *Vancouver Sun*, whose ranks were thinned with even more staff buyouts. The cost-saving measures included reassigning the newspaper's two legislative reporters in the capital of Victoria to general news. Morale at the newspaper reportedly plummeted as journalists struggled to report the news with fewer resources. "We're at the point now where cuts are going right into the bone, if not the marrow," said union official Mike Bocking, a *Sun* business reporter.[37]

Moves were made to generate additional revenue as well as to cut costs, but some critics questioned the ethics involved. Can-West put journalists on its Winnipeg news desk in charge of producing "advertorial" features designed to accompany advertising. In March of 2003, several of its dailies ran advertisements for the allergy drug Reactine alongside stories from CanWest News about hay fever and allergic asthma. A week later, the *National Post* ran an eight-page section devoted entirely to asthma and allergy stories and containing advertisements only for Reactine. Journalists became concerned that the move was a breach of the "church-state wall" that traditionally separated news and advertising at most newspapers. "It's appalling," said Klaus Pohle, a professor of ethics and law at Carleton's School of Journalism. "Where will this end? It's the health pages today but why not the news pages tomorrow?" CanWest executives denied they were pushing the boundaries of journalism ethics in the pursuit of profit. "We are not going to compromise anybody's integrity," said CanWest COO Rick Camilleri. Running stories to accompany advertising was something the company planned to do more of, added Gordon Fisher. "All we're saying is put the ad in an environment that we have sold to the advertiser. End of story," he said, "We're taking the things we have and selling them together."[38]

Vancouver stunned

The situation at the *Vancouver Sun* had, almost unbelievably, gotten even worse with the latest round of CanWest cost cutting. The largest daily west of Toronto was forced to rely on CanWest's newspaper in Victoria, the *Times Colonist*, for its coverage of BC politics. One result was that the *Sun*'s award-winning political columnist Vaughn Palmer often analyzed developments that had not been reported. "*Sun* readers are now frequently treated to Palmer's editorial-page take on a story before seeing the story itself," noted *BC Business* magazine. "If they see it at all."[39] Some saw a political motive to the coverage cuts in addition to a financial one. The Liberal party in BC tended much more towards conservatism than its federal counterpart. The Campbell government had been elected in 2001 on a platform of tax cutting and privatization, two Asper favourites. "Without taking sides on what's going on politically, everyone can agree what we have is an activist government that is doing a lot of things," observed Bocking. "Whether you support them or don't, those are things that should be discussed by people who will have to vote again.["](40)

The suspicion that Campbell was getting a free pass from the *Sun* grew when the newspaper downplayed news of the premier's 2003 drunk-driving arrest in Maui. "Here we are a paper that runs massive headshots, we get the headshot of all headshots, the Premier with the slate around his neck, and we run a little thumbnail," an un-named *Sun* reporter told *BC Business*. "We essentially tried to bury a front-page story on our front page. No one is going to convince me that that wasn't political."[41]

> If there were any doubts, the reporter says they were erased three days later when The *Sun* spiked a story reporter Peg Fong filed from Hawaii in which she interviewed the inmate who shared Campbell's Maui jail cell. Fong says she can't comment on the issue, saying only that she was told from the outset to direct all questions relating to the story to the senior editorial team.[42]

With the cutbacks and buyouts of senior staff members, the *Sun*'s newsroom was as hollowed out as its journalism. "Thanks to the 'synergies' arranged by CanWest Global, 40 percent of the desks in the *Sun*'s current newsroom stand empty," observed Richard Littlemore. "You could organize a swordfight at noon and endanger no one."[43] CanWest's harshest critic on the west coast, talk show host Rafe Mair, was effectively silenced with his 2003 firing from CKNW radio. A female assistant had complained, reported the *Globe and Mail*, that he "used the 'f' word inappropriately during a staff meeting and angrily referred to some newspaper stories as corporate 'blow jobs.'"[44] Mair moved his show to another station, but with nowhere near the audience he had attracted at market-leading CKNW. The *Georgia Straight* wondered if Mair had been the victim of a corporate conspiracy due to his declaration of war on CanWest a year earlier. CKNW was owned by Corus Entertainment, whose majority shareholder also controlled cable television giant Shaw Communications of Calgary, which linked it to CanWest. "The Asper family's CanWest Global, CTV, and other broadcasters rely on Shaw to distribute their television signals," noted *Straight* news editor Charlie Smith. "Did J.R. Shaw decide to silence Mair because the veteran broadcaster wouldn't toe the corporate line like the company's other 1,900 employees?"[45]

Senate hearings begin

When the long-awaited Senate hearings into the news media began in April of 2003, Tom Kent appeared as the first witness. The chair of the Royal Commission on Newspapers urged the prohibition of cross-media ownership, among other things. "Whether such convergence, in fact, benefits any investors is questionable," he testified. "What is certain is that it is a combination against the public interest." Kent, a former editor of the *Winnipeg Free Press* and a retired longtime Liberal bureaucrat, did not hesitate to prescribe government action. "It can be simply broken," he said of cross-media ownership. "It should be."[46] Russell Mills, by then dean of the School of Media and Design at Algonquin College in Ottawa, urged a return to the 1982 CRTC directive prohibiting cross ownership. "Ownership of both broadcasting and newspapers in the same city gives an owner too much control over the flow of news and opinion," he told the senators. Converged media companies such as CanWest should have to choose between their television and newspaper holdings, he added.

> Because media companies have made substantial investments in cross-ownership and convergence strategies ... they should be given reasonable time to adapt to the change. Legislation should be passed that would give media owners until the end of their current broadcasting licences to be in compliance with a ban on cross-ownership. In most cases, this would be about five years.[47]

The Asper position was just the opposite, and it had no stronger advocate than *National Post* columnist Terence Corcoran. He derided the Senate hearings as "the third media-control circus in three decades." To the *Post* business editor, Kent's view of press freedom was a "Marxist slogan that blotted the opening page" of the 1981 Royal Commission report.[48] ("Freedom of the press is not a property right of owners," the report read. "It is a right of the people."[49]) Corcoran claimed Kent had been "dead wrong on the issues of the day and the future of newspapers," yet "some of Canada's leading media gurus continue to use the Kent definition of press freedom."[50] Corcoran argued that government intervention was a more dangerous threat to press freedom than corporate control.

Now there are only two ways that "the people" exercise rights. One is in the market place, as buyers and owners of property and production, goods and services. The second route for people to exercise their rights is collectively through government. … It follows that the logical end-point of the Kent syllogism is the following: "Freedom of the press belongs to the government."[51]

CanWest critic converted

Given his undisputed compassion, perhaps it was sheer kindness that prompted Izzy Asper to promote Matthew Fraser. The *National Post* media columnist and Ryerson University professor had been a harsh critic of CanWest and Asper prior to them buying his newspaper. He once referred to Global Television as an "idea free zone," charging that its high profits had been reaped thanks to favourable federal regulations. "Even among his friends, no one denies that Mr. Asper owes his considerable personal fortune to the Global TV network's regulated profits," wrote Fraser. "Global's high margins have been made thanks largely to CRTC rules that allow television stations to rebroadcast low-rent American fare Canadians can watch on US networks anyway."[52] Fraser's criticism of Asper must have hit close to home, questioning as it did his patriotism.

> Mr. Asper's success seems to be based on a strict separation between profits and patriotism. He has deployed regulated income extracted from the home Canadian market to acquire broadcasting assets abroad — in Australia, New Zealand, Ireland, and perhaps soon in Scotland. But he has put back precious little into the country that generates those high earnings.[53]

Fraser also repeated verbatim in that column several of the criticisms Robert Lantos had made in a speech at Ryerson a few months earlier. Two days after Fraser's column appeared, Asper's lawsuit for libel was filed against Lantos. Despite repeating the alleged libel, which is legally equivalent to committing it in the first place, Fraser got off with a letter of protest from a CanWest spokesman.[54] A subsequent column, however, drew the ire of Leonard Asper. It suggested CanWest was a hand-picked federal "winner" and that Asper wanted more concessions from Ottawa to "help fatten Global

TV's profits."[55] The newly-installed CanWest CEO wrote in a letter to the editor that Fraser was "so far removed from the industry that he shouldn't be taken seriously."[56]

A PhD in political science from the Sorbonne in Paris, Fraser had written books on both broadcasting policy and popular culture. His 1999 book, *Free For All*, was harshly critical of Izzy Asper. "There can be little doubt . . . that Asper is the most prodigious agent of Americanization in the history of Canadian television," Fraser wrote. "Asper has made little effort to conceal his vision of Global as a cash machine whose profits are derived chiefly from importing American television shows."[57] After the Aspers bought half-ownership of the *National Post* in July of 2000, however, Fraser's coverage of CanWest became far more favourable. "What Fraser described less than a year earlier as CanWest's 'tentacles' had now become 'well-plotted acquisitions,'" noted the student-produced *Ryerson Review of Journalism*. "To him, CanWest was a model of how a conglomerate should work."[58]

According to Fraser, however, his change of perspective on CanWest pre-dated its takeover of his newspaper. "My views about CanWest Global . . . started changing in mid-1999, when Leonard Asper became chief executive and set the company on a new direction," he wrote in 2001. "That was before — not after — Izzy Asper purchased a stake in the *National Post*."[59] Fraser's attitude change on CanWest, according to some, was never more apparent than in a 2002 television interview with Izzy Asper. "Throughout the hour-long interview Fraser lobbed soft questions across a kidney-shaped coffee table and called his guest Izzy," noted the *Ryerson Review of Journalism*.[60] Antonia Zerbisias, who had co-hosted a weekly media show with Fraser on CBC Newsworld, gave his performance a thumbs-down. "Outrageously, he allows Asper to make unsubstantiated claims about CBC," she wrote. "What's more, when Asper does say something intriguing, Fraser cuts him off, interrupting him with his next question. Talk about missed opportunities."[61] Asper took full advantage of his chance to argue unchallenged on several issues. "We do not broadcast, never have from day one, one minute more American programming than CTV does," insisted Asper. "We spend more money, proportionate to audience, on quality Canadian programming, drama, children's programming and variety programming than CTV does."[62] As for the contentious national editorials, Asper called "mischaracterized garbage" the criticism CanWest had received for them.

> When we bought, I said, of course, we intended to make our point of view heard. That's one of the joys of being a publisher-in-chief ... So we're not doing anything we didn't forecast and we're not doing anything close to what the blind, one-eyed critics are saying we're doing: threatening freedom of the press, ordaining censorship. That is just utterly misguided, and deliberate in some cases.[63]

Fraser became the recipient of Izzy Asper's considerable compassion after the death of his wife, Rebecca Gotlieb. She was the daughter of Asper's old camp counselor, Allan Gotlieb. According to Fraser, Asper wrote him a "long letter of condolence in which he poured out his soul about his own personal tragedies." He also invited Fraser to spend some time at his Palm Beach vacation home. "At first, I balked at his invitation," recalled Fraser. "But Izzy kept insisting, so a couple of weeks later I got on a plane and flew down to Palm Beach." There, according to Fraser, he saw another side of Izzy Asper. "It didn't take long for me to realize that Izzy had already planned, and was ready to orchestrate, an entire program to cheer me up."[64] During the day, Fraser recalled, Asper took him around town in his vintage Mercedes, touring Palm Beach's most opulent mansions. At night, they sat up drinking until the small hours of the morning while Asper chain-smoked and told stories about his escapades in business and politics. "After my stay in Palm Beach, I could scarcely recognize the man behind the myth," recalled Fraser. "For me, Izzy was a warm, generous, funny and tremendously engaging man with big appetites and an even bigger heart."[65]

Surprise appointment

Fraser insisted he had no idea at the time that, just weeks later, he would be named editor-in-chief of Asper's *National Post*. His surprise appointment in May of 2003 was part of a major shake-up at the money-losing *Post*, the second in two years under CanWest. The *Post*, which had lost an estimated $200 million under Black, continued to be a financial drain on CanWest as a deepening recession dropped advertising revenue sharply. A projected loss of $20 million for 2003 led to predictions the Aspers would close the newspaper. What resulted instead was a mixed blessing for *Post* staff. While making a commitment to keep the newspaper alive for another three years, CanWest CEO Leonard Asper announced that its top editors were being replaced.[66]

Fraser was named editor-in-chief, to the astonishment of most *Post* watchers. David Asper joined him atop the newspaper's masthead as its chairman, a position previously held by Conrad Black. "The Asper name is on the line," Leonard Asper told assembled newsroom staff. "We're here to praise the *Post*, not to bury it."[67] Some expressed amazement at the appointments of Asper and Fraser to the newspaper's top two positions. "If either of them have any experience running a newsroom, nobody I consulted last week was aware of it," wrote Zerbisias. "More significant, neither is known for his people, management or editing skills."[68] For his part, Fraser promised to continue the newspaper's rightward direction. "The paper ... will remain faithful to its editorial mission as a cogent conservative voice on a media landscape cluttered with left-liberal chatter," he pledged on the next day's front page.[69] As for the *Post*'s anti-Chrétien crusade, Fraser declared it dead. The start-up phase of the newspaper, he told Chris Cobb, had ended up looking "stupid" with Chrétien's landslide re-election in the fall of 2000.

> It was a rude awakening. The prime minister outsmarted Conrad Black, the wind in the *Post*'s sails began to lose strength, and the founding religion of the *Post* began unraveling. We are still a conservative newspaper, but we're not trying to get anyone elected or unite any particular party. We are not torquing the news to pull down the prime minister.[70]

The subsequent exodus of *Post* columnists, including Mark Steyn, David Frum, Paul Wells, and Christie Blatchford, came as a shock to the Aspers, according to Cobb.[71] Fraser, however, made few changes at the *Post*, observed the *Ryerson Review of Journalism*. Its slogan became "Your Canada, your *Post*," which reflected a more patriotic tone than the anti-Canadian rhetoric that marked it under Black. Its cheeky nature changed little otherwise. "Fraser has kept faith with the Aspers' main concerns: the health of private television operators, lower taxes for rich people and unquestioning loyalty to the State of Israel," the *Review* observed in a profile of the elusive editor. "One of Fraser's few innovations — CBC Watch, which invites readers to complain about bias at the public broadcaster — is itself open to accusations of bias in favour of the first Asper concern."[72]

The black news hole

The Lincoln Committee on broadcasting policy did not issue its report until June of 2003, after more than two years of study and public hearings. The *National Post* got the scoop on its release by a full 48 hours. It reported that Alliance members of the committee would issue a dissenting opinion calling for "greater openness to foreign ownership and less stringent Canadian content regulations."[73] NDP committee member Wendy Lill, a noted nationalist, would not. That, the *Post* reported, was "as telling about the report's philosophical direction," as its title, "which includes . . . the words 'Cultural Sovereignty.'"[74] By the time the 872-page report was officially released, the *Post* had derided it for two days. It received little additional media attention, despite being described by one expert as "the most comprehensive review of Canadian broadcasting in 20 years."[75]

Titled *Our Cultural Sovereignty: The Second Century Of Canadian Broadcasting*, the report made 97 recommendations. They addressed many concerns expressed in Canada since the multimedia mergers of 2000. The recommendations called for an immediate moratorium on new broadcast licences for companies that also owned newspapers, pending a government review of the issue. "The danger is that too much power can fall into too few hands," warned the report, "and it is power without accountability."[76] A firm government policy on cross ownership was badly needed, it added, recommending Ottawa implement one within a year. "The potential problems with cross-media ownership are sufficiently severe that the time has come for the federal government to issue a clear and unequivocal policy on this matter."[77] The report also urged that foreign ownership restrictions be maintained to help preserve Canadian culture, and that funding for the CBC be increased for the same reason. Yet despite the report's scope and significance, it "virtually fell into a black news hole," according to Zerbisias.

> You'd think that, when five pounds of government reportage about broadcasting in this corporately merged and converged Canada hit all the desks in Media Land, the thud would be deaf-

ening. Instead, the mediaocracy has been strangely silent. Or, maybe not so strangely. . . . These are not proposals that some media barons wanted to see.[78]

The industry journal *Playback* labelled it the "under-reported" report. "Sank like a cinder block. Gone the way of the dodo. Fell off the radar. These are the phrases that spring to mind when one thinks of the Lincoln report." Public broadcasting advocates complained that the report was "unfairly buried" and "shouted down by big media."[79] CanWest dailies in particular ignored the report, noted Arnold Amber of the Newspaper Guild media union. "In fact, The *Gazette* in Montreal didn't even run the story. The *Vancouver Sun* dismissed it with a 71-word brief. The *StarPhoenix* in Saskatoon did only slightly better, carrying 138 words. It was a shameful and chilling demonstration of the problem at hand."[80] The *Vancouver Sun*'s brief was buried deep inside its business section on page D8 under the headline "More cash for CBC recommended." Editor-in-chief Patricia Graham explained that the report's release was considered a business story by editors at her newspaper. It got the placement it did, she said, because the *Sun*'s business coverage was primarily local.[81]

A few last words

The heat Izzy Asper took for using his newspapers to influence public opinion did not stop him from expressing his views prominently on their pages. In July of 2003 he gave a speech that was excerpted in at least seven CanWest dailies, including the *National Post*. In it, he urged Ottawa to use Canada's influence on the world stage to assist Israel in its dispute with the Palestinians. He called on the CRTC to refuse a licence to the Qatar-based Al-Jazeera news network, which was providing an Arabic perspective elsewhere around the world. He also attacked his favourite media target. "The refusal of the CBC to call a terrorist a terrorist, the taken-as-given that there is a 'cycle of violence,' when there is no such thing, and the call for only 'proportionate response' to terrorist activities is nothing short of odious."[82] When Mazen Chouaib of the National Council on Canada-Arab Relations was granted a reply on the *National Post*'s opinion page, Asper's rebuttal was printed alongside. "To believe the assertion that Israel is a Western democratic oasis situated in a scorching desert of Arab dictatorship is delusional," wrote Chouaib.

"One cannot claim the virtues of democracy while actively participating in the ethnic cleansing of an entire culture."[83] For Asper, the problem again came down to one of semantics. "Mr. Chouaib's view is that the Arab-Israeli conflict arises because of what he wrongly terms the 'Israeli occupation' of 'Palestinian land.' In fact, there never was — and there isn't now — any such thing as 'Palestinian land.'"[84]

In August, an Asper interview with Melissa Radler, the daughter of Hollinger's David Radler, was published in the *Jerusalem Post*. "In all our newspapers, including the *National Post*, we have a very pro-Israel position," he boasted. "So much so that the Canadian Islamic Congress has declared a boycott of all of our newspapers and TV stations." Radler asked him if that policy had any effect on government action or public opinion. "I'm told it has," Asper replied. "Certainly, we've raised issues that no other media has." On whether anti-Semitism was a reason for the criticism he had received in the Canadian media for setting editorial policy, Asper offered an analogy.

> When you own a newspaper, the inmates of the asylum don't run the asylum. I'm the last guy to be paranoid; on the other hand, in almost every criticism of our 'interference,' every example they use it always comes back to Israel. If I were being pro-something that they liked, they might not be as tough with us. . . . There's nothing they despise more than owners who say this is the position we want to take, and if you don't like these views, take yourself somewhere else.[85]

Then, suddenly in October, the larger-than-life media mogul was gone, felled by a heart attack two decades, almost to the day, after his first. Most obituaries focused on Asper's media empire building and philanthropy, but some journalists with whom he had crossed swords offered mixed remembrances. "Canada needs more Izzy Aspers," wrote Russell Mills. "The newspaper industry, however, does not."[86] His funeral in Winnipeg drew a "blue-chip crowd" of 1,600 mourners, according to *Maclean's*, including Jean Chrétien, Paul Martin, and Stephen Harper.[87] Leonard Asper, his voice breaking with emotion, eulogized his father. "Thank you for what you gave to the world and to your family," he said. "We have your checklist. We know what's left to be done. We will not let you down."[88]

CHAPTER 10

Like Father, Like Children

Each of the Asper heirs arguably inherited some, but not all, of their father's notable attributes in addition to a share of his ten-figure fortune. Leonard, who succeeded him as head of the CanWest Global empire, obviously got the greatest portion of I.H. Asper's business acumen. David, who showed creative talent but also a penchant for ruffling feathers, may have inherited Izzy Asper's famous ferocity. Gail, who kept the lowest profile of the three, was mostly involved with the Asper Foundation's charitable work, thus perpetuating Israel Asper's philanthropy. All were trained as lawyers at their father's insistence. All seem to have adopted their father's strongly-held beliefs about Canada and the world in general, which they expressed to a greater or lesser extent publicly. Izzy Asper's sons in particular showed the same verbal assertiveness as their father. They also both aspired to build CanWest into something much larger. "David and Leonard see their future in international markets," reported *Maclean's* in 1996, "or 'global domination,' as Leonard puts it."[1] By 2002, the new CanWest CEO had a more specific goal in mind. "Our ambition is to be one of the top five media companies in the world within 10 years," he told *Maclean's.*[2]

Although he was the youngest, Leonard showed at an early age the affinity for business that would make him the logical successor to his father. According to Gail, he was reading the *Wall Street Journal* and carrying around the Canadian securities handbook in pre-adolescence.[3] His brazen nature was demonstrated by the story she told about how, when Leonard was three, he picked flowers from a neighbour's front yard and tried to sell them back to the owner. The inclination toward larceny, reported Ric Dolphin of the *Calgary Herald*, ran in the family. "She and David used to do the same thing with crabapples."[4] One of Leonard's earliest lessons in business, he told another *Herald* reporter, came after he hurried home with

$6 in revenue from a lemonade stand he set up at age four. "Dad said, 'What did the gas cost for your Mom to go to the store to buy lemonade? What did the lemonade cost? What did you pay for the Dixie cups? How much was the wagon and how about that little tray you used for a cash register?'" The realities of entrepreneurship thus sank in at an early age. "By the time he'd finished," Leonard recalled, "I'd lost $600 and was a puddle of tears."[5]

Named CanWest CEO at 35, Leonard's youthful looks and pleasant disposition often led business associates to find him out of place in the cutthroat corporate world. "Leonard lacks the imposing presence, the growl and glare, of his ... father; in its place is an apparent willingness to please," noted *Report on Business* magazine in 2000. "A dimpled grin and a boyish enthusiasm gives the young CEO a somewhat merry air that, were he anyone else, might make it difficult for him to gain serious attention in a boardroom."[6] Those who made the mistake of underestimating him, however, were missing some important traits Leonard inherited from his father, noted *Maclean's*. "People talk so much about Izzy Asper's smarts that they miss or ignore similar qualities in Leonard, whose polite, soft-spoken manner and enormous devotion to his family belie his toughness."[7] Some saw Leonard as more like his mother, who was a calming influence on her children in contrast to the volatility of their father. There were definite paternal similarities, however, noted *Maclean's* editor Anthony Wilson-Smith.

> While father and son have different management styles, there are parallels in their personal tastes. Both love convertibles — Izzy his Mercedes-Benz, Leonard his exotic British TVR sports car. Music means a lot to them — Izzy devoted to jazz and Gershwin, Leonard still clinging to a boyhood enthusiasm for Rush, the thinking hoser's heavy-metal band, and newer hard-hitting groups like Soundgarden.[8]

The youngest Asper offspring attended college in the US because, he admitted, it was "a good way to get out of my family's shadow."[9] He studied political science at Brandeis University in suburban Boston, which was founded in 1948 as the first nonsectarian Jewish-sponsored university in the US. When he was 20, he worked a summer for federal Liberal cabinet minister Lloyd Axworthy, who had been one of his father's few Manitoba MLAs.[10] He studied tax law at the University of Toronto, where he later admitted to having introduced classmates to the US college tradition of the "kegger."[11] A collector of sports memorabilia and a hockey player even into his 40s, he went straight from law school into the family business. "Leonard's apprenticeship was systematic," reported *Maclean's*. "Working in the Toronto headquarters of Global Television, he spent a few months in each of the major departments, from production to finance, between 1991 and 1993."[12] In 1994, he returned to Winnipeg, where he became CanWest's executive vice-president of corporate development and then Chief Operating Officer in 1998. Leonard's elevation to CEO the following year came ahead of schedule and resulted from his finally getting WIC into the CanWest fold.

By the time Leonard moved into company headquarters in Winnipeg, his sister had already joined the CanWest corporate team. Gail came aboard in 1989 after working in corporate and commercial law for five years in Halifax. She served first as CanWest's in-house legal counsel and then as its corporate secretary. David Berman of *Canadian Business* magazine found her "as friendly, forthright and personable" as her father, but noted that she talked at four times the speed.[13] Described by Dolphin as "a born performer" who once aspired to the opera, she appeared in several Manitoba Theatre Company productions. "Most enthusiastically, though," noted Dolphin, "she is the public face of CanWest, the company's 'lieutenant governor,' as she puts it with a twinkle in her brown eyes."[14] Gail devoted most of her attention to the Asper Foundation's philanthropic endeavors, rather than to corporate affairs. "I didn't want to be on the roller-coaster my brothers were on," she told Gordon Pitts. "I wanted to be on the carousel, the steady ride."[15]

While he was the last to join corporate CanWest, as the eldest sibling David had actually been the first to get involved in the family business. He worked on the creative side, helping film such early CKND programming as *Polka Warehouse*.[16] Notably temperamental, he was described by the *Toronto Star* as "a man you don't want to cross on days when he can't find anyone to remove the thorn in his paw."[17] A student politician in both high school and university, in his younger days David was "a rebel, a good-time guy, famous hard-partier," according to Pitts.[18] His rebelliousness was reflected in the fact he dropped out of school briefly at age 16, only to relent and be bundled off to boarding school in BC.[19] "David's reputation is that he is charismatic, emotional, immensely likeable," noted Pitts, "but a bit of a loose cannon, a guy subject to enthusiasms, a personality of extremes."[20]

While he acceded to his father's insistence on law school, David "quietly resisted" Canadian corporate and tax law, according to Pitts. Instead, he studied criminal law in California. Back in Winnipeg, he bolted from his father's old law firm midway through articling there, preferring to strike off on his own and make a name for himself. He thus turned out to be the only Asper offspring to achieve reknown in the legal world. It came with his role in freeing David Milgaard, who had been wrongly imprisoned for murder in 1969. The campaign to have Milgaard's case reviewed began in 1986, with Asper handling the media in an attempt to pressure Ottawa to re-open the case. Six years later, the Supreme Court of Canada finally cleared Milgaard after 23 years in prison.

A 'shit disturber'

David then joined CanWest and worked at Global's operations in New Zealand, managed its Regina station, and headed its failed bid for Channel 5 in the UK. He found his niche, however, on the creative side, showing a flair for programming that helped vault Global ahead of CTV in the ratings. In 1996, *Globe and Mail* writer Trevor Cole found significance in Asper's "enormous blue lava lamp," which the visitor saw as "confirming the proximity of creative-type thinking."[21] Asper even joked about the observation, telling Pitts: "I'm the lava lamp guy in the group."[22] But while their father once saw his sons running CanWest as co-CEOs, it soon became obvious that such a scenario was unrealistic, according to Cole.

> By the mid-'90s it was becoming clear that Leonard alone would wind up in charge. David is, by his own admission, less patient than Leonard, preferring the quick resolutions of the operations side to the years of nurturing required in corporate development. And he was by some measures too emotional to lead.[23]

David freely admitted his unsuitability for corporate leadership. "I'm a shit disturber and a bit of a dreamer and I probably focus too much," he told Pitts.[24] During his time in London heading CanWest's bid for Channel 5, he reportedly threw an unresponsive fax machine out the window of his Covent Garden office.[25] His intemperate tirades as head of the company's publications committee made him a lightning rod for media criticism during the Asper Disaster. His sometimes erratic behavior also periodically made headlines. A former female reporter for the *National Post* filed a wrongful dismissal lawsuit in 2004 seeking $405,000 in damages from CanWest. It claimed David Asper "made an obscene and lewd gesture to [her] by unzipping the fly of his pants, sticking his finger out of his pants towards [her] to make it look like he was sticking out his penis and wiggling it."[26] He also made the news in 2005 with his resignation from the Blue Bombers board of directors after he was escorted out of Winnipeg Stadium by security guards. Asper was ejected after a close loss to Saskatchewan when he reportedly got into "heated confrontations" with team officials and players.[27]

One result of Izzy Asper's long years of experience in estate planning was his insistence on devising a method for his children to run CanWest after he was gone. He was particularly concerned about resolving any conflicts that might arise between them. Second-generation disputes were legend in the business world for ruining in a few years empires that had often taken decades to build. It was a phenomenon that had been exploited by the Aspers in building CanWest, most notably in their conquest of WIC. "We all know very well that married couples who love each other on Day 1 are ready to cut each other's throat 10 years down the line," Gail told *Canadian Business* in 1995. "There is no naiveté whatsoever about what can happen." Their father, she said, made sure the message got through. "Every week we get articles — we circulate them among ourselves — every article on any family dispute for the past 10 years."[28]

Izzy Asper also ensured that his children could buy each other out if they disagreed on how the company should be run or just wanted out. The original plan was for professionals to manage CanWest when he was gone. "I made a rule that no kids came into the business," he told an interviewer. "I don't believe in nepotism and I don't believe in dynasties."[29] But after Gail joined CanWest as in-house counsel, his sons also wanted in. "Leonard said, 'I thought we had a rule here, no kids in the business.' . . . So Leonard said he was coming in. And David said, 'Wait a minute' . . . so I couldn't say no. . . . They really liked what I was doing and they said 'that's what we want to do.'"[30] Asper insisted the nepotism wasn't his idea. "It was they who got this dynastic glaze in their eyes," he told *Canadian Business*.[31] That changed everything, Asper admitted to the *National Post*.

> I never trained my children, or caused them to be trained, to run this company. I trained them to own the company. There's

a huge difference. You don't get friction when three owners are sitting in a room, it's only when one of them is CEO and he gets defensive about what he did last week or dividends have to be cut because we bought Company X and your sister or brother or your nieces and nephews are mad at you.[32]

The result was a plan to which Izzy Asper devoted considerable thought. "Succession is an obsession with the Aspers," noted David Olive after discussing family matters with Leonard. "It was important, he said, for the family to maintain a united front in the face of criticism."[33] Leonard revealed the family "code of conduct" to *Maclean's* in 2000. "Outsiders watching for cracks to appear in the Asper clan's solidarity may have a long wait," predicted John Geddes.[34] The code was extremely specific about sibling relationships, he added, but it was not expected to be foolproof. "Leonard swears by the rule book as a way of avoiding disputes, but [he] puts more emphasis on warm relations among the three siblings," noted Geddes. "'If you have a jerk in your family,' he says, 'there is no piece of paper that can stop that person from doing something mischievous and self-centred.'"[35] The plan even provided a blueprint for succeeding generations of Aspers to continue operating CanWest according to its founder's wishes.

Among the terms of this 'very, very detailed' set of written rules: no talking publicly about family disagreements, and no jobs for spouses in the family company. When it comes to Izzy's grandchildren . . . guidelines for what sort of qualifications they will need if one day they want to get into CanWest are already set down in black and white.[36]

Asper values

Rather than turn their guns inward, the Asper heirs proved adept at following their father's example in promoting their views publicly. Gail did her part in taking the Asper message to the masses. "The Liberal government is to be applauded for a number of steps taken in the most recent budget that encourage charitable giving," she wrote in a 1997 letter to the *Globe and Mail*.[37] In her capacity with the Asper Foundation, Gail's fingerprints could be found on some of CanWest's national editorials, according to Ric Dolphin. "When you read one of those Southam editorials slamming the federal government for not providing better tax breaks for private donations," he noted, "Gail's the one behind them."[38] She shared her father's feelings about media bias against Israel, and she went public with them before he did in 2002. She gave a speech that February urging a letter-writing campaign to media outlets protesting anti-Israel bias in the media. She had already written several such letters to the *Globe and Mail*, Gail noted, that had not been printed. "We Jews have become incredibly complacent," she told a luncheon at Winnipeg's Shaarey Zedek Synagogue. "We can't even get 200 people out to a rally in Winnipeg to show our solidarity with Israel here."[39]

David Asper also had favourite topics on which he weighed in publicly, including reporting of the legal system. He blasted the press at a symposium on media and the courts in 2005. "Ill-informed commentary, in my view, and, to put it bluntly, bad reporting, pose, in my opinion, a serious threat to public respect for a system of law that I have to tell you overwhelmingly works." He gave the example of a recent Supreme Court decision on private health care insurance in Quebec. Some in the press, he pointed out, saw it as the beginning of the end for Canada's universal health care system. "It would be an interesting exercise for you to see precisely how Canadians were threatened with the end of the teddy bear of medicare," he said. Asper stressed the importance of reporting fairly and accurately because of the influence Canadian media had over the public mind. "The media — in our role as journalists and how we treat these big issues — is as much a subject of national interest as is what the courts are deciding, because of the effect that we can have on society."[40]

David's ire was also raised in 2003 by media criticism of his predecessor as chairman of the *National Post*. Shareholders in Conrad Black's Hollinger International had been questioning his management of the newspaper company. "Non-compete" fees paid to Black and other insiders upon the sale of Hollinger newspapers, they felt, should have gone instead to the company. Such payments, which unlike capital gains were tax-free at the time in Canada, had been included in CanWest's purchase of Southam. In exchange for a reduction in the purchase price, payments were made by CanWest directly to Hollinger executives. "Something needs to be said in order to put this issue in proper context," wrote Asper in a *National Post* column a week after the newspaper ran a lengthy recap of the controversy. He pointed out that non-compete payments were standard corporate practice. "If Lord Black ever decided to sell his interest in Hollinger, it is he — and not Hollinger — with whom we did not wish to compete. So what's the fuss?"[41]

Asper made it clear that, like Black, he was no fan of the safeguards shareholders had been clamouring for in the wake of high-profile corporate scandals like Enron. "The jackals who are madly barking at Lord Black's door are evoking principles of 'corporate governance' to justify their allegations against him," wrote Asper. "Lord Black has correctly characterized his critics as 'zealots.'"[41A] Some analysts noted that CanWest was hardly a bastion of corporate governance, having few independent directors lacking ties to ownership or management. With only nine directors, including three Aspers and three CanWest executives, the company's board was described by an Australian journalist as "not exactly a model of independence."[42] The inclusion of family members as "independent" directors was permitted by Toronto Stock Exchange rules, noted *Canadian Business* magazine. The loophole hardly allowed for good governance, however. "While those relatives may not be involved in the day-to-day operations of the company," noted writer John Gray, "can anyone conceive of them voting to oust the person who carves the turkey on Thanksgiving?"[43]

In his father's mold

Of all the Asper offspring, Leonard took after his father most in the public arena, proving to be almost as prodigious and pugnacious. "He has his father's quick tongue and love of a good fight," noted Pitts.[44] According to Olive, Leonard also had "his litigious father's intolerance for dissenters, both in and outside the camp."[45] His verbal fractiousness was evidenced by the insults he often dispensed toward critics of CanWest. In appearing before the Lincoln Committee hearings in 2002, for example, he ridiculed critics of media concentration. "Canadian media are more fragmented and less concentrated than ever before," he testified. "I submit that people who believe otherwise are not looking at the facts and they also probably believe Elvis is still alive."[46] In 2003, while announcing changes at the *National Post*, he derided the *Post*'s competition in Toronto as an "axis of snivel."[47] He seemed to love the public pulpit as much as his father did, and he used it often to dispense his opinions.

In 2001, he set out his economic prescription for Canada in a speech that was reprinted in the *National Post*. It closely matched what his father had been urging for decades. He judged Canada an economic "failure" due to its weak currency, high taxes, and its national debt. "Our dollar is . . . low because of our tax regime," he argued. "No rational economic being would want to earn profits in Canada when they keep only 50 cents of every dollar." He also claimed high income taxes were contributing to a "brain drain" out of Canada. It was an argument right off the pages of *The Benson Iceberg*. "This contributes to an exodus of our most highly skilled talent and a refusal of foreign talent to relocate here," he said. "The entrepreneurs and the risk takers are on a train south."[48]

Leonard proved as adept as his father at vilifying critics of CanWest. After witnesses blasted the company before the Lincoln committee, he shot back with "vintage Asper irreverence," according to Hugh Winsor.[49] In a speech to the Canadian Club in Ottawa, he labelled the critics "disgruntled and opportunistic journalists." He said that Southam writers had been "the subject of editing for many years — long before CanWest ever came along," and that the company's critics had abused the word "censorship." "I guess we no longer need editors," he said, "because they're all just censors."[50] Asper described other CanWest detractors as "anti-business academics."

> They just can't come to terms with the notion that the media is a business and that owners of those businesses must treat them as such in order to attract capital. But their warped sense of this also leads them to conclude that everybody but the owners has the right to run the business.[51]

Far from having grown too dominant, Asper argued that media companies were losing ground to fragmentation. Rather than discouraging growth, government should encourage media companies to grow in size and strength, he added. "Strong companies reinvest in content, people, capital and their communities. Weakened companies slash costs and stop investing. Breaking media up into uneconomic fragments will result in business failures and less, not more, diversity of outlets." Far from reducing diversity, he argued that cross ownership had done exactly the opposite. "Cross ownership strengthened Canada's media companies and created more diversity, and more sources of information because it has created better-financed companies," he said. He pointed out that CanWest had been absorbing millions of dollars in losses by the

National Post. "Large media companies have the staying power, the courage and the vision to nurture new media outlets to viability."[52]

Market share wasn't the only thing CanWest was trying to get back, according to Asper. He charged the CRTC with "overbilling" broadcasters by $71 million during the previous year in what he called "a pure tax."[53] CanWest alone, he pointed out, paid $7 million that year in so-called Part II fees. While Part I fees covered the CRTC's expenses — $23.1 million in 2003 — Part II fees charged broadcasters for using Canada's public airwaves. The 1.365-percent levy on revenues netted Canadians $92.6 million that year. In 2004, CanWest and other companies, through their lobby group the Canadian Association of Broadcasters, sued the federal government for the return of Part II fees. CanWest had been eager in 1995 to pay an annual fee of £36.2 (Cdn. $79.5 million) for the privilege of starting a new network in the UK. It objected, however, to sharing with Canadians a penny more than it had to of the hundreds of millions in advertising revenues it pocketed annually in its own country. *Toronto Star* media columnist Antonia Zerbisias found herself stonewalled when she sought comment on the story, which had gone largely unreported. "The CAB would like to fly this way below the public radar," concluded Zerbisias. "This is one issue that should result in a lot of paper being used in reporting on it. But, oh, wait a minute. The broadcasters own all the other major papers, don't they?"[54] In 2006, a federal court judge ruled the levy unconstitutional, and ordered that $790 million collected since 1998 be refunded to broadcasters.[55]

Asper also urged Ottawa to drop the country's one-third limit on foreign ownership to encourage the US to do the same. "We've always believed that the freer the market the better," he said. "We would like access to the US and other markets."[56] The following year he renewed the call to lift foreign ownership restrictions. "We ... have to make this a North American market in its entirety," he said. "We can't do it piecemeal."[57]

More CBC attacks

Like his father, Leonard Asper reserved his strongest attacks for the CBC. He echoed in 2003 his father's argument that the CBC should not be competing with private networks by airing popular programming such as news and sports. "Because private broadcasters can now afford to, and are showing a willingness to, invest in news and information programming, the CBC should produce only that kind of programming that is not commercial," he said. "That's why we say they shouldn't carry *Hockey Night in Canada* using your money, my money, taxpayers' money to outbid private broad-casters for something that private broadcasters would do just as well." *Hockey Night in Canada* was one of the few programmes that actually earned a profit for the CBC. Asper argued, however, that the public broadcaster should instead spend taxpayer dollars on unpopular programming. "Where it's uneconomic to invest is [in] what they call indigenous drama: kinds of things that don't unfor-tunately get the ratings but are deemed to contribute to Canadian culture or help Canadian artists," he said. "That's what the CBC should do though, is try new programs, try the drama, the variety programming and the arts programming that doesn't make it on an unsubsidized broadcaster."[58]

Leonard Asper also saved his harshest criticism of the CBC for its coverage of the Middle East, in particular that by correspondent Neil Macdonald. In a speech in September of 2003, Asper reprised his father's attack on the world media, accusing them of bias against Israel. He went one step farther, however, and attributed the bias to racism. "Racism is very difficult to prove, particularly when the accused do not openly state the reason for their attacks or their bias," he told an audience at the Gray Academy of Jewish Education in Winnipeg. "No reporter screams: 'I hate Jews.'" The racism of news media was instead an "institutionalized" bias against Israel, according to Asper. "Knowingly or not, the media who cover Israel do not recognize it as either a homeland or a fortress for the protec-tion of Jews both within Israel and for Jews living everywhere." He saw the reporting as resistance to making Israel a Jewish homeland. "Therefore to them Zionism is racism," noted Asper, "and some reporters condemn all Jews for the existence of what they deem to be a racist state." Terrorism in support of displaced Palestinians, he pointed out, had resulted in wild conspiracy theories.

> The reversion to the "blame the Jews" solution for terrorism everywhere is prevalent among the intelligentsia, including journalists. The Jews and therefore Israel are to blame for 9/11; they are to blame for the attacks on the United States and UN installations; they are to blame for the war in Iraq, and even economic decline.[59]

Attacking the media

Asper's speech, which was excerpted in the *National Post* and other CanWest newspapers, echoed his father's attack on lazy, stupid, ignorant journalists for dishonest reporting. Part of the problem, he said, was that unlike its early underdog years, Israel had come to be seen as the agressor in the conflict. "Many news journalists are either doctrinaire socialists or hold political views left of centre," he said. "They are generally supportive of anyone who they deem to be oppressed, victimized or otherwise aggrieved by a stronger party." The problem was made worse, he added, by the fact that Israel was "unprepared for propaganda wars." It was thus losing the battle for hearts and minds in the television age. "Journalists, some of whom are even Jewish, complain openly that they generally receive only an official government statement from Israel, often post-deadline, while from the Arabs they are granted interviews with whomever they want — Hamas, Hezbollah, al-Aqsa or Islamic Jihad."

> They get instant access to wild funerals, replete with bug-eyed youths chanting "death to Israel and America" and they are given packaged home videos from Arabs. These home video shots are either fabricated or edited to paint Israelis in the worst possible light. Professional ethics have fallen by the wayside in the interests of good raw video and deadlines.[60]

Some reporters covering the Middle East, he added, were "fooled by the openness of Israeli society" and the debate in that country over treatment of the Palestinians. Disagreement between politicians from the Labour and Likud parties, he said, as well as between Israeli newspapers, led many journalists to an incorrect conclusion. "The raging debate," he said, "confirms in many journalists' minds that Israel does bear at least some blame for the deaths that occur on both sides of the conflict." The biggest problem, he added, was that many journalists covering the Middle East simply lacked the background to do so competently. "Many reporters sent to the Middle East are unqualified for complex war coverage," he said. "They know nothing about the history but worse, they do not bother to make their own inquiries." Most journalists, he said, did not know that "the terrorist and weapons-infested Jenin refugee camp is run by the United Nations and has been for more than 50 years." Sympathy for Palestinian refugees seemed to Asper undeserved and due mostly to the ignorance of journalists. "Most do not have any clue that the so-called Arab refugees became refugees because they were urged to leave by Arab leaders when they were attacking Israel in 1948." Asper singled out only one media outlet and one journalist by name in charging "hints of anti-Semitism" in the Canadian media.

> When Hezbollah, the well-known terrorist group, was finally banned in Canada, Neil Macdonald of the CBC pompously, but dangerously, suggested Hezbollah was a "national liberation movement victimized by unfair smears cast around by supporters of the Jewish state." No reference to Israel, just "the Jewish state."[61]

Pointing out that while some journalists were "neither Marxists nor anti-Semites," Asper lamented that "they have little help." Fortunately, he told his audience, CanWest had been working toward a solution. "There is some hope, as we have found in observing the results of various programs to educate journalists," Asper continued. "With fair-minded journalists, who actually do care more about the truth than their own ideologies, there has been a positive response once the hard facts are known. But for some, their work must be done for them." In addition to training programs, proper media hiring practices were important to ensure the correct coverage of news, he added. "Media proprietors and managers must ensure that the people they hire do not bring their ideology into their newsrooms, and that journalists do proper research before filing stories." He echoed his father's call to action by urging his audience to hold the media's feet to the fire and point out anti-Israel bias where they detected it. "The media must be held accountable, just as they purport to hold others accountable. Respond to bias when you see it. Demand informed, objective and accurate reporting."[62]

A 'dreadful accusation'

Unfortunately for him, Asper had engaged in just the kind of sloppy reporting he accused many journalists of when he quoted Macdonald. It was a mistake that prompted him to issue a correction the following day. "A typographical error did occur in my speech, in which a quote was misplaced," admitted Asper. He added that he otherwise stood by his comments and had six years' worth of material to support him. "The sentence should have read Macdonald suggested Hezbollah was a 'national liberation movement,' with the rest not in quotes. In fact, the second part of the sentence was the quote from the *National Post* editorial on May 16, 2003, commenting on Neil Macdonald's remarks."[63] Even his correction wasn't entirely accurate, however.[64] Macdonald demanded an apology, but added that he didn't expect one.

> I expect more bullying, more bombast, more ideological, anti-journalistic nonsense. I used to work for the newspapers they now own. Several of my ex-colleagues, still there, tell me they find the Aspers' approach to journalism an embarrassment. But they cannot speak publicly. Thank heavens I can.[65]

Macdonald called Asper's charge of anti-Semitism "a dreadful accusation, one loaded with hateful, historical baggage." He pointed out that the complete sentence in question of his contentious CBC report had actually read: "Is Hezbollah a national liberation movement, or, as Israel and its supporters maintain, a murderous global menace?" Factual fidelity, Macdonald pointed out, seemed to matter little when it came to accusations of media bias. "In Mr. Asper's crusading hunt for Marxists and anti-Semites in the media, the accuracy of the quote hardly mattered," Macdonald wrote in the *Globe and Mail*. "He repeated what he wanted to believe I'd said." Macdonald pointed out that his report from Beirut the previous December had been an attempt to confirm a *National Post* front-page story. It had quoted Hezbollah chief Hassan Nasrallah as advocating the export of suicide bombing worldwide. "The Canadian government had been considering banning Hezbollah based on the Nasrallah quotes," noted Macdonald. "But Hassan Nasrallah, I discovered in Beirut, had said no such thing. Canadian embassy staff in Beirut came to the same conclusion."[66] The *National Post* story had been written by a London-based freelance writer named Paul Martin who, under the pseudonym Sayed Anwar, also wrote for the *Washington Times*. That conservative daily was owned by the Unification Church, or Moonies. "I watched the videos. I watched the speeches," Macdonald told Antonia Zerbisias on the phone from Jordan. "He [the London freelancer] came up with three quotes, one of which, to be charitable, was a gross mistranslation, and the other two were never even uttered." Martin later named as his source Walid Phares, a Florida State University professor who made the claim on the Fox News Network.[67]

Some in the media were outraged that Asper had spoken out on such contentious issues. Christopher Dornan, director of Carleton's journalism school, thought it was entirely appropriate for Asper to give his opinion on an issue of concern to journalists. After all, he was the CEO of Canada's largest private-sector news media company, reasoned Dornan. "No, the problem is not that he spoke out," he wrote in the *Globe and Mail*. "It is what he said." The CanWest leader's criticism of Canadian journalists, according to Dornan, was not only ill-advised and ill-founded, but worse. "Here's what's wrong with Mr. Asper's position: It's dumb as all get out." While Asper had prefaced his remarks by stressing they were his personal views and not CanWest editorial policy, Dornan found his disclaimer "either disingenuous or naive." The resulting influence on CanWest journalists was unavoidable, he noted. "When the person in charge of a national media corporation offers his deep-down opinion on what he hopes for in news coverage, the people who work for him cannot help but take notice." The extreme nature of Asper's comments betrayed his own ideology, noted Dornan.

> Journalists are all too often constitutionally Jew-hating Marxists who are intellectually dishonest and therefore morally bankrupt? Pardon? Mr. Asper takes a complicated matter that merits serious attention and reduces it to baiting and name-calling. He should know better, but apparently, he doesn't. This guy hasn't the foggiest idea how journalism works, but for the moment, much of Canadian journalism works for him.[68]

Izzy Asper died a few days later, leaving Leonard as "as Canada's most important media magnate," according to the *Globe and Mail*.[69] According to Gordon Pitts in 2002, he was already the country's most important media executive, because "unlike his rivals at BCE and Bell Globemedia, Leonard actually owns the shop."[70]

Fox News Canada

The Canadian Cable Television Association applied to the CRTC in 2003 to include the Fox News Network on its member cable systems, but was turned down. That was because CanWest had since 2000 held a CRTC licence to partner with Fox and broadcast its content north of the border. CanWest planned a "hybrid" digital

all-news television network, but Fox balked at the CRTC's requirement of 35 percent Canadian content. Global's news division was
by far the thinnest of all the Canadian networks. Even if Fox had
agreed to carry that much Canadian content, it would have meant
a major infusion of resources to meet CRTC requirements. CanWest
dropped the hybrid idea, but the Aspers harboured long-term plans
for breaking into the all-news business. "There comes a time of
maturity in the CanWest organization when we can focus on news,"
Leonard Asper said in 2002. "We desperately want a CBC News-
world-type channel."[71]

The CCTA also applied in 2003 to air satellite news network al-
Jazeera and 15 other ethnic networks. The Arabic-language network had been founded in 1996 with US$150 million in funding
from the emir of Qatar. It soon established itself as an international
news force, with 70 correspondents in 35 bureaus, and it regularly scooped Western media. Al-Jazeera also often drew the ire
of the White House by airing tapes from al-Qaeda leader Osama
bin Laden. With an audience estimated at 30–40 million, including viewers in the US, the UK, and France, it soon became the
world's most-watched television network.[72] The application to air
al-Jazeera in Canada was opposed, however, by the Canadian Jewish Congress and B'nai Brith. They claimed al-Jazeera violated the
country's strict laws against "hate speech." B'nai Brith described
it as "virulently" anti-Semitic. The CRTC allowed broadcast of al-
Jazeera in 2004, but required its distributors to edit out any "abusive
comment."[73] As a result of the extra cost of monitoring and editing,
no Canadian cable system added al-Jazeera.

When Fox News decided it wanted to broadcast its own service in
Canada rather than partner with Global, its application drew hundreds of supporters. Included were REAL Women of Canada, Focus
on the Family, and B'nai Brith. "It is our firm belief that a diversity of
viewpoints and perspectives which stay within the bounds of Canadian law is an essential part of freedom of expression as enshrined
in our charter," wrote Frank Dimant, B'nai Brith Canada's executive vice-president.[74] With only 82 parties opposing the application
and 531 in favor, the CRTC approved Fox News without restriction.[75]
Some commentators questioned the decision due to Fox commentators promoting contempt against an identifiable group — Canadians. Bill O'Reilly regularly mocked Canada and had referred to it
as a "nanny state" with an "entitlement culture."[76] While al-Jazeera

provided some of the best reporting from the Middle East, Canadians were prevented from viewing events in the region from its perspective. Fox, meanwhile, was criticized even by many Americans for its biased coverage. The liberal group moveon.org filed a false advertising complaint in the US over the network's "Fair and Balanced" slogan.[77] The irony, to some, bordered on hypocrisy. "No Al-Jazeera anchor would do what Fox network anchors did and wear Old Glory lapel pins on air," noted Ryerson journalism professor John Miller. "No Al-Jazeera executive would censor content at the request of a government, as Fox owner Rupert Murdoch did when he said, 'We'll do whatever is our patriotic duty.'"[78]

Convergence continued

While Izzy Asper had been a "notable skeptic" about convergence, he indulged his youngest son's passion for the concept. According to David Olive, who was by then writing for the *Toronto Star*, he "risked driving Leonard out of the family business if he didn't embrace the young man's fascination with multi-media endeavours."[79] Leonard's zeal for convergence was obvious from the outset of CanWest's venture into print. "He loses himself in enthusiasm when he talks about his convergence strategy, stepping from side to side as he answers the reporters' questions," noted one observer. "When he says 'multiple media,' his hands fly apart. When he says 'reaggregating,' they swoop back in to compact the air in front of him. Izzy Asper watches the performance in bemused silence."[80]

Rick Camilleri had been brought in by CanWest in 2002 as its point man on convergence, and he also proved an ardent proponent of the concept. He announced in 2004 that CanWest planned to merge its television and newspaper content by year's end and charge fees for online access. A multimedia digital newspaper he described as "PowerPoint on steroids" would provide access to both television and newspaper content. "Our publications put out reams of content every day and we plan to deliver that," Camilleri told a Royal Bank investors' conference. "This is how we see television and newspapers fusing together." Advertising would be revolutionized, he said, by adding multimedia capabilities such as classified ads with digital video clips. "It's a platform that will enable us to engage in one-to-one dialogue with our readers, subscribers and viewers."[81] A month later, he said advertisers were "thoroughly embracing" the new business model, which would create "huge employment opportunities." CanWest's convergence strategy would succeed where AOL-Time Warner's had failed, he claimed.[82]

After rebounding above $14 early in 2004, however, CanWest's stock price had fallen to a 52-week low of $9.15 by mid-September. CTV had overtaken Global in the television ratings war since 2002 due to the deep pockets of BCE. Its new corporate parent provided strong leadership for the network of twenty-three stations and took full advantage of its buying power over Global's group of ten. By 2004 CTV had boosted its audience share by 50 percent and boasted eighteen of the twenty most popular network programs. Included among Global's programming blunders had been luring late-night comedy show host Mike Bullard away from CTV in 2003. David Asper was appointed his executive producer. "We're passionate about star building and then basically taking those stars and exporting them to the world," said Camilleri. "We look at Mike as another potential multimedia star."[83] After watching Bullard beaten more than 3–1 in the ratings by his replacement on CTV, *The Daily Show With Jon Stewart*, Global pulled the plug. Bullard had lasted only thirteen weeks on Global.[84] That fall, CanWest moved to bolster its convergence team by importing five multimedia executives from the US. They included former vice-presidents at Time Warner, the *New York Times*, and *Playboy*, along with two former AOL executives. CanWest also restructured its operations, grouping its Canadian newspapers and Global Television into CanWest MediaWorks and its foreign properties into CanWest MediaWorks International.[85]

Camilleri, who was named president of CanWest MediaWorks, said the company hoped to boost its integration strategy by "repurposing" content across multiple media. "We have an archive some 100 years deep, in some cases," he said, "and it's really about mining that archive."[86] But Global's ratings continued to languish, and the *National Post* struggled to break even. As its convergence strategy stumbled, CanWest profits fell. In June of 2005, the company axed Camilleri and reached into its past to replace him, naming loyal warrior Peter Viner as the new head of CanWest MediaWorks. Viner had been a CanWest executive since 1980 and had headed Global Television, run its Australian operations, and even been publisher of the *National Post*. Leonard Asper denied the move signalled an end to CanWest's convergence plans. "This change in leadership in no way alters our Canadian multi-media strategy," he insisted.[87]

CHAPTER 11
Media Bias

The conventional wisdom on media bias reversed itself in the 1990s much as the consensus on media power had following World War II. Complaints from conservatives, backed by ammunition from media monitoring projects, created a new public perception of liberal media bias. Polls showed that the percentage of Americans who saw a liberal bias in the media rose from 12 percent in 1988 to 43 percent in 1996. The perception was a departure from decades of academic research that had shown news media to be quite conservative. The shift was attributed by some observers to Republican politicians in the US adopting claims of liberal media bias as a "core rhetorical strategy."[1] A content analysis of 632 newspaper articles from 2000-2001 found sixteen times as many claimed a liberal media bias than argued for a conservative bias. Stories that alleged liberal bias, moreover, tended to present the situation in absolute terms. Only 35 percent were "balanced" in tone, the study found, compared with 95 percent of those claiming conservative bias.[2] The tactic worked in part because it resonated with public perceptions of news media and journalists as elitist. It succeeded because it also utilized the best-proven of all propaganda techniques. "Like any classic Big Lie, the one about the so-called liberal media is based on strategic calculation," noted the *Columbia Journalism Review*. "Calling the media liberal *works*."[3] As the left fired back with accusations of conservative media bias, a growth industry emerged in books on both sides of the debate.[4]

What clinched the matter for many was *Bias*, a 2001 book by former CBS News reporter Bernard Goldberg. Unloading a career's worth of complaints, Goldberg claimed most news coverage made Democrats look good and Republicans look bad. "I could be wrong, but I think homelessness ended the day Bill Clinton was sworn in as president," he wrote. "Which is one of those incredible coincidences, since it pretty much began the day Ronald Reagan was sworn in as president."[5] At first turned down by several mainstream publishing houses, Goldberg finally found a taker in Regnery, the leading US publisher of conservative polemics. A first printing quickly sold out as *Bias* struck a chord with the American public. It became a surprise hit, racing to top spot on the bestseller list. The *New York Times* books editor called *Bias* "perhaps the most astonishing publishing event in the last twelve months."[6]

Bias confirmed to many Americans what they had long suspected — those in the news media were not like them. Goldberg's broadside pointed up an uncomfortable fact. The attitudes of most journalists on such hot-button issues as abortion, gun control, capital punishment, and gay rights did not reflect those of most voters. That fact had been long-established. Studies had repeatedly shown journalists at major media outlets to be more liberal than Americans as a whole.[7] Their politics, however, were explained in large part by simple demographics. Journalists tended to be younger, better educated, and more urban than the population at large — all factors associated with liberalism. Despite their politics, many journalists argued that reporters were constrained by ethics such as objectivity, balance, and fairness. They thus had limited power to influence the news. In fact, reporters were subject to several levels of "gatekeeping" by editors who had the power to alter or even "spike" their copy.

Liberal leanings?

A survey of Washington correspondents in 1937 showed that 64 percent had voted for Democrat Franklin D. Roosevelt. Another 6 percent supported the socialist candidate.[8] A replication of the research a quarter century later confirmed the press had liberal leanings.[9] One of the first major attempts to impute liberal bias from media content was Edith Efron's study of the 1968 presidential election, *The News Twisters*. Her analysis of television network news claimed it ran almost 9-1 against Republican Richard Nixon on ABC, almost 10-1 on NBC, and more than 16-1 on CBS.[10] A more scientific replication, however, found Efron's data to be highly subjective and unreliable.[11] A survey of 240 journalists in 1980 found that 54 percent described themselves as left of centre, while only 19 percent said they were on the political right. In the previous four presidential elections, at least 80 percent of surveyed journalists had voted Democrat.[12] Later published as *The Media Elite*, the study by Robert Lichter and Stanley Rothman was dismissed by sociologist Herbert Gans. He claimed it was biased because it was designed to advance a political agenda. Gans's classic study of news media found them to be neither liberal nor conservative but instead "reformist." He found the study by Lichter and Rothman to be "less than scientific" by engaging in such tactics as guilt by association.[13]

Scholarly studies had long seen news coverage as more determined by the conservative influence of media owners. Critical research had shown news organizations strongly supported the status quo and thus perpetuated social control.[14] Long-running media monitoring initiatives like Project Censored in the US and NewsWatch in Canada noted an absence of coverage critical of corporations.[15] The either-or focus on journalists versus owners, however, recognized only two of a number of influences that had been found to compete in shaping the news. These included news gathering routines, the power of sources, audiences, and advertisers, and the growing influence of pressure groups and public relations.[16]

Perceptions of media bias could also be explained in large part by differences in audience perspective, as demonstrated by the well-documented "Hostile Media Effect." A 1985 Stanford study surveyed pro-Israeli and pro-Arab audiences for their reactions to news stories. Members of both groups saw reports of the 1982 massacre of Palestinians in Lebanese refugee camps as biased in favour of the other side.[17] Replications consistently confirmed the phenomenon and showed how selective perception and selective retention could create polarized worldviews.[18] Media content, however, was hardly identical across news outlets. Increasingly it was presented from dueling perspectives, creating additional differences in worldview due to selective attention and selective exposure to media.

The increasingly polarized perspectives were due in part to a change in marketing that accompanied the fragmentation of audiences in the 500-channel universe. For most of the 20th century, media outlets sought to attract the largest audience possible to sell in turn to advertisers. That encouraged them to be as inclusive as possible, and thus liberal. The media's strategy under fragmentation, however, became more exclusive in the form of target marketing.[19] Satellite broadcasting, cable television, and specialized magazines allowed the targeting of specific demographic and even "psychographic" groups. Psychographic research revolutionized marketing by grouping people not by crude categories such as age, race, gender, education, and income. Instead it considered shared beliefs and interests as determined by sophisticated surveys and focus group research. One group coveted by advertisers was political conservatives, as they tended to be of higher income. Tailoring journalism to appeal to them, however, only resulted in increased polarization and political partisanship.

An example was the *National Post*. By appealing to readers who felt alienated from the country's more liberal media outlets, it aimed at attracting upscale readers. Their disposable income would in turn attract advertisers, or so the theory went. Building an advertising base in a falling economy, however, was difficult enough for any start-up newspaper. As a national publication, the *Post* was forced to rely solely on national advertising. That category was traditionally the scarcest in any newspaper, well behind classified and local retail advertising. As a result, even Gannett's *USA Today* lost money for fifteen years despite a circulation of more than two million.[20] Black's sale to the Aspers of a half-interest in the *Post* only two years after its debut — and the other half to follow less than a year later — was in hindsight a brilliant business move.

Monitoring media bias

The media-monitoring initiative of the Fraser Institute began in 1987 as the National Media Archive (NMA), which published the monthly newsletter *On Balance*. The vast majority of articles in *On Balance*, according to one study, "expressed concern about the left-wing bias of journalists and the effect this had on coverage of issues."[21] The NMA claimed coverage of the 1988 Free Trade Agreement by the CBC and the *Globe and Mail* portrayed it in an overwhelmingly negative light.[22] *On Balance* criticized news media outlets in 1994 for uncritically accepting Ottawa's assertion that 1.5 million Canadian children lived in poverty. "It just irritated me that there was so much unanimity among all the stories," NMA director Lydia Miljan explained. "I thought, well, I just don't see any poor waifs selling pencils on the streets."[23]

Miljan's 2003 book *Hidden Agendas: How Journalists Influence the News* was the Canadian equivalent of Goldberg's *Bias*. Unlike that journalistic hit job, however, *Hidden Agendas* came shrouded in a cloak of scholarly authority. It was short-listed for the $35,000 Donner Prize, awarded annually to the best book on Canadian public policy. *Hidden Agendas* was co-authored by University of Calgary political scientist Barry Cooper and based largely on Miljan's dissertation research under his supervision. It backed up its accusations of media bias with hard data. A survey of 270 journalists concluded they leaned left more than a control group of 804 members of the public. Liberalism was most obvious at the CBC, the authors found, where journalists displayed a "strong left-of-centre position on economic issues."[24] Their content analysis found news tended to be left-biased because it focused on negative events that reflected poorly on the capitalist system. For example, stories would appear about increased unemployment, but not about increased employment. Reports would focus on the adverse short-term social consequences of tax cuts, but not on their possible long-term trickle-down benefits.

To Miljan and Cooper, this tendency indicated a media bias in favour of "statism" over free-market capitalism. They concluded it was the left-wing bias of journalists that resulted in the news being slanted to the left. "Owners do not provide the labour for the product, therefore the product does not necessarily reflect the owners' values," they wrote. "Instead, because journalism is in essence a human endeavour, it must reflect the values and political orientations of those who do it."[25] Canadians concerned over increased concentration and convergence of news media in the hands of right-wing owners thus had nothing to worry about.

> When we examine the larger context of ownership and convergence we can offer some reassuring comments to those who fear this trend. Because, as we demonstrate, news stories are dependent on individual journalists writing, selecting sources, and so forth, the issue of who owns the company, or how large those holdings may be, does not have a large — or even small — impact.[26]

The problem with the study, aside from the extent to which its conclusions tortured logic, was in its methodology. Ascertaining the political leanings of survey respondents was supposedly achieved by asking some very loaded questions. One inquired whether communism was "evil and unworkable" or "a good idea but wrecked by bad leadership." A review of the literature ignored most of the voluminous research into influences on news media content. The lone reference on totalitarian propaganda in the 1930s was not a scholarly study but instead Hitler's *Mein Kampf.* It was cited not as an example, but as an authority.[27] The study's biggest and fatal flaw, however, was its admitted objective of proving a point, to which end the authors gathered data. "We think the opinions of journalists are important because those opinions influence the news they produce," Miljan and Cooper stated at the outset. "The burden of this book is to prove it."[28]

Given the authors' undisguised mission, with little pretence of disinterested inquiry, their findings may have largely been determined by their method. This habit of proceeding from conclusion to method to data collection to analysis was the major complaint scholars had with most Fraser Institute studies. It left the *Hidden Agendas* authors open to accusations of bias even more egregious than the bias they ascribed to journalists. To their credit, Miljan and Cooper admitted the stale nature of their data. Most of it was gathered over the winter of 1997-98, before Black founded the *National Post* and sold Southam to CanWest. Those developments, the authors admitted, signalled that their work was "far from over." They promised a replication that would provide "necessary corroboration" of their results within a few years.[29] No doubt it will.

A 'cultish adoration'

A study of the CanWest national editorials noted that they appeared infrequently in 2003, with the first not running until March. It urged the invasion of Iraq.[30] "No further resolutions or hand-wringing is necessary," argued the editorial, invoking a favourite phrase. "The indictment against the Iraqi regime has been proven over and over."[31] Enthusiasm for invasion was most marked in the *National Post*, which prompted one of its columnists to quit in a public display of disgust. "At times, the *Post*'s hostility to critics of the war was simply childish," wrote Patricia Pearson in the *Globe and Mail.*

> There wasn't a peace movement. There was a "peace" movement, quote unquote. There wasn't a valid argument that UN inspectors be given more time to find Iraq's weapons of mass

destruction, or that pre-emptive invasion should be seriously hashed out in light of precedents in international law, or that an alternative to force might be imagined.[32]

Pearson, who was one of the *Post*'s few liberal columnists, found that the newspaper's openness to dissenting opinions disappeared after the Aspers acquired it. "A kind of Political Correctness, so excoriated as a disease of the left, began to prevail," she noted. "When CanWest, controlled by the Asper family, acquired the paper from Conrad Black, I no longer dared to express sympathy for Palestinians." The granddaughter of former Liberal prime minister Lester B. Pearson said she quit her column in protest of what she felt was actual political peril on the *Post*'s pages. "What finally provokes a journalist to resign in protest of bias? The answer is when she begins to feel that that bias is doing her nation harm." The biggest threat to Canada from the *National Post*, according to Pearson, came from its constant push for American values. "I cannot sit back and watch this nation attacked, relentlessly and viciously, by a newspaper that would trash so much of what we believe in, from tolerant social values to international law." The *Post*'s post-9/11 editorials displayed what she called a "cultish adoration" of American power unleashed. "This vision of America blatantly favours the rich, displays a breathtaking indifference to the environment, crushes civil liberties, manipulates patriotism by stoking fear, insults its allies, and meets skeptics with utter contempt."[33]

Lawrence Martin, whose political column had been dropped by CanWest, judged the *National Post* similarly. Its coverage of the run-up to invasion, he concluded, showed the newspaper favoured American values over those more Canadian. "The *National Post* is so American it should come in a holster," Martin wrote. "Canadians, as opinion samplings suggest, haven't migrated rightward in big numbers, only their printing presses." The move to the right was obvious, he argued, in uncritical coverage of the US case for invasion. "Some wonder, for example, how there could be almost as much of a drumbeat for an Iraq war in the Canadian media as in the United States. Some wonder how President George W. Bush's allegations can be reported at face value by Canadian journalists."

If it were in years gone by, my guess would be there would be reams of articles on how he has been humiliated by United Nations weapons inspectors finding nothing, on how he has plummeted in the polls, on how he gets away with such assertions that Saddam Hussein has the capacity to invade America and ruin its economy. But that would be in the Canada of years gone by, before the media forces of the right gained the predominant place.[34]

Insurgent or terrorist?

Any reassurance Canadians may have taken from Miljan and Cooper that it was the left bias of reporters that coloured the news would soon be shaken. CBC radio reported in September of 2004 that some CanWest newspapers had taken to inserting the word "terrorist" into wire service stories out of the Middle East. A Reuters dispatch from Jerusalem referred to the al-Aqsa Martyr Brigades, but the *National Post* added the description "terrorist group." Reuters editors asked CanWest to remove the bylines of their correspondents when such changes were made. "When you change the meaning of the story — and we feel that changes the meaning of the story, because terrorist is an emotive term that we don't use in the way that they used it — what we would suggest is that they just take the name of our reporter off," said Reuters spokesman Stephen Naru.[35]

A review of earlier CanWest editions found the *Ottawa Citizen* to have inserted the label into wire service stories numerous times. An Associated Press report on the battle for Fallujah in Iraq that was published in the *Citizen* on September 9 used it seven times.[36] A comparison with the original AP version found the word "insurgent" used each time instead. *Citizen* editors admitted the changes contravened the newspaper's policy, which was to use "terrorist" only to describe someone who deliberately targets civilians.[37] An AP story on the *Citizen*'s front page was also corrected. That account of an Israeli attack that killed ten Palestinians on the West Bank referred to six of them as terrorists. After a complaint from the AP, the *Citizen* admitted a switch from the original "militants" had been due to an "editing error."[38]

The Canadian Arab Federation and the National Council on Canada-Arab Relations complained of "biased reporting against Muslims and Arabs." They asked the Ontario Press Council to investigate. Mazen Chouaib, executive director of the NCCAR, claimed in a *Globe and Mail* column that the wording changes by CanWest were helping to give Muslims a bad name. "For many Arab Canadians, this is another example of what they have long complained about: CanWest seems to make every effort to demonize them and their culture," Chouaib wrote. "No one has suggested that Osama bin Laden and his killers are not terrorists. They are. But to blame an entire race and culture is wrong."[39] *Ottawa Citizen* editor-in-chief Scott Anderson admitted to the *New York Times* that the changes had been "silly errors," but he took umbrage with the charge of vilification.[40] "Mr. Chouaib suggests that using 'terrorist' besmirches the Muslim community," Anderson wrote in the *Citizen*, "but I would argue that it is those who kill in the name of Islam who besmirch the Muslim community."

> On the charge that CanWest uses "terrorist" in some stories and editorials, we plead guilty. Those who blow up family pizza parlours in Jerusalem or commuter trains in Madrid are terrorists. . . . It is alternatives such as "gunmen" or "activist" that betray a bias, insofar as they sanitize what groups such as al-Qaeda are really about.[41]

Journalists at the *National Post* were even more unrepentant. "The [Reuters] agency's use of euphemisms merely serves to apply a misleading gloss of political correctness," an editorial declared. "And we believe we owe it to our readers to remove it before they see their newspaper every morning."[42] Kelly McParland, the editor who added the terrorist tag, explained why. "The killers at the al-Aqsa Brigades are terrorists, by any definition," wrote McParland. "If the CBC can't figure that out, it needs a kick up its airwaves."[43] Columnist Robert Fulford also took issue with Chouaib's complaint. "Muslim publicists prey on terrified bureaucrats, anxious to show their multi-cultural credentials, and on dozy, gullible newspaper editors, like the *Globe and Mail* people," he fulminated. "Like a brain-dead human-rights commissioner or a university harassment officer searching for language crimes, Chouaib believes that you can make something better by giving it a nicer name."[44]

Convergence concerns

By 2004, cross ownership of media had been the subject of fierce debate around the world as companies pushed for deregulation. In Britain, a Communications Act in 2003 allowed newspaper owners to hold licences for stations of the Channel 5 network. The change had generated strong opposition in the House of Lords due to concerns over increased ownership concentration. An amendment was thus added to apply a "plurality test" to media mergers to ensure they did not unduly reduce local diversity. In Australia, Liberals gained control of the Senate in late 2004. That allowed the Howard government to finally abolish restrictions on cross-media ownership. The political concession made to achieve the objective, however, was agreeing to allow foreign media ownership. Concerns over increased ownership concentration were eased somewhat by the introduction of a "diversity test." Mergers and acquisitions would be subject to approval in order to ensure a minimum of five media owners in large cities and four in regional markets.[45] Across Europe, measures were similarly enacted to ensure diversity in the face of Big Media's urge to merge, according to UK media economist Gillian Doyle.

> Rules which limit media and cross-media ownership are generally seen as an essential means of sustaining and promoting pluralism. . . . To avoid allowing such power to fall into the hands of too few individuals or corporations, virtually all countries in Europe have adopted some special rules which restrict ownership of the media.[46]

Nowhere was the battle over cross-media ownership more intense than in the US. The 1996 Telecommunications Act directed the FCC to regularly review its ownership rules to ensure they were necessary. Such a review began in September of 2001. Under chairman Michael Powell, the FCC voted 3–2 along party lines in 2003 to lift the 1975 ban on ownership of a daily newspaper and a television station in the same city.[47] Hearings had been held around the country, but few Americans knew about the proposed change because of scant news coverage. A 2003 survey found 72 percent of Americans had heard "nothing at all" about a relaxation of media ownership rules. "As the FCC moved toward final action on a plan that would greatly benefit a handful of large companies," reported the *American Journalism Review*, "most newspapers and broadcast outlets owned by those companies barely mentioned the issue." The survey by the Pew Research Center also found that the more people knew about convergence, the more they opposed it. "In other words," noted Charles Layton, "Big Media had an interest in keeping people uninformed."[48]

Word of the change spread mostly over the Internet, and a storm of bi-partisan protest resulted. It came from groups as diverse as fundamentalist Christian organizations, the liberal moveon.org, and the conservative National Rifle Association. The populist uprising saw more than two million e-mails, letters and petition signatures presented to the FCC. The *Columbia Journalism Review* noted that the protest resulted from growing concern about the parochialism of US media following 9/11 and their role in the invasion of Iraq. The intensity of public outrage over the rule change, it reported, surprised even industry insiders. "It is not every day that the ideological lines get redrawn over an issue," observed Gal Beckerman. "Media had become a political issue, as deeply felt as the economy, health care, or education."[49] The protest resulted in a Senate review, which voted 55–40 in September 2003 to overturn the change. An appeal to the courts also saw it ordered reversed. A judge found that cross ownership was not in the public interest and that the FCC had ascribed excessive diversity to the Internet.[50] A planned appeal to the US Supreme Court was dropped by the Bush administration in January of 2005 and Powell resigned.[51] The FCC began to revisit the cross ownership issue in 2006 as part of its periodic review of ownership rules.

By 2005, however, the flood tide of convergence had turned to an ebb of deconvergence. The US media giant Viacom announced in March that it would split into two companies. One would own and operate its broadcasting operations, including the CBS network. The other would hold its cable television and movie divisions, including MTV and Paramount Pictures. According to the trade journal *Television Business International*, the move signalled a sea change in media management. "Focus has become the new mantra of the conglomerates," it observed. "Some might find themselves at risk from being wretchedly oversized and not well-coordinated."[52] That description fit CanWest, as its experience with convergence was proving disastrous. "It's tempting to say convergence is dead," observed *Maclean's* in 2005. "But it was never really alive. It was just a tune the industry hummed for awhile."[53]

Even journalism schools began re-thinking their enthusiasm for convergence. After eagerly jumping aboard the multimedia bandwagon early on, some schools gave up after a decade of struggling with the concept. Brigham Young University's journalism school was one of the first to embrace the convergence model in 1995, combining its student newspaper and television newsrooms. BYU introduced multimedia journalism in a bid to create "super reporters" who could move easily between print, broadcast and online media. After the expected convergence jobs failed to materialize, however, faculty members voted to drop convergence from the curriculum in 2006. "Convergence took away necessary depth in core writing skills," explained dean Stephen Adams. "Students knew a whole lot about a whole lot of things, but didn't know very much in depth."[54] Even as the concept of convergence foundered, however, media owners insisted it was the way of the future. They continued to press regulators to lift remaining restrictions on multimedia growth. In Canada, the convergence question would continue to be asked as Senate hearings into the news media ground toward a conclusion.

Vancouver 'heart-breaking'

The desperate state to which journalism had descended in Vancouver could not escape the notice of the Senate news media inquiry. It held two days of its cross-country hearings in the west coast metropolis in early 2005. On the first day, senators heard from a populace furious not only with its mouthpiece media, but also with Ottawa for allowing it. Days earlier, news had emerged that CanWest planned to publish a fourth daily newspaper there, in addition to its *Vancouver Sun* and *Province* and the *National Post*. The Swedish newspaper group Metro International had announced plans to begin publishing a free commuter tabloid in Vancouver and elsewhere in Canada. In response, CanWest announced it would print its own free sheet *Dose* in five cities.[55] A Competition Bureau spokesman told the *Georgia Straight* the extra CanWest title in Vancouver would not arouse regulatory concern if it was aimed at a new market. "A company expanding into other markets is not something that's explicitly prohibited in the Competition Act," said Tim Weil.[56] When *Metro* hit Vancouver streets, it was revealed that CanWest had acquired one-third ownership of it as well.[57]

One of the first witnesses to appear before the Senate hearings in Vancouver was David Beers. Since his 2001 firing by the *Sun*, Beers had put his disaffection with CanWest's dominance of Vancouver's news media into action. With funding from concerned local groups, including the BC Federation of Labour, Beers had helped found an online publication called *The Tyee* in 2003.[58] "Vancouver is a heart-breaking place to be a dedicated news reporter, news editor or news reader, because a single company owns the big papers, the big TV news station and so many other media properties," Beers told the senators. "There is simply not enough competition to keep that owner honest." Cutbacks in political coverage by the city's largest newspaper, he added, resulted in *Sun* headlines being taken "verbatim off the press releases of the Liberal Party." After he and others had been purged from the *Sun*'s payroll, Beers noted, it began pouring money into regular "advertorial" sections titled "Believe BC." Beers said his sources inside the newspaper told him the *Sun* expected to lose $1.5 million on the features. They blurred the line not only between news and advertising, he pointed out, but with politics as well.

> Nothing like it had every been seen before, and it was stuff cranked out by the public relations department singing the praises of various business sectors of the company, not particularly labelled as advertisement. . . . The whole idea being to supposedly bump up everybody's enthusiasm about the potential of the economy and this new government and to make their advertisers feel good.[59]

Beers urged the inquiry to recommend breaking up Big Media by reinstituting the 1982 cross-media ownership ban. He also asked it to seek disclosure by owners of their profit margins, and to recommend measures to assist alternative media. If such steps were not taken, he warned, the kind of news media Vancouver suffered from would eventually be seen across the country. "Most of Canada will soon enough look the same," he said, "if your panel is not successful in crafting and winning a different way forward."[60] The senators heard from many witnesses on the first day of their Vancouver hearings. Most lamented the meagre, biased journalism fed to them daily by the CanWest news monopoly. The senators seemed aghast. Asked Senator Jim Munson: "How does a newspaper get away with not having a political reporter at a [legislative] assembly?" A former CTV reporter, Munson noted an absence of working journalists in attendance. "We'll wait and see tomorrow whether this meeting was covered," he said.[61]

Not only were the Senate hearings in Vancouver not reported in either local CanWest daily, they didn't make the newscast of Global's BCTV either. While there was some coverage by Vancouver radio stations, the only newspaper article appeared the following week in the neighbourhood *Westender* free sheet. It was published by Victoria-based Black Press, owned by David Black (no relation to Conrad). According to its account, the senators "looked surprised by the forceful delegations."[62] A second day of hearings convened across town at the University of BC, where not all witnesses agreed the news media situation in Vancouver was serious. "It is not as dire as a lot of people seem to think," testified Donna Logan, director of the UBC graduate school of journalism. "I mean, we do have the CBC here, and they have a very respectable audience. We do have CTV here and they have a very respectable audience."

> There is an incredibly strong ethnic press here that serves a huge population. And we have two — three television stations and numerous radio stations, I mean, too numerous to count almost, and they command huge audiences. So, the situation is not as simplistic as it is often portrayed here. It is much more complex.[63]

Logan had testified before the Senate inquiry once already, flying to Ottawa to appear as part of its first round of witnesses in 2003. On their visit to Vancouver two years later, she was able to brief the senators on the findings of the Canadian Media Research Consortium's first study. In producing a "report card" on news media, the CMRC surveyed more than 3,000 Canadians on their news consumption habits and on the credibility they lent it. "We came up with several ways to measure credibility," Logan told the senators. "These included: accuracy, bias, fairness and balance, accountability, sensationalism and trust, independence, and finally, consolidation and ownership."[64] On the controversial topic of bias, Logan told the senators Canadians were "quite cynical" about the news. "A surprising number of Canadians do not think the news is impartial," she said, clicking through a PowerPoint presentation. "Almost 80 percent of Canadians think that reporter's bias influences news often or sometimes. The finding of reporter bias is very similar to results in the United States, as you can see on this slide."[65]

The only problem was that Logan was engaging in a classic "apples and oranges" comparison. The US survey had asked whether "news organizations" were politically biased in their reporting. The CMRC survey question was loaded from the outset, laying any possible bias at the feet of individual journalists. It asked: "How often do you think reporters let their own political preferences influence the

way they report the news?"[66] The CMRC study had been released the
previous June at the Banff Television Festival. Logan there attributed the "disturbing" findings on news media credibility to almost
everyone except media owners. "I think the media has to do a much
better job of demonstrating its independence," she said. "Canadians... feel that reporters are influenced by government officials, by
bureaucrats, by powerful groups and people with money."[67]

Joel Connelly covered BC for the *Seattle Post-Intelligencer* and
found in Vancouver's news media a cautionary tale for his readers.
"News coverage has been so slanted that Vancouver's daily papers
should be read at a 45-degree angle," quipped Connelly in 2005.
"With its blatant biases and recent cuts in staffing, CanWest dem-
onstrates the perils of having a daily paper monopoly." In advance
of a provincial election, he noted, CanWest papers "trumpeted
good economic news with Page 1 stories and TV features on the 2010
Winter Olympics building boom." News that reflected poorly on the
Campbell Liberals was hardly mentioned. "The opposition's case —
75 medical facilities closed or downsized, 113 schools shut, slashes
in environmental protection and the highly suspicious BC Rail deal
— has been given short shrift." When the NDP began to gain in the
polls, noted Connelly, the *Vancouver Sun* and BCTV reported that
the BC Teachers Federation planned a strike vote just after the election. As a result, Campbell "basked in the opportunity to talk and
look tough."[68]

The environment was another issue on which the CanWest newspapers seemingly did not share the public's concern, especially on
the west coast. Canadian media carried so much misinformation
on global warming, claimed the *Georgia Straight*, that "it might best
be described as journalistic malpractice." *Vancouver Sun* columnist
Michael Campbell, the premier's brother, "regularly holds forth on
climate change," it noted. "Last year, he scolded the scientific community for its slapdash work." One particularly prolific climate-
change denier, retired University of Winnipeg professor Tim Ball,
seemed to be a CanWest favourite. "Over the past five years, he has
published no less than 39 opinion pieces and 32 letters to the editor
in 24 Canadian newspapers," calculated the *Straight*. "Fifty of these
pieces ran in papers owned by CanWest MediaWorks. These efforts
totalled an incredible 44,500 words."

> Papers like the *Vancouver Sun* and the *National Post* published
> articles by such deniers as Patrick Michaels and S. Fred Singer,

who some might remember from his previous efforts to allay public fears about the dangers of secondhand smoke. And the *Province* promoted right-wing columnist and environmentalist-bashing Jon Ferry to the position of editorial-pages editor.[69]

Even normally conservative pundit Norman Spector went off on the BC media for letting Campbell get away with "murder" on the environment. "It used to be that the media in British Columbia were the most aggressive in the county," he told talk show host Joe Easingwood on CFAX radio in Victoria. "You have to wonder now whether they're asleep." When the premier sparked angry local protests in 2006 by announcing two coal-fired power plants, Spector noted, no mainstream media outlets questioned it. When the NDP opposition announced its plans for curbing BC's greenhouse gas emissions in 2007, the story was buried deep inside both Vancouver dailies. "It's the issue on which the federal government could fall, on which we could have an election," Spector pointed out. "And Carole James and the NDP come out with a serious proposal . . . and it's nowhere! Where is it on the front page of our newspapers?"[70]

Social control in the newsroom

Concern over the growing power of media owners to shape news content was expressed in the US as early as the 1947 Commission on Freedom of the Press. Its report proposed self-regulation of news media in the form of "social responsibility" to stave off growing calls for government action. The Hutchins Commission, as it was known, concluded that the content of US newspapers was "twisted" by numerous influences, including the interests of owners.[71] The growing newspaper chains, it warned, were becoming "great concentrations of private power."[72] It concluded that the "exaggerated drives for power and profit which have tended to restrict competition and promote monopoly" had worked against the public interest. If media owners did not act responsibly, the commission added, government regulation might be required despite strong First Amendment guarantees of US press freedom.

Sociological studies of influence on news content dated to the 1950s and had shown ownership to be a deciding factor. Warren Breed had been a reporter for the *Oakland Post Enquirer* before studying for his doctorate. He thus well understood the unwritten rules of journalism.[73] His classic 1955 study "Social control in the news room" showed how reporters were socialized into adherence with a newspaper's unwritten editorial policy. News gathering, noted Breed, had to be delegated by media executives who set policy to reporters who were often younger and more liberal. His interviews with a random sample of 120 journalists at mid-sized newspapers illuminated the process Breed had experienced. He deemed publishers the true source of power to shape the news, but he found their power was not unlimited. Instead, it was offset to a degree by professional ethics, newspaper unions, and journalism education. The enforcement of a newspaper's policy, Breed concluded, was achieved indirectly through what a publisher could directly control. That included hiring, firing, and promotion — in short, power over the "mobility aspirations" of journalists. Editorial policy, Breed noted, was consequently reflected in the "slanting" of news stories.

> "Slanting" almost never means prevarication. Rather, it involves omission, differential selection, and preferential placement, such as "featuring" a pro-policy item, "burying" an anti-policy item in an inside page, etc. . . . Policy is covert, due to the existence of ethical norms of journalism; policy often contravenes these norms. No executive is willing to risk embarrassment by being accused of open commands to slant a news story.[74]

Pamela Shoemaker and Stephen Reese noted in their compendium of influences on news content that the power of ownership was subtle but effective. "The absence of visible attempts at control does not mean that none are being made," they noted. "Whenever media workers deduce what their supervisors want and give it to them, *de facto* control has been exercised."[75] Herbert Gans studied major US media outlets in the 1960s and 1970s and found that media executives had "virtually unlimited power" over news content. They were not able to exercise their power continuously, however, due to time constraints and their location outside the newsroom. According to Gans, that made their occasional intervention in news content all the more influential. "Their role in story selection and production is intermittent," he noted. "They do not exercise their power on a day-to-day basis. Perhaps because they do not do so, the journalists pay close attention to their periodic suggestions, and at times, they overreact."[76]

Such overreactions were occasionally observed anecdotally, often with lamentable results. Sometimes over-eager newspaper editors and television producers went well beyond the wishes of ownership in their zeal to toe the company line. The result at Fox News, according to its former media critic, sometimes bordered on the absurd. One 2002 story even displayed the address and phone number of a California judge who ruled unconstitutional the words "under God" in the Pledge of Allegiance. The Fox newroom chief said he had included the information onscreen so viewers could contact the judge directly. "To their credit," noted Charles Reina, "the big bosses recognized that their underling's transparent attempt to serve their political interests might well threaten the judge's physical safety and ordered the offending information removed from the screen."[77] *Maclean's* editor Anthony Wilson-Smith saw the same sort of thing in Canada. "More often than not, it is the middle-managers and career climbers among reporters and editors who are at the root of journalistic evils," he noted. "They hear their bosses cough, and whip their interpretation of that into a hurricane by the time it hits the newsroom."[78]

The 2006 firing of columnist Vivian Smith by CanWest's *Victoria Times Colonist* may have been such an example. Smith, a 30-year journalist who doubled as the *Times Colonist*'s writing coach, wrote a column that summer critical of the city's high-priced tourist attractions.[79] After a meeting with industry officials, the newspaper printed a rare front-page correction of a minor factual error in the column and informed Smith of her dismissal. The story was reported by the Internet blog *Public Eye* within days and circulated for weeks online before CanWest head office even heard about it.[80] Smith's firing was never reported in the mainstream BC media, but on the Internet it reached as far as the media blog of the *Guardian* newspaper in the UK.[81] *Times Colonist* columnist Lynne van Luven, who also taught journalism at the University of Victoria, and feature writer Janis Ringuette, quit in protest.[82] The Canadian Association of Journalists issued a statement of concern over advertiser influence on the news.[83]

When new CanWest Publications president Dennis Skulsky learned of Smith's firing, he ordered her reinstated. He issued a statement assuring CanWest journalists they were "free to cover any organization or subject with the full support of their Editor and Publisher."[84] According to CanWest spokesperson Dervla Kelly, Skulsky was "very upset" when he learned about Smith's firing.[85] *Times Colonist* publisher Bob McKenzie apologized, calling it an "error in judgment."[86] Ringuette and van Luven were invited to rejoin the newspaper, but both declined. Ringuette deemed the front-page correction misleading. Van Luven rejected the offer after editor-in-chief Lucinda Chodan banned her from writing about the incident. "I would feel somewhat hypocritical," explained Van Luven about why she refused to resume her column. "I feel that, as a community newspaper, [the *Times Colonist*'s] job is to air the news, and this is a piece of important background to the news."[87]

Self-censorship

Many independent-minded journalists scoffed at the notion of owners exercising control over the volumes of news they hurriedly produced daily. The process, however, had been well documented. Instead of censoring them, media owners relied on their journalists to censor themselves. A study reported by the *Columbia Journalism Review* found that direct censorship of the news by owners was not a problem, but self-censorship was "pervasive." A survey of 287 journalists, including 81 senior editors and executives, was conducted by the Pew Research Center. "Journalists say that, typically, they do not decide on their own to avoid newsworthy stories," it reported.

> More than half of those who think stories are sometimes ignored say that journalists either get signals from their bosses to avoid such stories or ignore them based on how they think their bosses would react. Of those who believe newsworthy stories are being avoided to protect corporate interests, fully three-quarters say journalists get signals or anticipate negative reactions from superiors.[88]

A 2003 study at McGill University found that 56 percent of Canadian journalists believed the views and interests of a newspaper's ownership affected its news coverage. The perceived influence of owners was highest at CanWest dailies, according to the survey of 361 journalists across the country. The percentage was lowest at the French-Canadian dailies *La Presse* (16 percent) and *Le Devoir* (31 percent). Ownership influence was felt most at the CanWest dailies *Ottawa Citizen* (67 percent), *Montreal Gazette* (76 percent), and *Vancouver Sun* (83 percent). The infamous Atkinson principles were thought by 71 percent of journalists to influence content of the *Toronto Star*. The results, according to political scientists Stuart Soroka and Patrick Fournier, could be cause for concern. "Greater concentration of ownership is not in and of itself a bad thing," they noted. "One can imagine an individual owning all of Canada's newspapers and letting a diverse group of editors freely determine news content and commentary. The situation may become problematic if the views and interests of a limited number of owners regularly affect newspaper content."[89]

The intermittent interventions of the Aspers in the content of Can-West newspapers fit perfectly the process described by Breed and Gans. By occasionally signalling what they wanted to see in print, they set the parameters for content. Publishers who didn't go along with what they wanted, like Russell Mills, risked being fired. Editors who wanted to be retained or promoted would tend to enforce compliance with the perceived wishes of ownership. Journalists who did not fall into line, like those at the *Montreal Gazette*, were threatened with suspension or firing. Those who got out of line anyway, as in Regina, were disciplined. Others would take note and become less likely to protest. As a result, journalistic independence was eroded. Among managers of former Southam newspapers, deviating from what their new owners wanted to see published became known as a "career-limiting move."

As Breed noted, there were countervailing influences on the power of ownership, such as professional ethics and union power. After CanWest quashed the 2001 byline strike in Montreal, the Newspaper Guild grieved the threatened sanctions.[90] Almost two years later, an arbitrator sided with the union, ruling that under their collective agreement *Gazette* reporters had an absolute right to withdraw their bylines.[91] The decision was hailed as a landmark victory for press freedom, yet ironically it went unreported in the *Gazette*. "Indeed," as *Toronto Star* columnist Antonia Zerbisias noted, "no CanWest paper saw fit to inform readers that, for the past couple of years, they might have been getting less than all the news fit to print."[92]

CHAPTER 12
The Press We Deserve

Its final report on the news media in Canada was not released by the Standing Senate Committee on Transport and Communications until June of 2006. It had been more than three years since the study was conceived. The committee had been disbanded twice while Parliament was dissolved for federal elections. It began its work by going on hiatus for more than a year while the Lincoln committee on broadcasting completed its hearings. In the end, its report was released by a new committee chair, Lise Bacon, as the term of Senator Joan Fraser had expired. The political landscape onto which the report landed had also been radically altered by the federal election earlier that year. After more than a dozen years of Liberal rule, a minority Conservative government came to power that February under Stephen Harper. A deregulationist regime in Ottawa meant that any prescription the senators might have had for legislative reform of Canada's news media ownership had little hope of being adopted. Perhaps that was why they proposed such modest measures to curb the growing power of Canada's media giants.

The likelihood of any limits on media ownership being enacted by Ottawa also lessened for another reason. While Izzy Asper had been tight with the Liberal party, his heirs moved almost as close to the Conservatives. One senior editor at Global Television even ran in the 2006 election as a Conservative candidate in Toronto with Asper blessing. A new chairman of CanWest's corporate board came directly from Tory ranks and aligned the Aspers uncomfortably close for some critics with the new party in power. A CanWest executive was discovered helping to fundraise for the Conservative cabinet minister in charge of broadcasting. Then, in early 2007, CanWest expanded its communication empire by another quantum leap amid protests over increased ownership concentration.

The acquisition also pushed Canada's foreign ownership limits, which the Aspers had long opposed.

A revelation made during the 2004 election campaign should have set off alarm bells that a political sea change was under way with a new generation of Aspers. CanWest's corporate helicopter had been used to ferry a harried Harper above Toronto traffic on his way to an appearance in Hamilton. As *Toronto Star* media columnist Antonia Zerbisias noted, the favour was business as usual for broadcasters. CanWest and CTV contributed generously to the campaign funds of both the Liberal and Conservative parties, she pointed out. "Perhaps it's not so surprising that some very controversial media issues are not being discussed during this campaign."[1]

Power of the press

The political problem of media regulation had been identified more than three decades earlier by Keith Davey, the first senator in Canada to take on Big Media. He became concerned in the late 1960s about increased control of Canada's press by large newspaper chains like Southam, Thomson, and FP Publications. Davey first considered Parliament the appropriate body to conduct an inquiry into press concentration. He soon realized, however, that appointed senators would be better insulated from editorial pressure brought by publishers. His concerns had been confirmed, Davey noted, by the easy passage in 1969 through an elected US Senate of the Newspaper Preservation Act. The NPA exempted from US anti-trust laws newspapers that had been illegally sharing facilities, fixing advertising rates, and pooling profits in arrangements similar to Vancouver's Pacific Press. President Richard Nixon's flip-flop on the issue, according to Davey, justified his concern. "Politicians looking to re-election," he noted, "must depend substantially upon the mass media in the very real world of practical politics."[2] Nixon was re-elected in 1972 with the highest modern level of newspaper endorsements despite his government's unprecedented press censorship in the Pentagon Papers case and a simmering Watergate scandal.[3]

Davey struck a Special Senate Subcommittee on Mass Media that forced media companies to open their books. It described what it found as "astonishing." Media owners were making enormous profits. The secrecy surrounding their financial success, Davey's committee declared, was delicious in its hypocrisy. "An industry that is supposed to abhor secrets is sitting on one of the best-kept, least-discussed secrets, one of the hottest *scoops*, in the entire field of Canadian business — their own balance sheets."[4] From 1958 to 1967, before-tax profits at Canadian newspapers ranged from 23.4 percent to 30.5 percent. After taxes, they were 12.3–17.5 percent, compared to 9.2–10.4 percent in other manufacturing and retailing industries. "Owning a newspaper, in other words, can be almost twice as profitable as owning a paper-box factory or a department store," observed the senators.[5]

The profit measure preferred by Davey's senators was "return on equity," which compared earnings — before or after taxes — to the market value of a company. Another quite different profit measure was preferred by companies, which they used to publicly report their earnings. "Return on revenue" instead counted how much they ended up keeping of every dollar they took in. It usually resulted in a figure several percentage points lower than after-tax return on equity. At Canadian newspapers from 1958–67, for example, return on revenue ranged between 11.8 percent and 15.6 percent.[6]

In broadcasting, according to CRTC data, profits were also robust, particularly in the emerging medium of television. Before-tax return on equity from 1964 to 1968 ranged between 21–26 percent in radio and 36–64 percent in television.[7] These disclosures proved a revelation, and not just in Canada. According Ben Bagdikian, dean of journalism at the University of California, they helped expose the "best kept secret" about the newspaper business — its profitability.[8] Despite such good money to be made, noted Davey's Senate committee, by 1970 there was "genuine" newspaper competition in only five Canadian cities.

> This tendency could ... lead to a situation whereby the news (which we must start thinking of as a public resource, like electricity) is controlled and manipulated by a small group of individuals and corporations whose view of What's Fit to Print may closely coincide with ... What's Good For Business ... There is some evidence, in fact, which suggests we are in that boat already.[9]

Davey's committee proposed several steps to remedy the problem of media concentration. A Press Ownership Review Board would have had the power to block newspaper sales or mergers that increased concentration. Federal subsidies would have encouraged the founding of alternative publications. Despite generating considerable debate for a number of years, however, neither proposal was enacted. It was disappointing for senators who had hoped to stem the tide of ownership concentration before economic forces overtook the media irreversibly. "We had to conclude that we have in this country not the press we need, but rather the press we deserve," recalled Davey in his memoirs. "The sad fact is that the media must self-regulate because most Canadians are not prepared to demand the press they need."[10]

Income trusts and political loyalties

CanWest's struggle to reduce debt led it to get in on the income trust boom of 2005. A report that September estimated such corporate arrangements had cost the federal government $300 million in tax revenues the previous year. Companies feared the loophole would be closed and rushed to place assets in trusts that periodically paid shareholders most of their earnings, not just a portion in dividends. CanWest decided to place its dozen major dailies — not including the *National Post* — into an income trust and sell shares in it. By creating the CanWest MediaWorks Income Fund, the company hoped to raise $700 million to pay down its debt, which stood at almost $3 billion. Before the trust was launched, however, Liberal finance minister Ralph Goodale hinted he might change the tax rules. The uncertainty cooled the market for CanWest's income trust. Instead of raising $700 million, it brought in only $550 million, and the yield it paid had to be increased to make it more attractive. In all, Goodale's musings cost CanWest an estimated $300 million.[11] The government announced the following month it would not tax the trusts, but stock prices had already soared amidst charges of insider trading. A 14-month RCMP investigation cleared Goodale, instead blaming a Finance department official for the leak.[12] The finding was not in time to save the Liberals from defeat in the 2006 election, in which the income trust debacle was a major issue. Ironically, the Conservatives subsequently moved to tax income trusts.

CanWest was able to reduce its debt to $2.59 billion by selling shares in its income trust, but its annual cost of servicing that debt still stood at $157 million by late 2005.[13] When the CanWest Media-Works Income Fund announced its first results in early 2006, revenues at its dailies were up five percent. Profit was up by almost half from the same period a year earlier. That performance was not as good as analysts had expected, however, which led the company to promise more cost cutting. "Payroll is higher than we're comfortable with," CEO Peter Viner admitted to analysts on a conference call.[14] A writer for *Maclean's* identified the problem created for journalism when the former Southam dailies were placed in an income trust. The move rightly brought a "new era of anxiety" for journalists, noted Steve Maich. "The analysts saw only rising costs, and results that didn't live up to their lofty expectations."[15] The public sale of newspaper company shares, studies had shown, made the chains directly subject to the short-term whims of the market. Investors sold stocks they felt were underperforming, encouraging companies to cut costs.[16] Placing them in an income trust made the former Southam dailies even more vulnerable to quarterly financial performance, according to Maich in *Maclean's*.

> They are now assets to be drained of value, like oil wells on the downside of their productive life, and quality is a luxury to be sacrificed on the altar of immediate cash flow. It's not like we weren't warned. Critics said the real danger of the income trust boom was not lost tax revenue, but the distortion of long-term business incentives, and now we have a case in point.[17]

The financial health of CanWest itself, which retained 74 percent ownership of the income trust, took a nosedive in early 2006 as its earnings fell by 14 percent. The downturn was due most of all to the slumping fortunes of its Global Television operations, not to mention the continued losses of its *National Post*. The dismal results led one stock analyst to question the viability of both the *Post* and the free commuter tabloid *Dose*, which had been included in the income trust. "We would not be surprised to see the *Post* and *Dose* publications close down in this fiscal year," wrote BMO Nesbitt Burns analyst Tim Casey in a research note to clients. "We ascribe no value to either asset in our valuation."[19] *Dose* had been distributed in Vancouver, Edmonton, Calgary, Toronto and Ottawa since April of 2005, but it had failed to turn a profit. In an effort to further pay down its debt, CanWest announced it would sell its investment in the Irish television network TV3.[20] It denied, however, that the *National Post* would be sold, folded, or converted to a private family venture, as had been rumoured. "It has had the best first quarter it's had in some years," said Viner. "We continue to give it our unqualified support."[21]

Drastic moves were required, however, to keep the *National Post* — and CanWest itself — afloat. In May of 2006, its *Dose* publications were killed off after little more than a year because they were losing a reported $9-10 million annually.[22] Two days later, CanWest announced that its TV3 holdings in Ireland had been sold to a private equity firm for $198 million.[23] The proceeds did not go toward paying down debt, however, because they were needed just to keep the company in the black. As advertising sales slumped, the company's earnings fell to $13 million in its second quarter, down from $50 million during the same period in 2005.[24] If not for the TV3 sale, CanWest would have posted a $9.1-million loss in its third quarter.[25] The poor results prompted Moody's Investor Service to cut CanWest's liquidity rating and place its credit under review for a possible downgrade.[26] The Dominion Bond Rating Service also placed CanWest's credit rating under review due to the company's weak financial results. It warned that they "may lead to a breach in the company's financial covenants."[27]

Despite the company's tight financial situation, CanWest somehow continued to make strategic investments overseas. In February 2006, it bought another radio station in New Zealand and two more in Turkey. In April, CanWest announced it would invest another US$61 million in Turkish radio, causing some analysts to question the company's international expansion.[28] Radio was a consolation prize as its entry into the mostly-Muslim country, Leonard Asper told *Television Business International*. "The reason we are in radio in Turkey and not TV is because we were outbid for the TV station," he said.[29] One overseas acquisition that slipped from the Aspers' grasp, however, was the *Jerusalem Post*. It had been sold by an imploding Hollinger to Israeli businessman Eli Azur for US$13.2 million in 2004. CanWest announced it had acquired an option on half of the newspaper their father had long coveted.[30] Azur, however, balked at selling to the Aspers for political reasons, according to a major Israeli conservative daily. "Azur says that one reason for the dispute is [Leonard] Asper's desire to give the newspaper an extreme right-wing orientation," reported *Ha'aretz*. Asper denied that he wanted to replace the *Post*'s editor and publisher, or that he intended to use it for political purposes. "We are only planning to bring the precise and objective facts to public attention," he said in an interview, "against the false propaganda of those who are anti-Israel."[31] The dispute was referred to an arbitrator in New York. He ruled in 2006 that the only binding agreement entered into by the parties had been to negotiate in good faith on a joint venture.[32]

The income trust controversy seemed to mark a turning point in CanWest's relations with the federal Liberal party. The company's annual meeting, which was held during the 2006 federal election campaign, vented displeasure against the party in power. "It would be an understatement to say that we are not pleased with the government of Canada," said interim CanWest chairman David Drybrough.[33] As if to underscore the company's annoyance, David Asper issued a statement criticizing Liberal prime minister Paul Martin for a series of campaign ads he called "insulting."[34] Asper endorsed the candidacy of Global executive Peter Kent, who had taken a leave of absence to run for the Conservatives in Toronto. Kent made news during the campaign with allegations of media bias in favour of the Liberals, for which he urged journalism schools to monitor election coverage.[35] When he failed to win a seat, Kent returned to CanWest Global as its deputy editor-in-chief in charge of current affairs coverage and documentary development.

As if to announce a break with tradition, the *National Post* declared: "The Asper family is known for its past support of the Liberal party."[36] Izzy Asper's sons, however, were never as enthusiastic in their support of the Liberals as their father had been. David Asper had even pointed out the "political diversity" within the family in 2001. "I happen to carry provincial and federal political party memberships which are not of the Liberal party," he told an audience in Calgary. "Not one of my critics cared, for example, to take note of my publicized endorsement of former Manitoba Conservative premier Gary Filmon." Asper noted that CanWest had spread its political contributions well beyond just the Liberals, following ideological rather than party lines. "We have provided significant financial support for both the Alliance federally and the Progressive Conservatives provincially," he said. "We are driven by policy ideas and innovation and not by some blind political loyalty."[37]

As the 2006 federal election approached, David Asper dispelled any doubt that might have remained about his political leanings. At a campaign rally, he joined Harper onstage to endorse his candidacy. The open declaration of personal support for a politician was something new for media owners in Canada. "You have to wonder about the wisdom of Mr. Asper's endorsement of Mr. Harper," noted Christopher Dornan, director of Carleton University's journalism school. "Not from Mr. Asper's point of view, but from Mr. Harper's. Why invite accusations of having the press in your pocket?"[38] After the Conservatives came to power, the links between CanWest and the new ruling party became apparent. That August, CanWest named Derek Burney as its new chairman, finally filling the vacancy created by Izzy Asper's 2003 death. Burney had been Conservative prime minister Brian Mulroney's chief of staff in the 1980s and later his ambassador to the US. It was his most recent position, however, that some felt made Burney a bit too close to the new government.

As head of Harper's transition team to power, they pointed out, Burney provided the perfect conduit for CanWest to the new ruling party in Ottawa. "There is no question that his extensive experience and connections within government are exceptionally valuable," noted *Maclean's*.[39] The political coziness was questionable, however, due to CanWest's reliance on federal regulators for the profitability of its billion-dollar business. "It ties the media companies into an existing social network of decision-makers that affect policy and government regulation in Canada," noted Carleton communication professor Dwayne Winseck. For some critics of CanWest, Burney's appointment was going too far. "I don't think that people prominently associated with one political party are appropriate people to chair the board of a media company," said Russell Mills. "It doesn't create the right climate for journalists to work in."[40]

The Senate report on news media

The final Senate report on news media stopped short of urging the breakup of Big Media by recommending that the ban on cross ownership be re-instituted. Instead it suggested that the Competition Bureau review future news media mergers to prevent dominance by one owner in any market. It recommended automatic review above a certain concentration threshold, mentioning an audience share of 35 percent. It also proposed allowing Cabinet to review any news media merger that government ministers considered questionable. Press freedom provisions in the Charter of Rights and Freedoms, the senators reasoned, should only go so far. "The media's right to be free from government interference does not extend ... to a conclusion that proprietors should be allowed to own an excessive proportion of media holdings in a particular market, let alone the national market."[41]

The inquiry had declined to do what the Davey committee had done when it made media companies open their books and disclose their profits. Instead, the 21st century senators relied on published numbers, from annual reports of newspaper companies and from CRTC data for broadcasters. The figures were thus confined to return on revenue and not the return on equity yardstick favoured by Davey's committee. They showed, however, that media companies were still extremely lucrative. Owning newspapers was even more profitable than in Davey's day. Radio profits were about the same, and television's were lower. In 2005, return on revenue for the major newspaper chains varied from 17–24 percent, with CanWest coming in at 20.7 percent.[42] In radio, return on revenue ranged from 18.5–22.7 percent between 1999 and 2003, and in television from 13.6–18.6 percent.[43]

Most news coverage of the Senate report did not focus on media profitability or ownership concentration, however. That angle was "well buried and buttressed by sneering editorial comment," noted *Ottawa Citizen* columnist Charles Gordon. "It was just another useless report, the critics said, like the Davey committee 36 years ago and the Kent commission a decade after that, and nothing good came from those."[44] The suggestion of federal oversight of news media mergers in particular did not sit well with publishers. According to an editorial in the *Toronto Star*, it would allow for political interference in the press.

> Giving elected politicians the power to make decisions that influence the survival and health of newspapers and other media would undermine public confidence in MPs and news gatherers alike. Would politicians approve mergers to curry favour with media owners? Would the media tailor coverage to win political approval for mergers?[45]

Most news coverage focused on the report's prescriptions for the CBC. Of particular interest to most media outlets, especially those owned by CanWest, was its recommendation that the government broadcaster "get back to basics." The report urged the CBC not to duplicate content provided by the private sector. It recommended the CBC drop coverage of professional sports and the Olympics and concentrate instead on providing quality, commercial-free programming. The *National Post* trumpeted the CBC angle on its front page and carried a package of articles inside emblazoned "Broadcaster Under Siege." Columnist Don Martin lamented that "full privatization will have to wait for Harper to land a majority government."[46]

The 600-word front page story in the *Post*, which also ran in other CanWest dailies, devoted exactly 23 words to the concentration issue. The topic was mentioned "almost as an aside," noted *Toronto Star* columnist Antonia Zerbisias. "As if the genesis of the report had nothing to do with CanWest's much-criticized moves to centralize editorial functions in 2003."[47] The CBC angle followed conveniently on the heels of a controversy that had erupted the previous day. The public broadcaster announced that its National newscast would be delayed by an hour on Tuesdays that summer to accommodate a US network reality show. Also announced that day, noted Zerbisias, was Quebecor's plan to cut 120 positions at its *Sun* newspaper chain, including political journalists.

> And yet, in all those thundering editorials and news stories about the Senate report, most of which dismissed its findings while chastising CBC, not a single one referred to the bloodbath that had just happened at the *Sun* papers. Which tells you all you need to know about Canada's giant media corporations.[48]

Big media off the hook

One week after the Senate report was released, CanWest announced it would pull out of the Canadian Press news co-operative to save its $4.6 million in annual dues. That money, it said, would go instead to bolster its own news service, which increasingly migrated back to Ottawa as another Asper innovation proved misguided. "This decision is about providing a unique and diverse array of content to our readers," said Scott Anderson. The *Ottawa Citizen* publisher doubled as vice-president of editorial for CanWest's newspapers. "We feel that by reinvesting the money we currently pay Canadian Press we can do a better job serving our readers."[49] Withdrawing from membership in CP required a year's notice, so CanWest's departure would not become effective until June of 2007. Ironically, a threat by Hollinger to pull the Southam dailies out of CP in 1999 first prompted Sheila Copps, then Liberal heritage minister, to consider a media inquiry.[50]

The CanWest-Conservative Party connection grew stronger when a company executive was discovered fundraising for the re-election of Heritage Minister Bev Oda. CanWest's vice-president of regulatory affairs, Charlotte Bell, had helped organize a $250-a-plate dinner for Oda, whose portfolio included responsibility for media. "Charlotte Bell's job is to meet ministers, is to sell the case of industry," NDP heritage critic Charlie Angus told the House of Commons, "and for her to be out there using her name selling tickets for the minister for a fundraising event, it stinks."[51] University of Ottawa law professor Michael Geist pointed out Oda's financial connections. Her re-election campaign, he noted, had attracted "enormous corporate support" from the broadcast industry. "Questions about Oda's fundraising activities," he observed, "could leave Canadians asking whether there is a hefty price tag associated with key government policies."[52] Oda, a former Rogers and Global executive and CTV vice-president, cancelled the dinner after its propriety was called into question.

As part of the CRTC's review of broadcast television, CanWest had sought the removal of limits on the amount of advertising allowed during a broadcasting hour. It also asked for an end to the 10 percent public benefits requirement of broadcasting takeovers, which Leonard Asper labelled "a tax."[53] In a 2006 speech, Asper also declared additional consolidation of Canadian media "inevitable." He called on the federal government to step aside and allow it. "There will never be concentration like there was before because you have Google, you have MSN and you have Yahoo," Asper told the Empire Club in Toronto. "You have all these companies that are competing against The *Vancouver Sun*s and the *Ottawa Citizen*s. So let us consolidate."[54] The timing involved in a CanWest executive fundraising for the government's minister in charge of regulating its business irked Angus. "The broadcast review happens in two weeks," he told the House. "The cash grab happens next week. Why is the minister using her office to trade political access for political contributions?"[55]

Mutual back-scratching

Another area of Asper family interest that Oda oversaw was the government's relationship with the long-planned Canadian Museum for Human Rights in Winnipeg. By 2006, the federal funding commitment to completing the project had grown to $100 million. The

Winnipeg and Manitoba governments had pledged $20 million each to supplement $61 million raised from the private sector, including the Asper family. Gail Asper, who spearheaded the fundraising drive, asked Ottawa to also provide $12 million a year in operating expenses. The Liberals had balked at that while they were in power. After the Conservatives were elected, however, the Aspers found that their entreaties to the federal government received a more favourable hearing. By designating the museum a national institution, Oda was prepared to provide the $12 million annually.[56] The sticking point was who would control the museum's operations, noted the *Globe and Mail*.

> The tricky part of the public-private relationship is the question of who dominates the museum board (and therefore who determines such ticklish issues as how the "national" human-rights museum might treat, say, Palestinian rights). Clearly, the board would include Asper family representatives — but would the government let them control it?[57]

The mutual back-scratching also saw CanWest come to the prime minister's aid in a long-running dispute with the parliamentary press gallery. His election platform had included promises of more government openness, but instead Harper tightly restricted press access to himself and other ministers. The strategy reminded some of the perception management tactics employed in the US by the White House. In Ottawa, the prime minister's office announced Harper would only take questions at Parliament Hill press conferences from reporters who put their names on a list. He would call on selected reporters to ask questions instead of answering them as before from those lined up at microphones. Reporters boycotted the new rules because they claimed they would allow Harper to "cherry pick" favourite journalists and freeze out those who might ask tough questions.[58] The dispute went unresolved for months until the CanWest News Service broke ranks and obtained exclusive interviews with the prime minister for two of its reporters after agreeing to go on his list.[59]

Consolidation continues

The media in Canada began another period of consolidation late in 2005, when Bell Canada Enterprises sold most of its controlling interest in Bell Globemedia. The dealmaking accelerated into 2007. The significance of BCE's divestiture lay in who picked up its shares in Bell Globemedia. Thomson had initially taken only 29.9 percent ownership to avoid triggering CRTC scrutiny, which became automatic at 30 percent. In late 2005, it increased its ownership to 40 percent. The Ontario Teachers Pension Fund came in for 20 percent, as did Torstar. That surprise move suddenly linked longtime newspaper rivals the *Star* and the *Globe and Mail* in a larger corporate web. The *National Post* gleefully published a mock "*Globe and Star*" front page. "My first reaction was to ask to change beats," wrote the *Star*'s Zerbisias. "My second was to contemplate suicide." Her role as the country's most strident media critic was suddenly on shaky, or at least smelly ground, admitted Zerbisias. "Media concentration has landed plop plop plop like a steaming pile of bad news on my very own front stoop," she quipped. "Which makes me worry that things might get very slippery for a media critic, if you get my drift."[60]

Co-operation between competitors in the formerly-cutthroat Toronto newspaper market was becoming more the norm than the exception. Some even saw things moving in the direction of Vancouver's complete co-operation. "Over time you'll see increased moves to joint distribution," Torstar CEO Robert Prichard told an investors conference that September. "Eventually I think we'll end up with a single distribution, but that's probably a decade out." Torstar had been printing editions of the *National Post* at its Vaughan production plant north of Toronto for some time, he pointed out. "The *Post* is going to be printed by someone," noted Prichard. "We might as well do that printing because of a much better utilization rate of our plant."[61]

Much like Southam's 1985 share swap, its connection to CTV also made Torstar practically unattainable as a takeover target. The Aspers had reportedly coveted Torstar for its dominance in Toronto, where CanWest lacked a major presence. David Asper spoke of a CanWest–Torstar alliance as "a virtual certainty," according to the *Globe and Mail*. "Putting Torstar and CanWest together would have created a national multimedia giant with a big presence in Toronto, the largest advertising market. CanWest now has to figure out its Toronto strategy. There is no easy answer."[62] Its $283-million investment also established Torstar, once one of the few major newspaper companies in Canada without a broadcasting arm, as a player in the convergence game. The company had sought an electronic presence since losing a 2002 bid for a new Toronto television licence to Craig Broadcasting.[63] Suddenly it was part of a well-funded multimedia consortium able to act on any opportunity, unlike debt-hamstrung CanWest. The next opportunity presented itself sooner than anyone expected.

In July of 2006, three weeks after the Senate report on news media was issued, another major merger again transformed the Canadian media. Bell Globemedia scooped up Toronto-based CHUM Ltd. for $1.4 billion. The sale had been triggered by the death of CHUM founder Allan Waters in late 2005. It was also encouraged by lingering indigestion caused by the company's $265 million takeover of Craig Broadcasting in 2004. Bell Globemedia got CHUM's 33 radio stations, twelve CITY-TV and A Channel television outlets, and 21 specialty television channels, including MuchMusic and Bravo. The purchase did not include the bulk of CHUM's news staff, however, which had been eviscerated the day before with 281 layoffs.[64] Bell Globemedia promised to sell the five A Channel stations to gain CRTC and Competition Bureau approval for the deal. Critics called for Ottawa to reject it entirely in the wake of the Senate report. "Somebody has to keep an eye out for the public interest in the continued diversity of news sources," said Senator Joan Fraser. Liberal MP Dan McTeague, who sat on the House of Commons industry committee, was even more emphatic in his reaction to the deal. "We've gone too far down the road of media concentration," he said.[65]

The rubber stamp

The regulatory climate in Ottawa, however, had turned tepid with the election of Harper's Conservatives. In June of 2006, Industry Minister Maxime Bernier urged the CRTC to rely on market forces to the "maximum extent feasible."[66] The change of control at CTV, television unions claimed, should have resulted in a public benefits package of at least $68 million. The company, however, claimed BCE's selloff resulted in the company not having a controlling shareholder, so no change of control had taken place.[67] In July, the CRTC agreed with the company and absolved it of providing public benefits. Its ruling was made over the objection of one commissioner. Stuart Langford claimed in a dissenting opinion that his CRTC colleagues "allowed themselves to be distracted from the simple facts." That was achieved, Langford claimed, "by a virtual avalanche of legal documents and legal opinions." According to Langford, a lawyer and former journalist, the torrent of legalese "either by design or by chance", confused the transaction. It was, he pointed out, "a straightforward transfer of control from one shareholder, BCE, to a group of four shareholders."[68] Later that year, the media conglomerate was renamed CTVglobemedia to signify the diversification of its ownership.

Before 2006 ended, Oda issued a response to the Senate report that confirmed there would be no government action even on its mild recommendations. "The government recognizes that convergence has become an essential business strategy for media organizations to stay competitive in a highly competitive and diverse marketplace," it read. The response quickly rendered moot both the Senate report and the Lincoln committee recommendations. Senator Jim Munson expressed frustration with the government's response. "I am very disappointed that they would have this attitude," he said. "We feel [the report] gives some creative ideas on how we should monitor massive media concentration." A union official was more pointed as to why the Senate report got such a cold shoulder. "Big media is in the driver's seat of big politics," said Peter Murdoch of the Communications, Energy and Paperworkers Union of Canada. "It's clear who the government is listening to. It's not just outrageous or appalling. It's scary." The only newspaper in Canada to even report Oda's response at the time was the *Toronto Star*. Zerbisias noted the Heritage Minister's relationship with the broadcasting industry.

> Last month, at a broadcasters' convention in Ottawa, Oda told her audience "I'm with you. I'm one of you." She also said she is "committed to more regulatory flexibility." Well, let me tell you, after covering this business for the better part of 17 years, I have learned that, when broadcasters talk about "flexibility," it's always Canadian artists, citizens, consumers who bend over. [69]

A Global alliance

The new Conservative government's *laissez-faire* approach to the continuing consolidation of Canadian media opened the door for an expansion of CanWest. It was an opportunity the Aspers seized on in early 2007 despite being deeply in debt. The strategy they employed, however, flouted Canada's restrictions on foreign media ownership. It also gambled control of the firm their father had built into a media powerhouse. "They're betting the personal farm," offered one unnamed analyst when details of the Aspers' bold venture emerged. "It's a risky move."[70] It was an ingeniously-financed expansion that defied Canada's foreign ownership limits, which the Aspers had long opposed. Ironically, their takeover target was a production company founded by Izzy Asper's arch-rival Robert Lantos. After excoriating CanWest Global as "toll collectors" on American programming, the Canadian company Lantos helped start was taken over by the Aspers and Americans. Alliance Atlantis was the largest production company in Canada, and one of the most successful in the world. It was a classic example of "vertical integration," not only producing content but also owning multiple avenues of distribution. It held CRTC licences for thirteen specialty television channels, including Showcase and History Television, and it owned the movie distribution company Odeon Films. Alliance Atlantis also enjoyed a certified hit in the television show *CSI: Crime Scene Investigation* and its spin-off series, a billion-dollar franchise in partnership with CBS.

The value of Alliance Atlantis to CanWest Global was obvious, but its debt load left the firm unable to make a play for it alone. Alliance Atlantis had been put on the market in late 2006 by its controlling shareholders, and several contenders lined up to bid. Included were Rogers, Quebecor, Montreal-based Astral Media, and the Shaw family's Corus Entertainment. CanWest was considered a long shot suitor due to its debt burden. It was able to work out an ingenious partnership, however, with New York-based investment bank Goldman Sachs. Their winning bid for Alliance Atlantis was $2.3 billion, the same price BCE had paid for CTV. In an innovative arrangement, CanWest contributed only $262 million in return for a 36-percent stake in a new subsidiary that would own the specialty channels. The extent of the company's participation was twice increased from an initial 17 percent. Goldman Sachs took the CSI franchise, the Odeon division, and the rest of the new specialty channel company. The twist came in a deal to merge that company with Global's television holdings, including its specialty channels and the CH network, in 2011.

A new company would be formed, in which CanWest and Goldman Sachs would be partners. Their division of ownership would depend on the relative earnings of each company four years hence. Global was forecast to earn $57 million in 2007, compared to $151 million for the Alliance Atlantis specialty channels. The possibility thus loomed of the Aspers being minority owners in 2011. At the predicted 2007 levels, CanWest Global would account for 53.5 percent ($57 million + 36 percent of $151 million = $111.4 million) of the combined revenues of $208 million. "The risk is really just performance," said Leonard Asper. "We've put it all on our own shoulders to perform." Another option in the complex deal allowed for a public sale of shares to instead finance the new entity in 2011. CanWest also held a first right of refusal from Goldman Sachs in the event it decided to sell.[71]

The Aspers had yet another ace up their sleeves in Australia's Network TEN. It had become even more valuable with the change in cross-media ownership rules set to come into effect there in 2007. Investment bankers had been shopping Network TEN around, with the most interested buyer said to be Rupert Murdoch. Analysts valued CanWest's Network TEN share at $1.5 billion, which nicely counter-balanced its new partnership with Goldman Sachs. Leonard Asper announced the company would keep its options open. "Australia has got a huge balance sheet capacity," he told the *National Post*. "We think Australia has a very good run ahead of it because the market is not as competitive as Canada, the advertising market is turning and we've got great ratings. So we could watch it grow."[72] When no buyer met their price, that's exactly what they decided to do.

The injection of so much American capital into the Canadian media concerned critics, who pointed to the 33 percent limit on foreign ownership. Leonard Asper insisted, however, that Goldman Sachs would be a passive investor in the deal. "In Canada, it's a control test," he pointed out, "and we are in control in every way, shape and form."[73] The argument echoed his father's strategy in circumventing Australia's foreign ownership limits. Peter Murdoch of the

Communications, Energy and Paperworkers Union (CEP), which represents many Canadian journalists, called on the CRTC to hold hearings into the arrangement and to reject the deal. He described it as "the thin edge of the wedge to the ceding of Canadian cultural institutions to American investment." He rejected Asper's assurances that Goldman Sachs would have no control over Canadian television. "Quite clearly the person, or in this case the company, putting out the cash, in one way or another, will have effective control over the company."[74] Others pointed to the likely effect of increased ownership concentration. CanWest would need to boost Global's earnings as much as possible in hopes of ending up with majority ownership in 2011. "So you can bet that Global, along with CTV and its specialty channels, will jack up ad rates, with CHUM and Alliance Atlantis gone," predicted Zerbisias. "Those costs will eventually reach consumers."[75]

The financial peril CanWest faced with the clock ticking toward 2011 was significant. The 33-percent profit margin that Global's television operations had enjoyed in their heyday of 2001 had dwindled to a mere 5 percent in 2006. With three hit shows in the fall of 2006, however, Global's financial fortunes blipped upward. Revenues increased by 11 percent in the first quarter of the company's 2006-07 fiscal year and earnings rose 30 percent. The improvement was cause for optimism that CanWest would ultimately emerge as majority owner of its new Global Alliance, according to Leonard Asper. Even if it didn't, he claimed CanWest would still be ahead of the game. "Whether it's 45 percent or 55 percent, we're still going to have a stronger net asset value, even on a present value basis, than we have today."[76]

Their bold move to expand against all odds left the Asper heirs more dependent than ever on federal regulators. Their innovative acquisition of Alliance Atlantis would need the blessing of the CRTC. Their Global Television operations would need every advantage they could get from Ottawa to keep them mostly Canadian. The bridges they had been building to the new Conservative government would thus be more important than ever to CanWest. That in turn suggested mutual admiration would continue to be expressed between the federal government and Canada's largest news media company. Whether the result would be the news coverage Canadians needed seemed less likely than Davey's prediction it would be the press they deserved for failing to demand better.

CONCLUSION

The Pictures in Our Heads

Walter Lippmann was considered by many Americans to have been the greatest journalist of their 20th Century. He began his career in 1910 as a "muckraker," that early breed of crusading investigative reporter. He was a founding editor in 1914 of the progressive *New Republic* magazine. By one account, it was a publication that "may well have introduced the term liberal in its modern sense into the American political lexicon."[1] When the US entered World War I in 1917, Lippmann was recruited into the propaganda machine that quickly persuaded Americans to support the cause. Many prominent journalists helped demonstrate the power of the press to whip up war sentiment by portraying the Germans as bloodthirsty "Huns." When the war ended, Lippmann drafted president Woodrow Wilson's famous "fourteen points" that aimed, but failed, to forestall future European conflicts. Lippmann served as editor of Joseph Pulitzer's *New York World* until it folded in 1931. He then began writing a column for the *New York Herald-Tribune* that was syndicated in more than 150 newspapers. The *Herald-Tribune* closed in 1967, after which Lippmann moved his column to the *Washington Post*. He died in 1974.

More than just a journalist, Lippmann was a scholar, media critic, and intellectual who authored more than two dozen books, his first at age 23. Some were the most influential of the day. In 1920, *Liberty and the News* deplored press corruptibility during wartime. His content analysis of *New York Times* coverage of the Bolshevik revolution in Russia, *A Test of the News*, showed it was neither accurate nor impartial. His 1947 book *The Cold War* gave a name to the post-World War II battle of ideologies between capitalism and communism. Lippmann's most enduring contribution to a theory of mass media, however, was his 1922 book *Public Opinion*. It was generally credited with helping found both the study of political communication and the science of public relations. In it, he appropriated the printing term "stereotype" as a metaphor to describe the oversimplified images of people and places that were conveyed by the press to the public.

Lippmann despaired for democracy in *Public Opinion*. His experience with wartime propaganda had shown him how easily public perceptions could be altered by distorted information. "It very largely succeeded, I believe, in creating what might be called one public opinion all over America."[2] His study of Freud's theories of social psychology convinced him groups respond not to logical ideas but to powerful symbolism and images, however inaccurate. The problem, according to Lippmann, was that "the world outside" was too vast and distant for people to experience directly. For their understanding of it, they instead had to rely on the media. The information people received, however, was insufficient to provide an accurate picture of reality, according to Lippmann. It was often slanted by journalists to serve various purposes. What people did receive was then shaped by their own prejudices and preconceptions to form stereotypes, or what Lippmann called "the pictures in our heads." The resulting distortions, he claimed, were incapable of informing citizens sufficiently to allow for an informed public opinion, or thus any real democracy.

The problem of propaganda

The secret to successful propaganda, according to Lippmann, lay in controlling information. That revelation proved a building block of the emerging science of public relations. "Without some form of censorship, propaganda in the strict sense of the word is impossible," Lippmann wrote in *Public Opinion*. "In order to conduct a propaganda [campaign], there must be some barrier between public and event."[3] By controlling information, he said, propagandists could literally "manufacture consent" for policies the public did not otherwise want and that were not even in its best interests. Lippmann concluded that the influence of propaganda on public opinion was a development that had not been foreseen by the founders of democracy.

> The creation of consent is not a new art. It is a very old one which was supposed to have died out with the appearance of democracy. But it has not died out. It has, in fact, improved enormously

in technic, because it is now based on analysis rather than on rule of thumb. And so, as a result of psychological research, coupled with the modern means of communication, the practice of democracy has turned a corner. A revolution is taking place, infinitely more significant than any shifting of economic power.[4]

Despite his insight into the problem, Lippmann's prescription for remedying it seemed by contrast hopelessly naïve. His faith in "science" as a means of knowing the social world led him to urge creation of a "central intelligence" agency. It would have supposedly been independent of the decision-making machinery of government. It would have gathered verifiable facts and disseminated them through the media to the public, which he likened to a "bewildered herd." Political scientists, he said, should organize the intelligence, which would be shared with other scholars and even taught at university. This system of experts was needed to illuminate what Lippmann called the "invisible environment," which would otherwise be hidden from public view. There, he argued, could be found the difference between news and truth. "The function of news is to signalize an event," argued Lippmann. "The function of truth is to bring to light the hidden facts, to set them in relation with each other, and make a picture of reality on which men [sic] can act."[5]

Lippmann's analysis was offered in the infancy of social science, when conceptions of one knowable "truth" were more absolute than they would become. It turned out that notions of truth varied greatly between individuals depending on their perspective and preconceptions. Lippmann's panacea would have amounted to "technocratic control" of media and politics by the kind of information elite Harold Innis saw as inevitable in any society. It was, however, a solution urged by other experts on propaganda, such as Harold Lasswell. He found propaganda to be a much more complex, sophisticated and long-term process than had been first assumed after World War I. As a result, Lasswell also concluded that only a cadre of experts could deal effectively with the pervasive problem of misinformation and disinformation. Under the technocratic control envisioned by Lasswell and Lippman, however, rule would inevitably devolve to the masters of the media. Some political theorists fear this may actually be the situation we have today, where a preferred view of the world is skilfully fed to the public through a well-controlled mass media and political control is thus achieved.

What goes around

The *New Republic* founded by Lippmann and others in 1914 was bought 60 years later by Harvard lecturer Martin Peretz. The influential weekly, dubbed the "in-flight magazine of Air Force One," continued its progressive stance under Peretz, except in one area — Middle East foreign policy.[6] By 1991, Jack Shafer counted 90 columns and articles Peretz had authored for the magazine. "Of them, 40 or so jab, revile, or otherwise shellac the 200 million people who call themselves Arabs," noted Shafer. "Peretz's view . . . is easily summarized. The Arabs are an undifferentiated mass, consumed by antique tribal hatreds, fated to fratricide, torn asunder by their religious sectarianism."

> When not whomping Arabs, Peretz whomps his enemies in the press — make that Israel's enemies in the press. Not that, to him, there's much difference. "Forgive me, but in the prestige media — except for the *New York Times* — it is 'blame Israel first,'" he wrote in 1987. For many years his bête noire was the *Washington Post*. . . . Later, the *Boston Globe* assumed the role of evil incarnate. These days, it's Peter Jennings, whose primary sin is to speak of Palestine as if it exists.[7]

Later that year, the *Progressive* magazine quipped that Peretz had "come up with a new marketing ploy to attract subscribers: race hatred." A mass mailing claimed Iraq's persecution of the Kurds should be a rallying cry for Israel. "If this is what the Arabs do to fellow Muslims," wrote Peretz, "what would they do to the Jews if they had half a chance?" The *New Republic*'s editorials, noted the *Progressive*, had recently focused on Israel's handling of the latest Palestinian uprising. "The dead, wounded, and displaced of Palestine would be interested to know that Israel has staged 'a relatively benign occupation' (January 29, 1990) and that Israeli soldiers and police 'have kept Palestinian casualties — and their own — very low' (January 1, 1990)." The Palestinian death toll, noted the *Progressive*, had been placed at 956 in less than four years, while 1,952 homes had been demolished or sealed by the Israelis. "A few years back, The *New Republic* mailed a subscription plea headlined 'Liberals with Guts,'" the *Progressive* pointed out. "Indeed, it takes guts to market oneself as a liberal while practicing compulsive bigotry. . . . The magazine has never looked less 'liberal.'"[8] A few years later, the *New Criterion* observed that on many issues the *New Republic* was "all but indistinguishable from certain aspects of neoconservative thought."[9]

In 2003, the *New Republic* came out in favour of the Iraq invasion, to the horror of many of its longtime readers. It later apologized after it turned out Saddam Hussein had no weapons of mass destruction, but to many a Rubicon had been crossed. By 2005, the magazine's circulation had dropped to 60,000 from more than 100,000 in 2000. In early 2006, CanWest Global Communications bought 30 percent of the *New Republic* for US$2.3 million.[10] According to the *Globe and Mail*, Leonard Asper met Peretz, a fellow Brandeis alum, through connections on Wall Street — "his new friends at Goldman Sachs."[11] Little more than a year later, CanWest bought the rest of the magazine for a reported US$5 million. Peretz remained as editor-in-chief, but CanWest moved in publisher Greg MacNeil, who had headed the extinguished Canadian icon *Saturday Night*. Plans were announced to expand the *New Republic*'s Internet presence, while print publication was cut back from weekly to twice-monthly. "It just seemed to me, given my own intellectual and moral synergies with Leonard J. Asper, a very good partnership," Peretz told the *New York Times*.[12]

'I had been bought'

In journalism lore, the press is charged with being the watchdog on powerful institutions such as government and big business. In a mass-mediated information society, however, the press and especially the electronic media have become some of the most powerful institutions of all. Who will be the watchdog on the watchdog? That should be a job for media scholars, but inevitably they come with their own biases and even conflicts of interest. In Canada, academics became increasingly reliant on corporate funding for both research and teaching, which raised serious questions about their independence. Higher education in Canada began suffering funding cutbacks in the early 1980s. Journalism schools, which ranked low in the estimation of other academics, often found themselves last in line for money. More and more they turned to corporate funding from the CRTC's public benefits program, which created potential conflicts.[13] Peter Desbarats perhaps put it best after, as dean of journalism at the University Western Ontario, he accepted a

$1-million donation from Rogers to fund an endowed chair. "When journalists subsequently asked me to comment on the Rogers take-over of Maclean Hunter, all I could do was draw their attention to the donation," he recalled. "They understood right away that I had been, to express it crudely, bought."[14]

Desbarats noted that the independence granted to faculty members by universities through tenure could be "rapidly eroded" through dependence on industry fundraising. "Unavoidably I gave up something in return," he concluded. "No one should pretend, least of all university presidents, that this experience, multiplied many times and repeated over the years, doesn't damage universities in the long run."[15] Funding schools of journalism and communication not only satisfied the CRTC demand for public benefits, it also gave corporate donors apparent influence in exchange. The problem arose from coast to coast in Canada. The Irving Oil family, which controlled all four daily newspapers in New Brunswick, paid $2 million to fund an Irving Chair in Journalism at St. Thomas University in Fredericton and a Romeo LeBlanc Chair in Journalism at L'Université de Moncton. "While the universities receiving the new funds are overjoyed," noted Mount Allison University sociologist Erin Steuter, "concerns have been raised about the possible implications of new generations of journalism students learning from the very outset of their training who is paying the piper."[16]

Desbarats well knew the tenuous place of journalism in the university curriculum, as the Western Ontario program he headed almost fell victim to cost cutting in 1993. It was saved by just one vote after a last-minute appeal to the university's board of governors. Even then it only survived after being merged with a much larger program in library sciences into a new Faculty of Information and Media Studies.[17] Desbarats also saw a shortage of needed research on the news media in Canada due to a lack of university-level journalism schools. In 2005, such programs numbered 458 in the US, while in Canda there were fewer than ten.[18] "The academic tradition in the United States ... produces a relatively abundant flow of writing about news media," Desbarats noted in 1989. "By contrast, public debate about journalism in Canada suffers from a constant shortage of historical perspective and reliable data."[19]

The use and abuse of academic research in pursuit of political and corporate ends has been a recurring theme in social science. Some of the worst examples have been seen in the field of mass communication. Questionable research was often used as ammunition in the battle for and against media regulation. Media research was most abundant in the US due to a post-World War II proliferation of large Midwestern land-grant universities. The academic expansion occurred at the same time the study of communication was emerging as a separate discipline.[20] Going back to the 1930s, one of the earliest cases of research manipulation in the battle over media regulation actually saw restrictions placed on content. The Payne Fund studies used to justify film standards had been initiated, historians later confirmed, by religious groups determined to sanitize content on moral grounds.[21] Perhaps the most egregious example of media research manipulation involved the US Surgeon-General's report on television violence in the early 1970s. Studies done for the report found a link between television viewing by children and anti-social behavior, but the finding was minimized in the report's final version. Some scholars suspected it was suppressed due to the intervention of Joseph Klapper, a leading proponent of "limited effects" theory who was employed by CBS.[22]

Research manipulation was also noted in the 21st century debate on cross ownership in the US. When senator Barbara Boxer of California charged in 2006 that a study had been suppressed by the FCC because its result did not favour deregulation, an inquiry was called. Boxer produced an unreleased 2004 "working paper" that showed locally-owned television stations aired significantly more local news than network-owned stations. A former FCC lawyer claimed commission officials who did not favour its findings ordered copies of the study destroyed.[23] A week later, Boxer produced another report that had been ordered by the FCC but not released. It showed that while there was a 5.9 percent increase in radio stations in the US between 1996 and 2003, there was a 35 percent decrease in the number of owners. Clear Channel Communications benefited most of all from the 1996 Telecommunications Act that loosened radio ownership limits, the report showed. It expanded from 62 stations in 1996 to 1,233 in 2003.[24] As a result of the embarrassing disclosures, the FCC posted all of its ownership studies on its website.[25]

The revelations about suppressed studies followed a spate of disputed FCC research on cross ownership. When an internal FCC study found advertising rates increased only 3–4 percent on average as a result of media consolidation, it didn't pass the smell test with Jon Mandel. The head of the Grey Global Group advertising agency commissioned his own study. It showed that radio advertising rates more than doubled in some markets as a result of increased ownership concentration. With consolidation and lower production costs for cheap "reality" shows, television stations were also reaping extraordinary profits, according to Mandel. "The only business that is doing better," he quipped to the FCC's 2003 cross ownership hearings, "is selling crack."[26]

Another set of FCC studies on cross ownership was done by academics, and some were criticized by the advocacy group Center for Digital Democracy. It claimed they served mostly to "ratify" the deregulationist views of FCC chairman Michael Powell. One study singled out for criticism was done by David Pritchard, a Canadian teaching at the University of Wisconsin–Milwaukee.[27] It examined viewpoint diversity at ten cross ownerships and found only half expressed similar editorial positions during the 2000 election campaign. Pritchard admitted that a first study on the subject had been funded by Quebecor. It found a diversity of editorial viewpoints in three cross ownerships studied. As a result, Pritchard concluded the FCC's assumption that "media ownership inevitably shapes the news to suit its own interests may no longer be true (if it ever was)." Its cross ownership ban, as a result, had "outlived its usefulness."[28] After his paper was published in an academic journal, Pritchard explained, additional research had been commissioned by the FCC. Pritchard defended his research as "rigorously objective" and insisted he would "never cater to the whims of a funder." The fact his findings provided empirical support for the FCC dropping its cross ownership ban was entirely coincidental, he insisted. "I got no pressure whatsoever from anybody at the Commission that the results should go one way or the other."[29]

The west and the rest

One ironic aspect of the Izzy Asper story is how he started out as a champion for western Canada against its domination by a Toronto-Ottawa-Montreal power bloc. Asper even lay claim to having coined the phrase "western alienation" in 1969.[30] Little more than three decades later, his company held a tight grip on the news media of Saskatchewan, Alberta, and British Columbia. With the leading newspaper and a television station in the largest two cities of each of those provinces, CanWest was the convergence king of western

Canada. Looking eastward from there, the new power bloc was Toronto–Ottawa–Winnipeg. The Asper grip on news media grew ever tighter farther west until it was almost complete on Canada's west coast. CanWest's domination of Vancouver's media included both of its local daily newspapers, its largest television station, and almost all of its community newspapers. In the nearby capital of Victoria, CanWest Global owned the only daily newspaper and the largest television station. Even the 2006 Senate report on news media expressed dismay at how such a situation could have come to pass.

> Media mergers before 1986 had what was, in practice if not in theory, a free ride. This explains the ability of Pacific Press to obtain a monopoly of the daily newspaper market in Vancouver. . . . Concentration on the broadcasting side is explained by the CRTC's priorities. News was ranked well below support for Canadian culture.[31]

In approving CanWest's acquisition of Southam in 2000, the Competition Bureau was empowered to consider only the market for advertising, not news. "There was no evidence that newspapers, the Internet and television compete directly for retail advertising normally found in newspapers," it noted.[32] The CRTC's 2001 renewal of Global's television licences, according to the Senate report, "in effect, approved the acquisition by CanWest of the Hollinger newspaper empire. . . . There were no special conditions for the Vancouver market."[33] Despite signing off on CanWest's stranglehold on Vancouver's media, the CRTC did adjust licence conditions elsewhere on the basis of competition for news. Following its 2004 acquisition of Toronto One TV, the CRTC absolved Quebecor of adhering to the strict code of journalistic separation it had volunteered at TVA. Its cross ownership with the *Toronto Sun* was instead allowed to operate under only the management separation agreed to by CTV and CanWest. "One argument cited by the CRTC for the differential approach is that the Toronto market for news is highly competitive," the Senate report noted. "If the CRTC is going to base its decisions on the extent of news competition in each market, one might wonder why the conditions of newsroom separation imposed on CanWest's Vancouver operations were not stronger."[34] This disparate treatment, according to the senators, was partly to blame for CanWest's unprecedented control of Vancouver's media.

The CRTC is capable of treating some stations in a group differently from others, depending on the degree of media concentration in the particular markets, but has puzzlingly chosen not to do so in some striking instances. The end result is that by late 2001 . . . CanWest had approval for its extensive media holdings in Vancouver.

Along with its market dominance in western Canada, and especially in British Columbia, came considerable power over public opinion. As media scholars have noted, that power is exercised through such processes as agenda setting and the framing of issues. A western Canadian could be forgiven for suspecting that such a degree of local political influence might not have been permitted in eastern Canada. By contrast, in Toronto four daily newspapers were published by four different owners in one of the most competitive markets in North America. "Vancouver is the most concentrated media market in the country," noted Norman Spector. "This amount of power is worrisome in a democracy, and, though it's long past time something were done about it, the Senate committee proposes no remedy."[35] Yet to a concentrated and converged mainstream media, there was no problem. "Media critics have been harping on this issue for 30 years or more," observed the *Montreal Gazette* in an editorial. It claimed the real problem for media was instead "staying alive and profitable in a world of blogs, podcasting," and other new technologies.

> In the case of Vancouver, the two dailies have been under common ownership for almost 50 years. With so much history, one might reasonably expect critics to offer detailed and specific examples of abuse, rather than continuing to issue vague warnings of the potential dangers of corporate ownership. The Senate report contains no such examples.[36]

The CRTC's focus on cultural considerations, the Senate report concluded, had resulted in the market for news being taken into account in Quebec, but not elsewhere. "The CRTC's treatment of Global's Vancouver stations, which were in a market of obvious media concentration, is also at odds with its treatment of Quebecor's television group in Quebec," it noted.[37] French-Canadian culture, it seemed, was deserving of protection, but not the culture of other regions in the country. "Because the CRTC was concerned about the diversity of views in French-language media in Quebec," the Senate report observed, "more stringent conditions with respect to cross-media ownership were set out."[38] The long-simmering disaffection of western Canadians seemed justified by Ottawa's benign neglect in failing to prevent corporate control of their news media. Western alienation came to national attention in large part due to Izzy Asper's complaints in the early 1970s, but it grew in no small part due to his domination of its cultural apparatus.

The news media, as cultural theorists have noted, play an important role in representing culture. The pervasive nature of media, claimed David Altheide and Robert Snow, made them "the most powerful entities in the world today."[39] The power of media, they observed, lay in "controlling the definition of the situation."[40] According to the late James Carey, news media not only transmitted information, but also "a portrayal of the contending forces in the world." Consuming news, in Carey's view, was a ritual more like attending mass, where "nothing new is learned but . . . a particular view of the world is portrayed and confirmed."[41] Cultural considerations were particularly important in Canada, David Taras noted. Its diverse regional nature, multi-ethnic populace, and proximity to the cacophonous US media machine made the country unique. "Canada must depend on its media system to be a cultural and information lifeline in a way that other countries need not," Taras pointed out.[42] Convergence of CanWest's television holdings with Southam's newspapers, he concluded, gave it "extraordinary power in most Canadian media markets."

> While Izzy Asper's many achievements are extraordinary and many observers cannot help but admire his warm and gutsy style, the question is whether one man or one company should have so much control over the information available to citizens. The capacity to intervene in and even alter public life and public consciousness is nakedly apparent.[43]

Public benefits — or private?

The public benefits program the CRTC adopted to encourage positive outcomes from broadcasting takeovers was abused almost from the beginning. When Rogers took over Maclean Hunter and its Selkirk broadcasting arm in 1994, more than half its $101.9 million public benefits package went to upgrading its cable infrastructure.

That expenditure, some critics pointed out, benefitted Rogers most of all and was money it would likely have spent anyway.[44] Catherine Murray of Simon Fraser University noted the public benefits process was "unwieldy, secret, and subject to the whim of the private broadcasters' largesse." The benefits that were supposed to accrue to the public, it seemed to some, were instead being redirected to benefit the growing media giants. "They are increasingly being privatized, that is, interpreted to mean direct industrial benefit on the screen in priority program categories," according to Murray. "There are no systems to monitor the performance of the public benefits."[45]

The funding of journalism schools through the public benefits program resulted in the unavoidable perception of conflicting interests. The $3.5 million BCE provided to the Canadian Media Research Consortium resulted first in a marketing survey. It fit perfectly the category of "administrative" research proposed by Dallas Smythe. Survey research to determine media use and credibility would be valuable most of all to media outlets, their owners, and marketers. The study thus fulfilled the CMRC's stated mandate to "focus on the development of Canadian data for use in media planning."[46] It did not, however, ease the shortage of "historical perspective and reliable data" from which Desbarats noted debate about media in Canada had long suffered. To do so, a more critical approach would be required to examine, through content analysis and institutional research, the news Canadians receive and how it is shaped. Such an approach might be more along the lines of the NewsWatch Canada media-monitoring project at Simon Fraser University that has by contrast suffered from neglect. That way, differences between the real world and the "pictures in our heads" conjured by mass media could at least be illuminated.

As such critical research would doubtless be frowned on by media owners, a more independent funding arrangement would be required. Rather than abolishing the public benefits program, as sought by Leonard Asper, a percentage of its proceeds could be automatically devoted to educational initiatives. This fund could be administered by a panel of scholars unbeholden to owners. The findings of such media-monitoring research could be ordered publicized through those media outlets. Programs of media literacy could also be established with public benefits funds to ensure students learn how media work. Such programs have been proposed to offset the growing power of concentrated media, not just in universities and colleges, but in high schools as well.[47] Providing students with the tools required to "read" media content between the lines would be preferable to the stated educational objectives of two generations of Aspers. Their avowed ambition to influence higher education in favour of their worldview — as if their accumulated news media empire was insufficient for that purpose — should be denounced as antithetical to the very ends of education.

Correspondence entered into evidence at Conrad Black's 2007 fraud trial in Chicago provided an inside glimpse of his media machinations with Izzy Asper. "This isn't the end of a deal, it is only the beginning of the real deal," Asper enthused in a fax to Black days after buying Southam. "The possibilities are truly awesome and infinite."[48] Under expansion plans outlined by Asper, the consolidation of Canada's media would have grown even tighter in 2000. Even before the deal for Southam was completed, Asper said he was discussing a "strategic alliance" with Rogers and Shaw. "Ted [Rogers] has offered us a proposal which would give us a meaningful position in Sportsnet if we would join him in a joint venture on certain sports franchises." Asper told Black. "You can appreciate that a strategic relationship between CanWest and Rogers, and possibly Shaw, would give us the most bulletproof media position in Canada — radio, cable, television, print, magazines, Internet, direct-to-home satellite, multilingual broadcasting." Asper said he hoped to start with "joined sales forces, limitless cross promote," and sports media synergies in Toronto and Montreal. "And the beauty of all this is that it could be done without any approvals being necessary from the Competitions or CRTC, or government."[49]

After they consummated the sale of Southam, Asper and Black discussed a co-ordinated assault on Thomson, which was selling off its Canadian newspapers. Despite being deep in debt as a result of swallowing Southam, Asper coveted the *Winnipeg Free Press* in his home town. He had already been in discussions with Thomson and hoped to swap it for minority ownership of CanWest. "I believe we are the only game in town with Thomson on the *Free Press*, but, of course, one never knows," Asper wrote to Black. Instead, Thomson sold the *Free Press* and the *Brandon Sun* to Vancouver lawyer Ron Stern and Winnipeg businessman Bob Silver in late 2001 for $150 million.[50] Black wanted to merge Thomson's *Globe and Mail* with the *National Post*. Asper offered his assistance. "We believe that you will get your best deal on the *Globe & Mail* by being seen, by them, to be the ticket to unloading the other papers," he told Black in August of 2000. "If you agree with this scenario, we would tell [them] that we are willing to purchase 'whatever.'" Instead, Thomson cast its lot with BCE-CTV.

Under Asper's plan, however, newspaper competition would have been severely curtailed. "I certainly agree that there should only be one national newspaper," he told Black. "In order to gain all the synergies of the merger, in effect, you might turn the Toronto *Globe & Mail* into merely a Toronto edition of the *National Post.*" Even Asper allowed that alarm bells might go off in Ottawa under that scenario.

> Although we claim no expertise in the newspaper business, we do have a concern, perhaps ill-founded, that there would be an enormous public reaction, and possibly political repercussions, if the two papers simply merged and one disappeared, causing Regulators to complain about a lack of diversity and choice, even though none existed a mere two years ago.[51]

Perhaps the most telling nugget of information contained in the Black-Asper correspondence, however, was buried deep inside a confidential early CanWest "concept document" proposing the purchase of Southam. It set out the political congruence between Black and the Aspers. "It is noted that the Asper Group, in general, endorses and shares the editorial and philosophic views that Conrad and/or Barbara A. Black have expressed over the past years."[52]

Counterbalancing power

The Royal Commission on Newspapers urged limits on media ownership in 1981 because of the political power it feared could result. It had to admit, however, that its concerns were to that point so far hypothetical. "The best defence, on the evidence of the principal corporate proprietors themselves," conceded its report, "is that they do not exercise that power."[53] Fifteen years later, Conrad Black's takeover of Southam brought for some citizens the nightmare scenario of a political activist controlling much of Canada's news media. The brazenness of the Aspers that suggested they could dictate — almost literally — to their newly-acquired journalists brought unprecedented national alarm. It triggered not one but two federal inquiries. If ever the facts supported long-sought limits on media ownership, reformers thought, this finally would. If the actions of the Aspers did not trigger a regulatory response, they feared, perhaps nothing ever would. If not, then another solution should be sought to the problem of growing media power.

A group of scholars from the University of Windsor began studying Canadian media and politics in 1972 and took a particular interest in concentration and convergence. Having together reached retirement age, the Windsor group offered a 2005 book-length study as its final contribution to the debate.[54] "The Aspers and CanWest Global have shown themselves woefully inadequate in their ability to treat their new-found newspaper ownership as a public trust as well as a business," it concluded.[55] Government intervention would not be an ideal way to solve the problem of media concentration, it added, but it might be the only way if owners could not act responsibly. "It is highly unlikely that Canadian governments will continue to tolerate current levels of concentration of media power in the hands of those who show their commitment to use that power to control content," the Windsor group predicted. "The potential downside is simply too great."[56] Others remained unconvinced. "No one is holding their breath," observed Stephen Kimber in 2003 after the Lincoln committee called for a re-thinking of convergence. "The Aspers have too many friends in high places."[57]

Being newcomers to newspaper ownership, the Aspers were perhaps unaware of the importance to journalism of news media owners maintaining a veneer of impartiality. Since Izzy Asper's death in 2003, his heirs have been much more circumspect, perhaps having learned from the controversies they provoked. They doubtless continue to harbour strongly-held, dogmatic views of the world, however, and quite likely still wish to promote them. The way they might do that, however, may well have changed. Abandoning their program of national editorials, as has been noted, does not mean the Asper family has ceased attempting to shape our view of the world. Doing so through slanting the news might actually be more effective — and less obvious — than by pushing editorial page opinions. To influence the news by direct newsroom intervention, as media sociologists have found, is not even necessary. By merely signalling their wishes and exercising their power over hiring and promoting journalists, they can eventually ensure the news is shaped to their liking.

As the early studies of voting behaviour found, all of us have core beliefs instilled in us through our upbringing and socialization. As a result, most people have ingrained attitudes that are difficult — if not impossible — to alter. A more effective way of influencing public opinion than through editorials might be to manipulate the news and thus alter "the pictures in our heads." By framing Palestinian insurgents as "terrorists," for example, public opinion may be influenced in favour of Israel treating them harshly. By portraying tax cuts for the wealthy as necessary to our economic well-being, while ignoring their social costs, the interests of one class are favoured over another. Ignoring or downplaying concerns over increased political power in the hands of large media owners conveniently ensures that the issue will never make it onto the political agenda.

The more powerful Big Media grows, however, the more concerned Canadians should be. The simple way to stem the rising tide of media power by constraining convergence lies with the CRTC. It is charged with acting in the public interest to ensure the licences it grants to broadcast on public airwaves are held by acceptable parties. In some countries, such as the US, newspaper owners are considered unacceptable applicants for a television licence in the same city. In other countries, such as the UK and Australia, that prohibition has been lifted, but not without a "diversity" or "plurality" test to safeguard the public interest. In granting or renewing television licences, the CRTC should similarly consider ownership diversity of a market's news media. An applicant should not be approved if that would result in it owning most of a market's sources for news.

A 50 percent limit, as measured by audience share, would far exceed the maximum cross ownership allowed in some countries. It would, however, at least be a limit, which has long been needed on media ownership in Canada. It would have its first effect in Vancouver, where CanWest's audience share of evening news viewers was measured at greater than 70 percent. Its ownership of local daily newspapers, of course, is 100 percent. Taking into equal consideration the medium of radio, in which CanWest has no holdings in Vancouver, would still result in it controlling most of the local market for news. This exempts from the equation the Internet, on which CanWest carries both its local newspaper and television content, and community newspapers, in which it dominates. The result would be to force CanWest to choose between its newspaper monopoly and its BCTV superstation, or to break up Pacific Press by selling either the *Vancouver Sun* or *Province*.

If the inexorable glacier of ownership concentration cannot be reversed, some mechanism should at least be initiated for monitoring its abuses. To insist that an independent watchdog on media be provided as a public benefit would seem a reasonable price for holding increased power. To require that the findings of such a "watchdog on the watchdog" be publicized through the media it monitors would at least be a limited counterbalance to such power. To provide media literacy skills as part of every Canadian student's education would ensure they are able to examine critically the media messages that increasingly inundate them. These would seem more worthy public benefits than the self-serving purposes to which the funds have been, at least in some cases, previously put.

The 'fatal enemy'

The Canadian media system, according to David Taras, is in the midst of a profound crisis. "We are witnessing not an abrupt execution," he claimed, "but a slow, lingering death."[58] His analysis was offered in 2001, before the full extent of Asper family influence on Canada's news media was seen. A tectonic shift in Canadian journalism began when Conrad Black took over Southam and it accelerated when he used its resources to foiund the *National Post*. From its inception, the *Post* turned on its head Chicago journalist Finley Peter Dunne's century-old observation that "the business of a newspaper is to comfort the afflicted and afflict the comfortable." Instead, the *Post*'s campaign for tax cuts and social service reductions aimed to comfort the comfortable and afflict the afflicted. Asper family ownership of the dailies not only continued Black's mission *in absentia* but took it to a new level across the renamed Southam chain.

The ongoing crisis in Canadian journalism reached its zenith in 2002 during the Asper Disaster, which quickly replaced the 1980 "Black Wednesday" newspaper closures as the darkest chapter in Canadian journalism history. It will stand as such until the next crisis of concentrated news media control inevitably arrives. A bigger problem is that such a crisis might not even come to public attention if vigilance is not exercised in monitoring media influence. For that, media scholars and media critics play a vital role in the public interest, but they seem to be an endangered species. "Apart from the occasional column by Antonia Zerbisias in The *Toronto*

Star, and the contributions of a small group of academics, largely working in obscurity," noted *Globe and Mail* television critic John Doyle in late 2006, "major Canadian media organizations, including newspapers, rarely face criticism or analysis."[59]

In mid-2007, Zerbisias bowed out of media criticism, shifting the focus of her column to general lifestyles. According to blogger Dennis Earl, Zerbisias told him the *Star* would not be replacing her as media critic. "You would have to be crazy to do that job," Earl quoted from her e-mail.[60] The departure of Canada's last remaining media critic in the mainstream press went unreported in the *Star* or any other major daily, but it did not go unnoticed online. "Poof, she was gone," lamented Brian Brennan of the sudden end to her media column and popular blog. "No farewell to the world of media criticism."

> Did Star management have a hand in the demise of her blog? We likely will never know. She obviously still cares about some of the issues she covered in her media column. In a posting this week to the listserv of the Canadian Association of Journalists, Zerby signs herself "No Longer the Media Columnist" and urges fellow journalists to call on the CRTC to establish rules curbing media concentration in Canada. Too bad she couldn't have written another piece on this topic for the *Toronto Star*. All of which now leaves media criticism a non-starter as far as Canada's mainstream media are concerned.[61]

It is essential to redress the deficit of media criticism if the potential excesses of Big Media are to be guarded against. Journalism educators are often reluctant enough to criticize the mainstream media due to their need to arrange internships, and ultimately jobs, for their students. Allowing them to become beholden to Big Media through funding relationships increases the risk from one of silence to one of complicity. Communication scholars usually have little or no real-world experience in media, often allowing industry advocates to easily dismiss their criticisms as uninformed. The ombudsman function that emerged at many Canadian dailies following the 1970 Senate report has largely vanished due to cost cutting. The low level of credibility most Canadians ascribe to their news media might be increased if they at least showed a willingness to monitor themselves.

The societal stakes are high. News media manipulation must be seen in the context of several larger debates that have been going on for decades, if not centuries. One is the battle for control of the world economy between those who believe market forces alone should rule and those who think government intervention is occasionally required.[62] Another is the ethnic and religious conflict in the Middle East that has boiled over into a global war on terrorism. The Aspers have pronounced strong positions on both conflicts, the outcome of which could help determine the course of the 21st Century. To allow them or anyone else to dominate the debate on these or any other issues of importance would be unwise of Canadians. Those who have shown they will use their accumulated media holdings to force their worldview on others are too dangerous to allow to hold such power. "Concentrated control of the media is not the most urgent danger facing society," observed Ben Bagdikian in *The Media Monopoly*. "But the ability to cope with larger problems is related to the peculiar industries we call the media, to their ownership and the nature of their operation. . . . In a world of multiple problems, where diversity of ideas is essential for decent solutions, controlled information inhibited by uniform self-interest is the first and fatal enemy."[63]

The Asper Agenda

Following is the partial text of a speech by David Asper to the Calgary Chamber of Commerce in 2001, which set out fifteen points the family intended to advance on the pages of its newspapers:

My father was indeed the leader of the Liberal party in Manitoba in the early 1970s. To say that he was and is forever a supplicant of all Liberal parties entirely misses the reality of his politics. As early as 1961, at a conference of the Canadian Tax Foundation, he was an advocate of the flat tax concept. Working in an environment of Manitoba's first socialist government, and the federal government blindly pursuing policies he described as the "free ride society," my father foretold and fought against the entirely predictable debt-ridden and over-taxed society in which we now live. As welfare got extended irrationally, he campaigned on a theme that "welfare doesn't mean we pay for people who won't work." He was definitely not your usual big government Liberal of the 1970s. I want to say on his behalf that the tag often used that he was a former Liberal leader pays very little respect to what he was saying.

It should not be surprising that one or more of us living in the house with him might be somewhat influenced by my father's activities. With all of the various political and media characters gracing our living room, it was also not hard to witness the good-natured and often intense thrust and parry of idea-mongering. While my brother, sister and I ultimately pursued our own careers, eventually we came back, and now, the process of ownership transition is under way. So, I want to leave you with a snapshot of some of the things we will be advancing to embellish the issues on our national agenda:

1. We believe Canada is still very much a work in progress and the work of the federal government is wrongly beholden to the popula-

tion centres in Ontario and Quebec, with their attendant govern-ment-forming power. We suffer from economic discrimination and there needs to be a better balance.

2. Smart immigration policies are needed to help Western Canada continue to build its population base so as to develop political power and redress federal economic policies, which support Ontario and Quebec and their population bases.

3. We believe in an equal, elected and effective Senate in order to provide one national legislative forum where all stand as equals in order to promote truly national policy-making.

4. We believe that Parliament must be reformed to allow for free voting so as to give real meaning to the role of individual members of Parliament and to thereby subject legislation to meaningful scrutiny.

5. We believe the Supreme Court should be populated with justices representing each province — one province one judge.

6. We are opposed to the granting of any constitutional vetoes to any province. The current amending formula is workable, but we also believe in the need for ongoing constitutional review and recommendation structures in order to give life to the occasional need for the consideration of constitutional amendments.

7. We believe it is essential that property rights be enshrined in the constitution because the right to own and hold property is fundamental.

8. We believe in the decentralization of federal government offices, institutions and agencies across the country so as to promote local economies with necessary national administrative expenditures and to promote a national perspective.

9. We believe that with appropriate safeguards, the people ought to be able to recall a member of Parliament who has lost the confidence of the electorate.

10. We believe that having a notwithstanding clause in a constitution defeats the purpose of having a constitution in the first place and that it ought to be removed.

11. We believe that there ought to be fair and comprehensive public scrutiny of the appointment of senior public officials including the judiciary.

12. We abhor the use of deficit and debt financing for government activities in the belief that it wrongly mortgages the future and allows government to create an illusory or false economy for itself.

13. We believe that Canadian history ought to be a mandatory subject taught to Canadians throughout their Grades 1 to 12 school years.

14. There needs to be a taxpayer's bill of rights, to protect Canadians from over-reaching by the state when, for example, it imposes odious "pay now, argue later" rules.

15. The government should heed the report of the Senate on reform of capital-gains taxes to make ours lower than those in the United States as a means of making Canada a welcome haven for capital investment and reward.[1]

NOTES

Preface

1. See Marc Edge, "Byline wars at the *Vancouver Province*: The battle was not the only problem at the paper," *Media*, Spring 2001, pp. 6–7. Available online at http://caj.ca/mediamag/spring2001/cover1.html.
2. Available for free download at http://www.ihh.hj.se/mmt/corporategovernance.html.
3. Journal articles available online at http://www.marcedge.com/articles.html.
4. Available online at http://cjms.fims.uwo.ca/issues/02-01/index.html.
5. Simon Tuck, "CRTC concerned about 'diversity of voices,'" *Globe and Mail*, March 14, 2007, p. B7.
6. Deirdre McMurdy, "For CRTC chairman, less is more in world of television policy," *Ottawa Citizen*, May 18, 2007, p. A5.
7. See Paul Boin, "The CRTC's chairman has to go," *Globe and Mail*, May 30, 2007, p. A21; Antonia Zerbisias, "Time Canadians stood up to the CRTC," *Toronto Star*, May 27, 2007, p. C6.
8. Stephen Ward, "*Pacific Press: The Unauthorized Story of Vancouver's Newspaper Monopoly* [book review]," *Journalism History*, Summer 2002, p. 98.
9. James Adams, "Will Newman pen Izzy's life story?" *Globe and Mail*, July 8, 2006, p. R10.
10. Morley Walker, "Hunting for Newman, feeling guilty about it," *Winnipeg Free Press*, July 15, 2006, p. C5.
11. James Adams, "Writers and publishers adjust to Dion's victory," *Globe and Mail*, January 16, 2007, p. R3.

Introduction: The Asper Disaster

1. Canada, Standing Senate Committee on Transportations and Communications, "Interim report on the Canadian news media," April 2004, p. 8.
2. Ibid.
3. Conrad Black, "My adversaries' joy is premature," *Montreal Gazette*, August 5, 2000, p. B5.
4. Larry Zolf, "Whatever happened to left-wing Liberalism," CBC News *Viewpoint*, November 25, 2002. Available online at http://www.cbc.ca/news/viewpoint/vp_zolf/20021125.html
5. CanWest sold its Vancouver affiliate CKVU after being ordered to do so by the Canadian Radio-television and Telecommunications Commission upon acquiring market-leading BCTV.

6. Ian Jack, "Ottawa drops study of media ownership," *National Post*, March 17, 2001, p. A6.

7. Bill Schiller, "Aspers flex their media muscle," *Toronto Star*, December 23, 2001, p. A1.

8. David Asper, "Who controls freedom of speech?" *Calgary Herald*, December 14, 2001, p. A29.

9. Kim Lunman and Shawn McCarthy, "Sacking of Ottawa publisher draws international criticism," *Globe and Mail*, June 19, 2002, p. A1.

10. David Estok, "Aspers' views won't disappear, but just take on subtle guise," *Globe and Mail*, October 2, 2002, p. B11.

11. Keith Damsell, "CanWest set to launch news hub," *Globe and Mail*, January 20, 2003, p. B13.

Chapter 1: Citizen Asper

1. Jennifer Wells, "Izzy's dream," *Maclean's*, February 19, 1996, p. 40.

2. Martin O'Malley, "Asper finally runs, and runs hard," *Globe and Mail*, May 10, 1972, p. 9.

3. Edward Greenspon, "New TV stations may lead to 3rd network," *Globe and Mail*, September 5, 1987, p. B1.

4. Edward Greenspon, "Bicker, bicker: The three-way partnership at Global TV is acrimonious, to say the least," *Report on Business* magazine, March 1988, p. 26.

5. Trevor Cole, "Revenge of the ratings king: How did CanWest Global evolve into the powerhouse that's eating CTV's lunch?" *Report on Business* magazine, May 1996, p. 58.

6. Patricia Chisholm, "Tycoon of the tube," *Maclean's*, November 27, 1995, p. 36.

7. Peter C. Newman, *Titans: How the New Canadian Establishment Seized Power.* (Toronto: Viking), 1998, p. 212.

8. Ibid., p. 210.

9. Ibid., p. 209.

10. Ed Pearce, "On leadership," *Ivey Business Journal*, November 2001, p. 82.

11. Gordon Pitts, *Kings of Convergence: The Fight for Control of Canada's Media.* (Toronto: Doubleday), 2002, p. 54.

12. Naomi Lakritz, "Politician, entrepreneur, media mogul: The life and times of Izzy Asper," *Calgary Herald*, August 15, 2000, p. D1.

13. Larry Zolf, "You can't take Winnipeg out of the boy," *Globe and Mail*, August 8, 2000, p. A15.

14. Newman, *Titans*, p. 213.

15. Lakritz, "Politician, entrepreneur, media mogul."

16. Oakland Ross, "Asper's school slogan: 'Izzy clever . . . Izzy ever': Media giant had what it takes to win even as a teen," *Toronto Star*, August 5, 2000, p. A1.

17. I.H. Asper, "Flat rate tax benefits deserve study," *Globe and Mail*, October 14, 1971, p. B2.

18. Ibid.

19. "Taxation, Royal Commission on," *The Canadian Encyclopedia*. Available online at http://www.thecanadianencyclopedia.com/index.cfm?PgNm=TCE&Params=A1ARTA0007884

21. Linda McQuaig, *Behind Closed Doors: How the Rich Won Control of*

Canada's Tax System — and Ended Up Richer. (Markham, Ont.: Viking), 1987, p. 65.

22. Ibid., p. 163.

23. I.H. Asper, *The Benson Iceberg: A Critical Analysis of the White Paper on Tax Reform In Canada.* (Toronto: Clarke, Irwin), 1970, p. 15.

24. Ibid., p. xiii.

25. Ibid., p. xv.

26. Ibid., p. 282.

27. Ibid., p. 282.

28. McQuaig, *Behind Closed Doors*, p. 174.

29. "Taxation, Royal Commission on," *The Canadian Encyclopedia.*

30. I.H. Asper, "Tax reform proposals at a glance — with comparisons," *Globe and Mail*, June 21, 1971, p. B5.

31. McQuaig, *Behind Closed Doors*, p. 65.

32. "Asper will run," *Winnipeg Free Press*, October 16, 1970, p. 1.

33. Paul Pihichyn, "Grit candidate snubs 'old political games,'" *Winnipeg Free Press*, October 16, 1970, p. 2.

34. "Asper image works," *Winnipeg Free Press*, November 2, 1970, p. 10.

35. "Advocates a better Confederation bargain, Asper elected head of Manitoba Liberals," *Globe and Mail*, November 2, 1970, p. 10.

36. Paul Pihichyn, "Asper leads 'New Liberals,'" *Winnipeg Free Press*, November 2, 1970, p. 1.

37. "Asper image works."

38. "Liberal leader," *Winnipeg Free Press*, November 2, 1970, p. 25.

39. Ibid.

40. Donald Newman, "By-election call injects some life into Manitoba PC leadership rally," *Globe and Mail*, February 27, 1971, p. 11.

41. John Zaritsky, "A man who sells politicians like tomatoes," *Globe and Mail*, April 26, 1971, p. 8.

41. Murray Goldblatt, "Liberal conference backs guaranteed income plan," *Globe and Mail*, November 23, 1970, p. 1.

42. Arlene Billinkoff, "Asper, Molgat caution govt. on CFI inquiry," *Winnipeg Free Press*, February 1, 1971, p. 1.

43. I.H. Asper, "Pundits might take a shower before giving instant wisdom," *Globe and Mail*, October 6, 1970, p. B4.

44. Saul Cherniak, "Tax expert with dual identity chided by Cherniak [letter to the editor]," *Globe and Mail*, June 1, 1971, p. 7.

45. "Asper spurns chance to run in Manitoba," *Globe and Mail*, October 13, 1971, p. 8.

46. Egon Frech, "Tough battle expected in PC stronghold," *Globe and Mail*, November 13, 1971, p. 8.

47. "New National Policy needed?" *Canada & The World*, February 1973, p. 4.

48. Nick Hills, "At last, a strong role in Canada for the West," *Vancouver Province*, January 9, 1973, p. 4.

49. "Neglect of west hit," *Winnipeg Free Press*, January 25, 1971, p. 1.

50. Ibid.

51. I.H. Asper, "The winter of discontent in the west" *Globe and Mail*, January 26, 1971, p. 7.

52. George Bain, "Them as has, gits," *Globe and Mail*, January 27, 1971, p. 6.

53. O'Malley, "Asper finally runs, and runs hard."

54. Katie FitzRandolph, "Asper scores runaway win," *Winnipeg Free Press*, June 17, 1972, p. 1.
55. "Asper not shy on his first day in the House," *Globe and Mail*, July 4, 1972, p. 8.
56. Tim Traynor, "'As far as I'm concerned the union has broken up,'" *Winnipeg Free Press*, January 27, 1973, p. 8.
57. Ibid.
58. Ibid.
59. Hills, "At last, a strong role in Canada for the west."
60. I.H. Asper, "Financial facts need more than a perusal," *Globe and Mail*, October 12, 1972, p. B3.
61. David Lewis, "Mr. Lewis replies [letter to the editor]," *Globe and Mail*, October 14, 1972, p. 6.
62. Paul Jackson, "Asper confident of new deal," *Winnipeg Free Press*, January 26, 1973, p. 6.
63. "Western Liberals plan debate on BNA Act changes," *Winnipeg Free Press*, February 19, 1973, p. 1.
64. Ralph Bagley, "State control a bad thing — Asper," *Winnipeg Free Press*, June 23, 1973, p. 23.
65. Quoted in Glenn Johnson, "He left his mark on city he championed," *National Post*, October 8, 2003, p. A11.
66. Frances Russell, "Manitoba vote foreseen as joust of ideology, not issues," *Globe and Mail*, May 29, 1973, p. 3.
67. Ibid.
68. "Asper working hard to upset predictions that Liberals doomed," *Globe and Mail*, June 27, 1973, p. 8.
69. Egon Frech, "Signs of Schreyer sweep in Manitoba," *Globe and Mail*, June 28, 1973, p. 7.
70. Ibid.
71. Frances Russell, "Once ignored by all but Liberals, riding may decide the future of Schreyer Government," *Globe and Mail*, March 31, 1971, p. 7.
72. Alice Krueger, "Asper reluctant to elaborate future plans," *Winnipeg Free Press*, October 10, 1974, p. 4.
73. Ralph Bagley, "No losers and no setback: Asper," *Winnipeg Free Press*, June 29, 1973, p. 20.
74. "Vote count finally over in Manitoba," *Globe and Mail*, September 7, 1973, p. 1.
75. O'Malley, "Asper finally runs, and runs hard."
76. Newman, *Titans*, p. 216.
77. John Stackhouse, "Izzyvision," *Report on Business* magazine, May 1990, p. 78.
78. Egon Frech, "Ripples of change on political pond," *Globe and Mail*, August 10, 1974, p. 8.
79. Newman, *Titans*, p. 211.
80. Ibid.
81. Quoted in Ross, "Asper's school slogan."
82. "Asper's days as party leader may be limited," *Winnipeg Free Press*, July 29, 1974, p. 3.
83. Alice Krueger, "Asper steps down," *Winnipeg Free Press*, August 3, 1974, p. 1.
84. Barry Mullin, "Man of many talents, Asper leaves imprint on politics,"

Winnipeg Free Press, August 3, 1974, p. 9.
85. Krueger, "Asper steps down."
86. Krueger, "Asper reluctant to elaborate future plans."
87. "Half the fun of being Liberal," *Globe and Mail*, March 30, 1981, p. 8.
88. David Stewart-Patterson, "13 directors of Air Canada bounced by Conservatives," *Globe and Mail*, March 16, 1985, p. 1.
89. Wells, "Izzy's dream."

Chapter 2: CanWest Rising

1. Allan Levine, *From Winnipeg To the World: The CanWest Global Story*. (Winnipeg: CanWest Global Communications), 2002, p. 85.
2. Edward Greenspon, "Bicker, bicker: The three-way partnership at Global TV is acrimonious, to say the least," *Report on Business* magazine, March 1988, p. 26.
3. Gordon Pitts, *Kings of Convergence: The Fight for Control of Canada's Media*. (Toronto: Doubleday), 2002, p. 61.
4. Hershel Hardin, *Closed Circuits: The Sellout of Canadian Television*. (Vancouver: Douglas & McIntyre), 1985, p. 164.
5. Ibid.
6. Peter Desbarats, "The truth about Global's near-demise and Izzy Asper's rise," *Ottawa Citizen*, August 5, 2000, p. A13.
7. Greenspon, "Bicker, bicker," p. 29.
8. Ibid., p. 31.
9. Roger Newman, "'Misunderstood' Asper knows exactly what he's doing," *Vancouver Sun*, July 13, 1979, p. D1.
10. Greenspon, "Bicker, bicker," p. 31.
11. Matthew Fraser, *Free For All: The Struggle for Dominance on the Digital Frontier*. (Toronto: Stoddart), 1999, p. 141.
12. Greenspon, "Bicker, bicker," p. 33.
13. Hardin, *Closed Circuits*, p. 165.
14. Barry Berlin, *The American Trojan Horse: US Television Confronts Canadian Economic and Cultural Nationalism*. (New York: Greenwood), 1990, p. 108.
15. Israel Asper, "Retrospective," in Levine, *From Winnipeg To the World*, p. xiv.
16. Ibid., p. xvi.
17. "Top Liberal dismisses allegation about CDC," *Globe and Mail*, December 16, 1976, p. 9.
18. Asper, "Retrospective," p. xvii.
19. Levine, *From Winnipeg To the World*, p. 43.
20. Dan Westell, "Aristar sale by CanWest gets closer," *Globe and Mail*, May 13, 1983, p. B3.
21. David Olive, "Turnaround travails at the CDC," *Report on Business* magazine, May 1985, p. 28.
22. "Canwest exchanges diversity for flexibility," *Globe and Mail*, March 8, 1983, p. B17.
23. See Abraham Tarasofsky, *Something Ventured: The Canada Development Corporation 1972–85*. (Ottawa: Economic Council of Canada), 1987.
24. Quoted in Levine, *From Winnipeg To the World*, p. 58.
25. Roger Newman, "Asper takes over control of Canwest," *Globe and Mail*, February 15, 1984, p. B1.

26. Edward Greenspon, "The acquisitor," *Report on Business* magazine, October 1987, p. 27.
27. Peter C. Newman, *Titans: How the New Canadian Establishment Seized Power*. (Toronto: Viking), 1998, p. 212.
28. Mary Vipond, *The Mass Media in Canada*. (Toronto: Lorimer), 1989, p. 174.
29. Ellen Vanstone, "Simulcast sins drive up blood pressure and ad revenue," *Globe and Mail*, May 13, 1997, p. C2.
30. Fraser, *Free For All*, p. 139.
31. "Asper wins CKVU in court fight," *Vancouver Sun*, June 20, 1987, p. 1.
32. Paul W. Taylor, "Third service, third network: The CanWest Global system," *Canadian Journal of Communication*, 1993, p. 469.
33. Ibid.
34. Ibid.
35. Susan Gittins, *CTV: The Television Wars*. (Toronto: Stoddart), 1999, p. 278.
36. Terry Weber, "Asper's former partners must foot $1-million legal bill," *Winnipeg Free Press*, February 27, 1991. p. 3
37. Gittins, *CTV*, pp. 194-195.
38. Ibid., p. 253.
39. Trevor Cole, "Revenge of the ratings king: How did CanWest Global evolve into the powerhouse that's eating CTV's lunch?" *Report on Business* magazine, May 1996, p. 58.
40. Ibid.
41. Gittins, *CTV*, p. 277.
42. Richard Siklos, "CanWest programming, policies anger its rivals," *Financial Post*, July 25, 1991, p. 7.
43. Patricia Chisholm, "Tycoon of the tube," *Maclean's*, November 27, 1995, p. 36.
44. Peter C. Newman, "Thumbs down to Izzy Asper's national dream," *Maclean's*, December 23, 1996, p. 40.
45. Fraser, *Free For All*, p. 164.
46. Gittins, *CTV*, p. 309.
47. Ibid.
48. Fraser, *Free For All*, p. 168.
49. Gittins, *CTV*, p. 330.
50. Robert Lantos, "Only American TV is worth watching, you say?" *National Post*, November 25, 1998, p. A18.
51. Ibid.
52. Antonia Zerbisias, "TV titan sues movie mogul over speech," *Toronto Star*, February 5, 1999, p. 1.
53. "Clash of the celluloid titans," *Maclean's*, May 24, 1999, p. 10.
54. It was withdrawn after Asper's death.
55. Chris Cobb, "Canada won't get new TV network," *Ottawa Citizen*, February 7, 1998. p. A4.
56. William Boei, "Network TV status shunned at hearing," *Vancouver Sun*, November 4, 1997. p. D1.
57. Eve Lazarus,"Third network? No thanks!" *Marketing*, March 9, 1998, p. 28.
58. Vinay Menon, "CanWest buys TV producer Fireworks," *Toronto Star*, May 6, 1998, p. 1.
59. Fraser, *Free For All*, p. 217.

60. In 2006, Prime became a dedicated re-run channel re-christened TVtropolis.

61. Matthew Fraser, "Izzy Asper's Love Boat still cruising: CRTC protection has helped Global sail to profits," *National Post*, February 2, 1999, p. C7.

62. Fraser, *Free For All*, p. 217.

63. "Spending report fuels CTV-Global rivalry," *Montreal Gazette*, August 6, 1998, p. D7.

64. Antonia Zerbisias, "Broadcasters gear up for hearings," *Toronto Star*, September 16, 1998. p. 1.

65. Antonia Zerbisias, "Private broadcasters challenged," *Toronto Star*, September 24, 1998, p. 1.

66. "Global prez lashes out: Izzy Asper attacks producers, CTV for 'propaganda' campaign," *Halifax Daily News*, October 18, 1998, p. 52.

67. Antonia Zerbisias, "Is there an Irish anti-defamation league out there?" *Toronto Star*, October 16, 1998, p. 1.

68. Tony Atherton, "CRTC slashes red tape for broadcasters: Content regulation changes to boost prime-time Canadian content," *Ottawa Citizen*, June 12, 1999, p. A3.

69. Barbara Shecter, "CanWest buying NetStar, but it's not a done deal: ESPN could block it," *National Post*, January 21, 1999, p. C1.

70. Bertrand Marotte, "CTV outbids CanWest for TSN and Discovery," *Vancouver Sun*, February 6, 1999, p. F1.

71. Barbara Shecter, "CanWest called off NetStar race: Rejected conditions," *National Post*, February 8, 1999, p. C1.

72. Matthew Fraser, "What if Izzy Asper owned NetStar?" *National Post*, March 9, 2000, p. C6.

73. David Berman, "Life without father," *Canadian Business*, November 1995, p. 98.

74. "TV takeover tussle," *Maclean's*, November 12, 1995, p. 39.

75. "CanWest targets WIC," *Maclean's*, March 17, 1997, p. 33.

76. Jennifer Hunter, "Plotting a takeover: CanWest Global closes in on Griffiths family's control of WIC," *Maclean's*, January 19, 1998, p. 46.

77. "CanWest fires back at WIC," *Maclean's*, April 6, 1998, p. 53.

78. "WIC takes a poison pill," *Maclean's* April 13, 1998, p. 43.

79. "The battle for WIC," *Maclean's*, April 27, 1998, p. 53.

80. Barbara Shecter, "Couture lands in Asper-Lantos suit," *National Post*, February 21, 2002, p. FP6.

81. Antonia Zerbisias, "Vital Friends under attack by media giants," *Toronto Star*, May 2, 2001, p. D3.

82. Hugh Winsor, "CanWest ill-served by its own tactics," *Globe and Mail*, April 26, 2000, p. A4.

83. Sinclair Stewart, "CanWest orders study on critics," *National Post*, July 14, 2000, p. C4.

84. Winsor, "CanWest ill-served by its own tactics."

85. Trevor Cole, "Lenny: A mogul in the making," *Report on Business* magazine, October 2000, p. 66.

86. Pauline Couture, "Friends and conflict [letter to the editor]," *National Post*, July 20, 2000, p. C15.

87. Rob Ferguson, "CanWest is 'going to town' with WIC benefits package," *Toronto Star*, February 19, 2000, p. B3.

88. Drew Hasselback, "CanWest defends WIC split," *National Post*, April

25, 2000. p. C4.

89. William Boei, "Shape TV in Canada on WIC split, CRTC urged," *Vancouver Sun*, April 28, 2000, p. C1.

90. Dana Flavelle, "Regulator redraws TV industry blueprint: CanWest Global a truly national network," *Toronto Star*, July 7, 2000, p. F1.

Chapter 3: Going Global

1. Richard Siklos, "CanWest issue brings in half what Asper hoped," *Financial Post*, September 23, 1991, p. 16.

2. See Margie Comrie and Susan Fountaine, "Retrieving public service broadcasting: Treading a fine line at TVNZ," *Media Culture Society*, 2005, pp. 101-118.

3. Margie Comrie, "Television news and broadcast deregulation in New Zealand," *Journal of Communication*, 1999, p. 42.

4. John Partridge, "CanWest seals deal for N.Z. network," *Globe and Mail*, December 18, 1991, p. B7.

5. Trevor Cole, "Revenge of the ratings king: How did CanWest Global evolve into the powerhouse that's eating CTV's lunch?" *Report on Business* Magazine, May 1996, p. 58.

6. Allan Levine, *From Winnipeg To the World: The CanWest Global Story*. (Winnipeg: CanWest Global Communications), 2002, p. 85.

7. Harvey Enchin, "CanWest reaches deal to buy New Zealand radio network," *Globe and Mail*, July 1, 1997, p. B3.

8. Quoted in David Berman, "Channel changer," *Canadian Business*, September 1995, p. 46.

9. Mark Westfield, "Ten out of ten for a winning formula," *The Australian*, October 16, 2003, p. B3.

10. Quoted in Levine, *From Winnipeg To the World*, p. 90.

11. Levine, *From Winnipeg To the World*, p. 92.

12. Westfield, "Ten out of ten for a winning formula."

13. Lisa Murray, "Seriously fumbled: The great media carve-up of 2007 has left the Asper family in its wake, unable to sell its controlling stake in Ten," *Sydney Morning Herald*, June 23, 2007, p. 43.

14. "Asper: A life of sharp contrasts," *The Australian*, October 9, 2003, p. 22.

15. Ibid.

16. G. Easdown, "The man who would take," *Melbourne Herald Sun*, November 9, 1992.

17. John Partridge, "Australian TV deal 'big' for CanWest: Asper says sales, audience will double," *Globe and Mail*, October 21, 1992, p. B1.

18. Easdown, "The man who would take."

19. T. McCrann, "Ten Network ownership, and control, still a mystery," *Melbourne Herald Sun*, December 2, 1992, p. 67.

20. Malcolm Parry, "U.TV owner hopes to buy N.Z. network," *Vancouver Sun*, September 28, 1991, p. A2.

21. Robert Williamson, "Asper's global vision blurred: CanWest's stake in Australian television is raising eyebrows Down Under over the issue of foreign ownership," *Globe and Mail*, April 24, 1993, p. B3.

22. Australian Broadcasting Authority, "Investigation Into Control: CanWest Global Communications Corporation/The Ten Group Ltd.," November 1995, p. 128.

23. Ibid., pp. 183-184.

24. Mark Westfield, "ABA report on Ten unlikely to please CanWest," *The Australian*, June 12, 1995.

25. Levine, *From Winnipeg To the World*, p. 95.

26. Bruce McDougall "Succession: The McCains should have looked at other family businesses," *Financial Post* magazine, November 1994, p. 84.

27. Australian Broadcasting Authority, "Investigation Into Control," p. 185.

28. Levine, *From Winnipeg To the World*, p. 96.

29. Fiona Wingett, "Network axes news program," *Adelaide Advertiser*, June 19, 1993.

30. Richard Siklos, "CanWest gains 'whopping' dividend from Down Under," *Financial Post*, November 11, 1994, p. 9.

31. Deborah Brewster, "Parent details Ten leap to $103m," *The Australian*, November 22, 1995.

32. Robert Brehl, "CanWest chief must follow 'a great act,'" *Globe and Mail*, December 22, 1997, p. B1.

33. Levine, *From Winnipeg To the World*, p. 97.

34. Kim Sweetman, "Canadian TV boss sees 'no sense' in our media laws," *Adelaide Advertiser*, November 6, 1995.

35. M. Cole, "Canadian TV threat to beam from Fiji," *Brisbane Courier-Mail*, November 6, 1995.

36. Luke Collins, "CanWest's TEN stake safe after ABA ruling," *Brisbane Courier-Mail*, December 1, 1995, p. 27.

37. Australian Broadcasting Authority, "Investigation Into Control," p. 28.

38. Rebecca Lang, "TEN's owner in threat to pull out," *Melbourne Herald Sun*, December 7, 1995.

39. "CanWest bullish on float of Ten," *The Australian*, December 19, 1996, p. 21.

40. Australian Broadcasting Authority, "Investigation Into Control: CanWest Global Communications Corporation/The Ten Group Ltd. Second investigation," April 1997, p. 234.

41. Deborah Brewster and Clive Mathieson, "Ten's owner forced to cut stake," *Weekend Australian*, April 5, 1997, p. 1.

42. Bryan Frith, "Conflicting views put Asper at sixes and sevens on Ten," *The Australian*, May 22, 1997, p. 32.

43. Mark Westfield, "CanWest battles to maintain its grip on the box," *The Australian*, April 14, 1997, p. 24.

44. Clive Mathieson, "Mixed signals," *The Australian*, May 26, 1997, p. 36.

45. Deborah Brewster, "Right to sack board does not an owner make, Asper claims," *The Australian*, July 18, 1997, p. 27.

46. Anne Davies, "Canada weighs in to help TV mogul," *Sydney Morning Herald*, May 28, 1997, p. 29.

47. Clive Mathieson, "CanWest loses court appeal on Ten stake," *Weekend Australian*, August 9, 1997, p. 56.

48. Deborah Brewster, "Next move in CanWest snakes and ladders," *The Australian*, September 2, 1997, p. 22.

49. Andrew White, "Izzy's deal enough to make Singo cry," *The Australian*, September 12, 1997, p. 22.

50. Mark Westfield, "Coup of the decade goes to CanWest's one-off TEN deal," *The Australian*, January 28, 1998, p. 23.

51. Ivor Ries, "Izzy turns defeat into victory," *Australian Financial Review*, February 26, 1998, p.56.

52. Luke Collins, "CanWest set for $134 million windfall," *Hobart Mercury*, February 20, 1998.

53. Robert Brehl, "CanWest strikes gold with Australian investment," *Globe and Mail*, April 4, 1998, p. B3.

54. Westfield, "Coup of the decade goes to CanWest's one-off TEN deal."

55. Brehl, "CanWest chief must follow 'a great act.'"

56. Harvey Enchin, "CanWest setting up international office," *Globe and Mail*, January 21, 1993, p. B14.

57. Harvey Enchin, "CanWest buys interest in Chilean network: Hopes to put La Red in the black, use deal to gain toehold in Hispanic market," *Globe and Mail*, May 31, 1994, p. B1.

58. Ibid.

59. Andrea Mandel-Campbell, "Television South American-style," *Manitoba Business*, May 1995, p. 30.

60. Levine, *From Winnipeg To the World*, p. 107.

61. Paul Chard, "Too many gringos," *Canadian Business*, September 1995, p. 48.

62. Quoted in Levine, *From Winnipeg To the World*, p. 108.

63. Ibid., p. 109.

64. John Partridge, "Asper may have misplayed hand on British TV deal: Wanted too much control, consortium member says," *Globe and Mail*, July 9, 1992, p. B3.

65. Ibid.

66. "Talk radio airs in Britain," *Broadcaster*, March 1995, p. 8.

67. "Take Five," *The Economist*, May 6, 1995, p. 55.

68. Ritchie Gage, "Asper vision," *Manitoba Business*, December 1995, p. 5.

69. Raymond Snoddy, "CanWest set on Channel 5 bid," *Financial Times*, April 15, 1995, p. 4.

70. James Bethell, "City diary," London *Independent*, May 31, 1995, p. 34.

71. Quoted in Levine, *From Winnipeg To the World*, p. 111.

72. Ibid.

73. "Take Five," *The Economist*.

74. Nicholas Hellen and Matthew Lynn, "The odds against two rival companies making identical bids for Channel 5 were a billion to one," *London Sunday Times*, May 7, 1995.

75. Madelaine Drohan, "CanWest claims victory in U.K. bid," *Globe and Mail*, May 3, 1995, p. B3.

76. Hellen and Lynn, "The odds against two rival companies."

77. Alexandra Frean, "Experts surprised by size of bid for new franchise," *London Times*, May 3, 1995.

78. Paul Rodgers, "Costs may rule out Channel 5 bidder," *Independent*, May 7, 1995, p. 2.

79. Mathew Horsman, "UKTV fights whisperers in struggle for Channel 5," London *Independent*, May 30, 1995, p. 27.

80. Andrew Culf, "Channel 5 bidder attacks 'lies,'" *Guardian*, June 3, 1995, p. 10.

81. Raymond Snoddy, "Channel 5 contender appoints chief executive," *Financial Times*, June 5, 1995, p. 7.

82. Allison Vale, "Canwest loses UK bid," *Playback*, November 6, 1995, p. 4.

83. John Ivison, "UKTV favourite to win Channel 5," *The Scotsman*, October 20, 1995, p. 23.

84. Mathew Horsman, "ITC extends Channel 5 delay," *Independent*, October 24, 1995, p. 20.

85. Allan Laing, "Scottish dimension may have swayed television commission's vote," *Glasgow Herald*, October 28, 1995, p. 7.

86. Andrew Culf, "TV rivals attack fifth station 'deal': Fury as lower bidder wins," *Guardian*, October 28, 1995, p. 1.

87. Quoted in Levine, *From Winnipeg To the World*, p. 112.

88. Raymond Snoddy, "Channel 5 row deepens over ITC ruling on TV quality," *Financial Times*, January 5, 1996, p. 1.

89. Mathew Horsman, "Fresh Channel 5 bid evidence," *Independent*, January 17, 1996, p. 16.

90. Jennifer Wells, "Izzy's dream," *Maclean's*, February 19, 1996, p. 40.

91. Nicole McCormick, "Irish eyes smile for CanWest," *Broadcasting and Cable*, May 5, 1997, p. 64.

92. Ibid.

93. Emmet Oliver, "Canadians buy stake in UTV," *Irish Times*, October 10, 1997, p. 51.

94. "CanWest raises UTV stake to 30PC," *Irish Independent*, February 24, 1998, p. 15.

95. Eric Reguly, "Aspers eye greener pastures in Ireland," *Globe and Mail*, November 4, 1997, p. B24.

96. Susan Bourette, "CanWest gets nod on Irish TV network," *Globe and Mail*, November 3, 1997, p. B6.

97. Roisin Ingle, "Birth of a station," *Irish Times*, September 12, 1998, p. 62.

98. Hollie Shaw, "Canwest and Granada may extend ties," *National Post*, October 9, 2000 p. C1.

99. Kevin Dawson, "The TV empire of Israel Asper," *Dublin Sunday Business Post*, January 3, 1999, p. 18.

100. Farrel Corcoran, *RTÉ and the Globalisation of Irish Television*. (Bristol: Intellect Books), 2004, p. 208.

101. Ibid., p. 216.

Chapter 4: The Black Plague

1. After Rupert Murdoch's News Corp. and the Gannett chain in the U.S.

2. Richard Siklos, *Shades of Black: Conrad Black and the World's Fastest Growing Press Empire*. (Toronto: Minerva), 1996, p. 156.

3. Quoted in Peter C. Newman, "The inexorable spread of the Black Empire," *Maclean's*, February 3, 1992, p. 68.

4. Ibid.

5. Siklos, *Shades of Black*, p. 383.

6. Erwin Frenkel, *The Press and Politics in Israel: The Jerusalem Post from 1932 to the Present*. (Westport, CT: Greenwood Press), 1994, p. 168.

7. Jennifer Wells, "He styles himself as the editorial savior of papers, but critics label Conrad Black an editorial storm trooper," *Maclean's*, November 11, 1996, p. 56.

8. Conrad Black, *A Life in Progress*. (Toronto: Key Porter), 1993, pp. 453-454.

9. Sandra Olsen, "Banker slams media tycoon on PM claims," *Brisbane Courier-Mail*, February 12, 1994.

10. "Black distorts truth: Hawke," *Brisbane Courier-Mail*, April 12, 1994.

11. L. Clausen, "Hawke spy offer claim," *Brisbane Courier-Mail*, April 22, 1994.

12. Piers Akerman, "Black turns the tables," *Brisbane Courier-Mail*, April 22, 1994.
13. M. Cole, "Insults fly in Black-Hawke stoush," *Brisbane Courier-Mail* April 25, 1994.
14. "Flog Hawke, says press baron's wife," *Adelaide Advertiser*, April 26, 1994.
15. Stephanie Raethel, "Black mark against PM: Report hits link," *Brisbane Courier-Mail*, June 10, 1994.
16. Brenda Dalglish, "Black sells Fairfax stake," *Toronto Sun*, December 17, 1996, p. 37.
17. Edward Clifford, "FP board leaves offers to investors," *Globe and Mail*, December 13, 1979, p. B1.
18. See Marc Edge, "The good, the bad, and the ugly: Financial markets and the demise of Canada's Southam Newspapers,"*International Journal on Media Management*, Winter 2003, pp. 180–189.
19. Siklos, *Shades of Black*, p. 311.
20. Ibid., p. 307.
21. Ibid., p. 318.
22. Including the author, in February of 1993, after sixteen years at the *Calgary Herald* and *Vancouver Province*.
23. Michael Urlocker, "Southam feeling the heat from Black, Desmarais," *Financial Post*, April 25, 1995, p. 33.
24. Siklos, *Shades of Black*, p. 404.
25. Jonathan Ferguson, "Hollinger boss scoffs at critics," *Toronto Star*, August 10, 1996, p. F3.
26. Doulas Goold, "The pain of the obdurate rump," *Globe and Mail*, August 16, 1996, p. B8.
27. Brenda Dalglish, "Hollinger to pay special dividend," *Financial Post*, April 23, 1997, p. 5.
28. Brenda Dalglish, "Hollinger bids $923m for rest of Southam," *Financial Post*, May 1, 1997, p. 1.
29. Casey Mahood, "Black fails in quest to take Southam private," *Globe and Mail*, June 24, 1997, p. B1.
30. Quoted in John Miller, *Yesterday's News: Why Canada's Daily Newspapers Are Failing Us*. (Halifax: Fernwood), 1998, p. 62.
31. Casey Mahood, "Thomson's B.C. paper sale raises ire of rivals," *Globe and Mail*, May 21, 1998, p. B1.
32. Paul Waldie and Janet McFarland, "Black snaps up *Financial Post*," *Globe and Mail*, July 21, 1998, p. A1.
33. Casey Mahood, "Hollinger picks up big Southam block," *Globe and Mail*, August 20, 1998, p. B4.
34. Brenda Dalglish, "Hollinger bids for rest of Southam," *Globe and Mail*, December 3, 1998, p. B1.
35. Brenda Dalglish, "Southam accepts Hollinger's new bid," *Globe and Mail*, January 7, 1999, p. B10.
36. Barbara Shecter, "Hollinger swallows Southam: Raises stake to 97%," *National Post*, January 20, 1999, p. C1.
37. Anthony Wilson-Smith, "War of words: The gloves are off over circulation claims in the newspaper industry," *Maclean's*, February 8, 1999, p. 48.
38. See Chris Cobb, *Ego and Ink: The Inside Story of Canada's National Newspaper War*. (Toronto: McClelland & Stewart), 2004.

39. Robert Sheppard and Patricia Chisholm, "A strategic retreat," *Maclean's*, May 8, 1999, p. 30.

40. Alan Freeman, "Lord Black on hold: Ottawa blocks bid for peerage," *Globe and Mail*, June 20, 1999, p. A1.

41. Guy Abbate, "Court rejects Black's bid to sue PM over peerage," *Globe and Mail*, March 16, 2000, p. A1.

42. Gordon Pitts, *Kings of Convergence: The Fight for Control of Canada's Media*. (Toronto: Doubleday), 2002, p. 139.

43. Ibid., p. 139.

44. Ibid., p. 144.

45. See Robert Eringer, *The Global Manipulators*. (Bristol: Pentacle Books), 1980; Frances Stonor Saunders, *Who Paid the Piper? The CIA and the Cultural Cold War*. (London: Granta), 1999; Richard J. Aldrich, *The Hidden Hand: Britain, America and Cold War Secret Intelligence*. (London: John Murrray), 2001.

46. Hugh Wilford, "CIA plot, socialist conspiracy, or new world order? The origins of the Bilderberg group, 1952-55," *Diplomacy and Statecraft*, September 2003, p. 76.

47. Hugh Wilford, "Calling the tune? The CIA, the British left, and the Cold War, 1945-60," *Intelligence and National Security*, 2003, p. 47.

48. Rosalie Bertell, "An elite pow wow," *Peace*, July/August 1996, p. 6.

49. John Deverell, "Bilderberg secrecy lid intact," *Toronto Star*, June 1, 1996, p. E8.

50. Michael Macdonald, "Black hosts secret gabfest," *Windsor Star*, June 1, 1996, p. A11.

51. The deal was subsequently scaled back to $3.2 billion after CanWest encountered financing problems.

Chapter 5: Media Power

1. Dallas Smythe, "On the political economy of communications." *Journalism Quarterly*, 1960, p. 564.

2. Dallas W. Smythe and Tran Van Dinh, "On critical and administrative research: A New Critical Analysis," *Journal of Communication*, September 1983, p. 117.

3. See Werner J. Severin and James W. Tankard Jr., *Communication Theories: Origins, Methods, and Uses in the Mass Media*, 4th ed. (New York: Longman), 1997, p. 317.

4. Daniel J. Czitrom, *Media and the American Mind: From Morse to McLuhan*. (Chapel Hill, NC: University of North Carolina Press), 1982, p. 148.

5. Canada, Royal Commission on Newspapers, *Report*. (Ottawa: Minister of Supply and Services), 1981, p. 211.

6. See Scott Stossel, "The man who counts the killings," *Atlantic Monthly*, May 1997, pp. 86-104.

7. See James Shanahan and Michael Morgan, *Television and its Viewers: Cultivation Theory and Research*. (Cambridge: Cambridge University Press), 1999.

8. Donald L. Shaw and Maxwell E. McCombs, *The Emergence of American Political Issues: The Agenda-setting Function of the Press*. (St. Paul, MN: West Publishing), 1977, p. 5.

9. See Shanto Iyengar and Donald R. Kinder, *News That Matters: Television and American Opinion*. (Chicago: University of Chicago Press), 1987.

10. Jim Rutenberg, "Cable's war coverage suggests a new 'Fox effect' on television," *New York Times*, April 16, 2003, p. B9.

11. See Alexandra Kitty, *Outfoxed: Rupert Murdoch's War on Journalism.* (New York: Disinformation), 2005.

12. See Piers Robinson, *The CNN Effect: The Myth of News Media, Foreign Policy and Intervention.* (New York: Routledge), 2002.

13. Steven Kull, Clay Ramsay and Evan Lewis, "Misperceptions, the media, and the Iraq war," *Political Science Quarterly*, 2003-2004, pp. 569-598.

14. Harold Meyerson, "Fact-free news," *Washington Post*, October 15, 2003, p. A23.

15. See Stefano DellaVigna and Ethan Kaplan, "The Fox News effect: Media bias and voting," National Bureau of Economic Research Working Paper No. 12169, March 30, 2006. Available online at http://elsa.berkeley. edu/~sdellavi/wp/foxvote06-03-30.pdf

16. See Kitty, *Outfoxed*.

17. Charles Reina, "The memo," in Kristina Borjessin, ed., *Into the Buzz- saw: Leading Journalists Expose the Myth of a Free Press.* (Amherst, NY: Prometheus), 2002, pp. 45-46.

18. Ibid., p. 46.

19. Chris Cobb, *Ego and Ink: The Inside Story of Canada's National Newspa- per War.* (Toronto: McClelland & Stewart), 2004.

20. Gordon Fisher, "Have the journalists lost control?" *Canadian Issues*, August 2003, p. 11.

21. Cobb, *Ego and Ink*, p. 233.

22. Quoted in Ibid., p. 246.

23. Quoted in Ibid., p. 251.

24. Larry Patriquin, *Inventing Tax Rage: Misinformation in the National Post.* (Halifax: Fernwood), 2004, p. 4.

25. Ibid., p. 6.

26. Ibid., p. 30.

27. Ibid., p. 157.

28. Ibid., p. 164.

29. See Peter Ferguson and Cristine de Clercy, "Regulatory compliance in poll reporting in the 2004 Canadian election," *Canadian Public Policy*, September 2005, pp. 243-248.

30. See Ann Marie Owens, "Canadians split on creationism," *National Post*, November 25, 2000, p. A1.

31. Steven D. Brown and James Delodder, "When is a creationist not a cre- ationist? Appreciating the miracles of public opinion polling," *Canadian Journal of Communication*, 2003, p. 116.

32. Jonah Butovsky, "Phony populism: The misuse of opinion polls in the *National Post*," *Canadian Journal of Communication*, 2007, p. 91.

33. Ibid., p. 91.

34. Ibid., pp. 93-94.

35. See Daniel Yergin and Joseph Stanislaw, *Commanding Heights: The Battle for the World Economy.* (New York: Simon & Schuster), 2002.

36. See Diane Stone, *Capturing the Public Imagination: Think Tanks and the Policy Process.* (Portland, OR: Frank Cass), 1996, p. 136.

37. Andrew Rich and R. Kent Weaver, "Think tanks in the U.S. media," *Har- vard International Journal of Press/Politics*, Fall 2000, p.98-99.

38. Nurith C. Aizenman, "The man behind the curtain: Richard Mellon-

Scaife and $200 million of his money is the man behind the conservative revolution," *Washington Monthly*, July/August 1997, p. 28.

39. Trudy Lieberman, *Slanting the Story: The Forces that Shape the News.* (New York: New Press), 2000, pp. 26-27.

40. Lieberman, *Slanting the Story*, p. 3.

41. Reese Cleghorn, "The press shifts to the right, but slowly: Words have moved the center as journalists followed the pols," *American Journalism Review*, December 1994, p. 4.

42. Lieberman, *Slanting the Story*, p. 8.

43. Ibid., p. 18.

44. Ibid., p. 20.

45. Tasha Kheiriddin and Adam Daifallah, *Rescuing Canada's Right: Blueprint for a Conservative Revolution.* (Mississauga, ON: J. Wiley and Sons Canada), 2005, p. 92.

46. Ibid., p. 97.

47. Ibid., p. 55.

48. Barbara Kay, "Awakening the conservative within," *National Post*, November 30, 2005, p. A18.

49. Kheiriddin and Daifallah, *Rescuing Canada's Right*, p. 56.

50. Frum, who would become a founding columnist for Black's *National Post*, left the newspaper in 2000 to become a speech writer for U.S. president George W. Bush and authored his famous "Axis of evil" speech.

51. David Taras, "The winds of right-wing change in Canadian journalism," *Canadian Journal of Communication*, Fall 1996, p. 485.

52. Linda Goyette "Fraser Institute revealed: Right-wing think-tank brags about its media manipulation," *Edmonton Journal*, January 17, 1997. p. A16.

53. Clive Thompson, "Fraser attacks! A leaked document reveals the right-wing Fraser Institute's five-year plan for media domination," *This Magazine*. May/June 1997, p. 18

54. Donald E. Abelson, *Do Think Tanks Matter? Assessing the Impact of Public Policy Institutes.* (Montreal: McGill-Queen's University Press), 2002, p. 45.

55. Donald Gutstein, "Who funds the Fraser Institute?" *Straight Goods*, October 28, 2004. Available online at http://www.yourmedia.ca/library_articles/041028_gutstein_fraser.html.

56. Donald Gutstein, "U.S.-style media monitor comes north," *The Tyee*, February 10, 2005. Available online at http://www.thetyee.ca/Media-Check/current/FraserInstituteSpinCycle.htm.

57. Miro Cernetig, "Neo-cons young bucks of the new right," *Globe and Mail*, February 5, 1994, p. D1.

58. Krishna Rau, "A million for your thoughts," *Canadian Forum*, July/August, 1996, pp. 15.

59. Dalton Camp, "Politics, journalism of new right fueled by money," *Toronto Star*, February 5, 1997, p. A19.

60. Gutstein, "Who funds the Fraser Institute?"

61. "Donner Canadian Foundation," *Globe and Mail*, October 16, 1996, p. B6.

62. Gutstein, "Who funds the Fraser Institute?"

63. Charlie Smith, "Left-wing shtick: The *Vancouver Province* is criticized for removing a professor's comments from a story," Media, Spring 2000, p. 15.

64. See Marc Edge, "Byline wars at the *Vancouver Province*: The battle was not the only problem at the paper," Media, Spring 2001, p. 6. Available online at http://caj.ca/mediamag/spring2001/cover1.html.

65. He resigned when he was named chairman of the *National Post* in 2003.

66. Wendy McLellan, "Study paints journalists as left-wing, biased messengers," *Media*, Winter 2003, p. 31.

67. Campbell Clark, "Think-tanks changing their minds," *Globe and Mail*, August 20, 2005, p. F2.

68. Abelson, *Do Think Tanks Matter?*, p. 185.

69. Gutstein, "Who funds the Fraser Institute?"

70. Donald Gutstein, "Harperstein," *Georgia Straight*, July 6, 2006, p. 39.

Chapter 6: Convergence

1. Canada, Royal Commission on Newspapers, *Report*. (Ottawa: Minister of Supply and Services), 1981, p. xi.

2. No convictions were ever obtained against the companies.

3. Canada, Royal Commission on Newspapers, *Report*, pp. 215, 218.

4. Ibid., pp. 229-230.

5. See Allan Bartley, "The regulation of cross-media ownership: The life and short times of PCO 2294," *Canadian Journal of Communication*, Summer 1988, p. 56.

6. Only transactions involving 30 percent ownership of broadcasting companies or more required prior CRTC approval.

7. Bertrand Marotte, "Quebecor wins Videotron," *Globe and Mail*, September 13, 2000, p. B1.

8. David Asper, "To Chrétien's accusers: Put up or shut up," *National Post*, March 7, 2001, p. A17.

9. Graham Fraser and Allan Thompson, "Ottawa to study media ownership," *Toronto Star*, March 13, 2001, p. 1.

10. Ian Jack, "Ottawa drops study of media ownership," *National Post*, March 17, 2001, p. A6.

11. Heather Scoffield, "Ottawa and the Goliaths: Media giants are converging at breakneck speed. Will regulators be left in the dust?" *Globe and Mail*, March 26, 2001, p. R1.

12. "Putting up," *National Post*, March 7, 2001, p. A17.

13. Mark Steyn, "Shawinigan shenanigans: Our investigative reporting has been as solid as it gets," *National Post*, March 8, 2001, p. A14.

14. I.H. Asper, letter to Conrad Black, March 13, 2001. Available online at http://www.usdoj.gov/usao/iln/hot/us_v_black_exhibits/canwest_134.pdf.

15. Conrad Black, letter to Israel Asper, March 14, 2001. Available online at http://www.usdoj.gov/usao/iln/hot/us_v_black_exhibits/canwest_134.pdf.

16. Conrad Black, letter to Israel H. Asper, January 5, 2001. Available online at http://www.usdoj.gov/usao/iln/hot/us_v_black_exhibits/canwest_132.pdf.

17. Black, letter to Asper, March 14, 2001.

18. Lawrence Martin, "Turner betrays lack of class," *Montreal Gazette*, May 10, 2001, p. B3.

19. Lawrence Martin, "Liberals should beef up CBC and move it west," *Cal-*

gary Herald, January 18, 2001, p. A20.

20. Val Sears, "Offers $36 million public benefit package: Maclean Hunter 'sweetens' deal for Selkirk," *Toronto Star*, May 30, 1989, p. B1.

21. Ian Austen, "Does bigger mean better?" *Calgary Herald*, September 14, 1994, p. C4.

22. "BCE tables $230 million benefits package as part of CTV acquisition," Canada Newswire, July 18, 2000.

23. Matthew Sekeres, "Ryerson endowment hinges on BCE deal," *Globe and Mail*, August 15, 2000, A6.

24. Dana Flavelle, "BCE-CTV media giant given CRTC's blessing," *Toronto Star*, December 8, 2000, p. E1.

25. Which it was required by the CRTC to sell after acquiring TVA.

26. Allan Swift, "TVA will be independent, Quebecor says — Print, broadcast newsrooms to remain separate," *Toronto Star*, March 27, 2001, p. 1.

27. Antonia Zerbisias, "Ready or not, CRTC takes on media convergence," *Toronto Star*, April 14, 2001, p. 6.

28. Ibid.

29. Swift, "TVA will be independent, Quebecor says."

30. Leonard Asper, "Observations on the media, Canada and Winnipeg," *Canadian Speeches*, January/February 2001, p. 54.

31. Zerbisias, "Ready or not, CRTC takes on media convergence."

32. Canadian Radio-Television and Telecommunications Commission, "Transcript of proceedings," April 25, 2001, at 10325-10326.

33. Quoted in "Code of conduct the big debate at CRTC hearings," *Regina Leader-Post*, April 26, 2001, C2.

34. Peter Desbarats, "Get out of our newsrooms," *Globe and Mail*, July 11, 2001, p. A15.

35. Canadian Radio-Television and Telecommunications Commission, "Transcript of proceedings," April 19, 2001, at 4051.

36. Ibid., at 3667.

37. Ibid., at 3910-3911.

38. Antonia Zerbisias, "Networks balk at new rules for licences," *Toronto Star*, April 26, 2001, p. A21.

39. Chris Cobb, "Regulate media giants: nationalist culture groups," *Ottawa Citizen*, April 25, 2001, p. D3.

40. Heather Scoffield, "Ottawa and the Goliaths: Media giants are converging at breakneck speed. Will regulators be left in the dust?" *Globe and Mail*, March 26, 2001, p. R1.

41. Julian Beltrame, "Whose news?" *Maclean's*, May 28, 2001, p. 34.

42. Scoffield, "Ottawa and the Goliaths."

43. Heather Scoffield, "CRTC okays newsroom convergence," *Globe and Mail*, August 3, 2001, p. B1.

44. Susan Pigg, "CRTC okays merged newsrooms — But management of newspaper and TV operations must be separate," *Toronto Star*, August 3, 2001, p. A3.

45. Heather Scoffield, "Writers blast convergence decision: Using same staff for coverage by TV, newspapers will cut quality, group says," *Globe and Mail*, August 7, 2001, p. A6.

46. "In support of BCE/CTV union," [Excerpts from letters] *National Post*, May 29, 2001, p. C15.

47. "BCE tables $230 million benefits package as part of CTV acquisition."

48. Ibid.
49. Terence Corcoran, "At least they should send the money back," *National Post*, May 29, 2001, p. C15.
50. Anthony Westell, "Journalism foundation launched," *content* [Centre for Investigative Journalism], November/December 1990, p. 10.
51. Chris Cobb, "The Canadian Journalism Foundation: on the fast track to where?" *Media* [Canadian Association of Journalists], July 1994, p. 22.
52. Eric Reguly, "The lobbyists are everywhere — and they're multiplying," *Globe and Mail*, May 17, 2001, p. B13.
53. Richard Foot, "Conflict allegations force out journalism group's executive," *National Post*, May 28, 2001, p. A1.
54. Terence Corcoran, "At least they should send the money back," *National Post*, May 29, 2001, p. C15.
55. Ibid.
56. Murray Whyte, "CJF washes hands of donation," *National Post*, May 30, 2001, p. A2.
57. Donna Logan, "Re: CTV Television Inc. station group licence renewal," [letter to Ursula Menke, CRTC Secretary General], March 22, 2001.
58. Donna Logan, "Opportunities and challenges," *Media*, Fall 2000, p. 15.
59. Daniel Sieberg, "Beware automatic reaction against concentration, j-school head says," *Vancouver Sun*, April 27, 2000, p. D5.
60. See Marc Edge, "Balancing academic and corporate interests in Canadian journalism education," *Journalism and Mass Communication Educator*, Summer 2004, pp. 172-185.
61. Frank Luba, "CanWest invests in UBC," *Vancouver Province*, June 26, 2001, p. A2.
62. "CanWest pledges $500,000 to fund visiting journalist program at UBC," *Vancouver Sun*, June 26, 2001, p. B1.
63. http://www.journalism.ubc.ca/events_convergence.html
64. Ibid.
65. Katherine Macklem, "Can the Aspers do it?" *Maclean's*, April 8, 2002, p. 48.
66. Murray Whyte, "Newman quits broadcasting lobby," *National Post*, April 26, 2001, p. A4.
67. Antonia Zerbisias, "Vital Friends under attack by media giants," *Toronto Star*, May 2, 2001, p. D3.
68. Ibid.
69. Matthew Fraser, *Free For All: The Struggle for Dominance on the Digital Frontier*. (Toronto: Stoddart), 1999.
70. Matthew Fraser, "Broadcast watchdog has only one trick," *National Post*, May 1, 2001, p. A17.
71. Matthew Fraser, "A world of Friends," *National Post*, May 5, 2001, p. D11.
72. Richard Foot, "Conflict allegations force out journalism group's executive," *National Post*, May 28, 2001, p. A1.
73. Murray Whyte, "Foundation apologizes to writer," *National Post*, May 11, 2001, p. A13.
74. Kim Honey, "Southam drops national columnist," *Globe and Mail*, July 31, 2001, p. A8.
75. Cobb, *Ego and Ink*, p. 312.
76. David Parkinson, "Aspers vow to continue voicing opinions through media holdings," *Globe and Mail*, April 21, 2001, p. B4.

77. David Asper, "The Asper agenda: 15 points for Canada," *Calgary Herald.* April 24, 2001, p. A15.
78. Robin Raizel, "Mixed-up media," *Canadian Business*, October 15, 2001, p. 65.
79. David Olive, "Crazy, hazy days of convergence: U.S. experience exposes some flaws in media marriages," *National Post*, September 16, 2000, p. D.1.
80. Gillian Doyle, "Convergence: 'A unique opportunity to evolve in previously unthought-of ways' or a hoax?" in Christopher T. Marsden and Stefaan G. Verhulst, eds., *Convergence in European Digital TV Regulation.* (London: Blackstone), 1999, p. 151.
81. Ibid., p. 153.
82. Robert Sparks, Mary Lynn Young, and Simon Darnell, "Convergence, corporate restructuring, and Canadian online news, 2000-2003," *Canadian Journal of Communication*, 2006, p. 418.
83. Ibid., p. 419.
84. See Peter J. S. Dunnett, *The World Newspaper Industry.* (London: Croom Helm), 1988.
85. See Allan Brown, "Newspaper ownership in Australia." *Journal of Media Economics*, Fall 1993, pp. 49-64.
86. See Alec Klein, *Stealing Time: Steve Case, Jerry Levin, and the Collapse of AOL Time Warner.* (New York: Simon & Shuster), 2003; Nina Munk, *Fools Rush In: Steve Case, Jerry Levin, and the Unmaking of AOL Time Warner.* (New York: Harper Business), 2004; Kara Swisher with Lisa Dickey, *There Must Be a Pony in Here Somewhere: The AOL Time Warner Debacle and the Quest for a Digital Future.* (New York: Crown Business), 2003.
87. Martin Peers, "In shift, AOL Time Warner to de-emphasize 'convergence,'" *Wall Street Journal*, May 13, 2002, p. B1.
88. Keith Damsell, "CanWest Global's loss doubles: Ad decline, debt servicing cited," *Globe and Mail*, November 15, 2001, p. B9.
89. Dana Flavelle, "Deep cuts at *National Post*: Slimmed-down paper lays off 120, will focus on business," *Toronto Star*, September 18, 2001, p. A3.
90. Chris Cobb, *Ego and Ink: The Inside Story of Canada's National Newspaper War.* (Toronto: McClelland & Stewart), 2004, p. 313.
91. Aldo Santin, "Aspers ringing up jobs: Call centre, movement of key managers could bring 1,200 positions here," *Winnipeg Free Press*, May 31, 2001, p. A1.
92. Justine Hunter, "Loonie support surprisingly soft, poll indicates," *National Post*, June 16, 2001, p. SN7.
93. Terence Corcoran, "The top 10 taboos our rulers won't touch," *National Post*, June 16, 2001, p. SN2.
94. Chris Powell, "Royal buys into Global convergence," *Marketing*, June 25, 2001, p. 4.
95. Brett David Dolter, "Insert showed 'right-wing bias' [letter to the editor]," *Regina Leader-Post*, June 26, 2001, p. A7.

Chapter 7: The *Gazette* Intifada

1. Bertrand Marotte and Ingrid Peritz, "Gazette publisher steps down: differences with daily's new owners, CanWest Global, cited for Oct. 1 resignation," *Globe and Mail*, September 1, 2001, p. A8.

2. Ibid.

3. Aaron J. Moore, "A chill in Canada," *Columbia Journalism Review*, March/April 2002, p. 40.

4. "Canada must lead," *Montreal Gazette*, August 11, 2001, p. B6.

5. Moore, "A chill in Canada."

6. Gabriel Boutros, "Mideast editorial is dangerously shortsighted," *Montreal Gazette*, August 14, 2001, p. B2.

7. Clifford Krauss, "Canadian press freedom questioned in publisher's firing," *New York Times*, June 29, 2002, p. A3.

8. "Duopoly," *The Economist*, April 28, 2001, p. 38.

9. Rita Zekas, "New Gazette publisher backs CanWest's editorials," *Toronto Star*, January 11, 2002, p. A8.

10. Walter C. Soderlund, Ronald H. Wagenberg, and Walter I. Romanow, "CanWest Global's national editorial policy: Round one," in Soderlund and Hildebrandt, eds., *Canadian Newspaper Ownership in the Era of Convergence: Rediscovering Social Responsibility.* (Edmonton: University of Alberta Press), 2005, p. 121.

11. See Marc Edge, *Pacific Press: The Unauthorized Story of Vancouver's Newspaper Monopoly.* (Vancouver: New Star Books), 2001, pp. 69-70.

12. Benoit Aubin, "Back to the future," *Maclean's*, July 21, 2003, p. 32.

13. Elizabeth Church, "Southam's national editorial policy comes under fire," *Globe and Mail*, December 7, 2001, p. A12.

14. http://radio.cbc.ca/insite/AS_IT_HAPPENS_TORONTO/2001/12/7.html

15. Peggy Curran, "Journalists in a dangerous time," *Montreal Gazette*, November 28, 2001, p. D5.

16. Tu Thanh Ha, "Sparks fly in Quebec over changes at Gazette," *Globe and Mail*, December 21, 2001, p. A7

17. Bill Schiller, "Aspers flex their media muscle: 'Go CanWest young man or go to hell,'" *Toronto Star*, December 23, 2001, p. A1.

18. http://radio.cbc.ca/insite/AS_IT_HAPPENS_TORONTO/2001/12/7.html

19. Ibid.

20. Quoted in Schiller, "Aspers flex their media muscle."

21. Canadian Journalists for Free Expression, "Not in the newsroom!: CanWest Global, chain editorials and freedom of expression in Canada," April 2002. Available online at http://www.cjfe.org/specials/canwest/canwintro.html

22. Bernard Perusse, et al., "How CanWest is threatening press freedom," *Globe and Mail*, December 11, 2001, p. A27.

23. Ibid.

24. Antonia Zerbisias, "Aspers own their freedom — But what about that of their newspaper employees?" *Toronto Star*, December 16, 2001, p. 2.

25. Ha, "Sparks fly in Quebec over changes at *Gazette*."

26. Ibid.

27. Schiller, "Aspers flex their media muscle."

28. Ha, "Sparks fly in Quebec over changes at *Gazette*."

29. Don MacPherson, "A distinct voice: Anglo Quebec views need to be accommodated in editorial columns," *Montreal Gazette*, December 13, 2001, p. B3.

30. David Asper, "Southam editorials a sign of freedom," *Winnipeg Free*

Press, December 11, 2001, p. A13.

31. David Asper, "Who controls freedom of speech?" *Calgary Herald*, December 14, 2001, p. A29.

32. Ibid.

33. "CanWest submission to Heritage Committee," September 10, 2001, p. 5. Available online at http://www.canwestglobal.com/sep_10_heritage_review.pdf

34. Mark Harrison, Norman Webster and Joan Fraser, "Every community's newspaper needs a free voice," *Globe and Mail*, December 15, 2001, p. A21.

35. Graham Fraser, "Editorial harmony creates discord at *The Gazette* — Montreal paper fights to keep respect in face of new policy," *Toronto Star*, December 16, 2001, p. 2.

36. Lysiane Gagnon, "How the Aspers are shooting themselves in the foot," *Globe and Mail*, December 17, 2001, p. A15.

37. Charles Gordon, "The local voice is the issue," *Ottawa Citizen*, January 31, 2002, p. A14.

38. "Quebec Culture Minister to force 'plurality of opinion' in media," *National Post*, December 19, 2001, p. A2.

39. Mike Gasher, "Does CanWest know what all the fuss is about?" Media, Winter 2002, p. 8.

40. Alexander Panetta, "CanWest decision rapped: Quebec angered by move over editorials," *Toronto Star*, December 20, 2001, p. A1.

41. John Miller, "CanWest's assault on press freedom," *Toronto Star*, December 13, 2001, p. A32.

42. Peter Desbarats, "What it takes to make a great paper," *Globe and Mail*, December 31, 2001, p. A11.

43. Gasher, "Does CanWest know what all the fuss is about?"

44. Ibid.

45. "Bad news," *Marketing*, January 28, 2002, p. 38.

46. "Canadian chain editorials called bad for business," *Editor and Publisher*, January 7, 2002, p. 20.

47. Adam B. Kushner, "The bogus editorial voice in Canada," *Masthead*, Summer 2002, p. 22.

48. Ibid.

49. DeNeen L. Brown, "Canadian publisher raises hackles: Family is accused of trying to restrict local newspapers' autonomy," *Washington Post*, January 27, 2002, p. A25.

50. Harvey Enchin, "CanWest aims to reduce dependence on ad revenue," *Vancouver Sun*, January 31, 2002, p. D7.

51. "The Gazette Intifada," *Media*, Winter 2002, p. 6.

52. Ibid.

53. Stephen Kimber, "In the wonderful world of Iz, it's 1984 all over again," in David Dadge, ed. *Silenced: International Journalists Expose Media Censorship*. (Amherst, NY: Prometheus Books), 2005, p. 53.

54. Ibid., pp. 52-53.

55. Brown, "Canadian publisher raises hackles."

56. Kimber, "In the wonderful world of Iz, it's 1984 all over again," p. 53.

57. Ibid., p. 54.

58. Kevin Cox, "Halifax writer quits after column pulled," *Globe and Mail*, January 5, 2002, p. A6.

59. Stephen Kimber, "Why I won't write for the Aspers," *Globe and Mail*,

January 7, 2002, p. A15.

60. Ibid.

61. Australia, Senate Environment, Communications, Information Technology and the Arts Legislation Committee, "Report on the Broadcasting Services Amendment (Media Ownership) Bill 2002," June 2002, p. 61.

62. "Kimber on tour," *Maclean's*, May 27, 2002, p. 9.

63. Kimber, "In the wonderful world of Iz," p. 53.

64. David Swick, "From thrilled to the truth: Many weren't happy with newspaper owner's policies," *Halifax Daily News*, August 14, 2002, p. 2.

65. Nicolaas van Rijn, "Defying Southam 'a career disaster': Editors urged to get head office advice on columns, letter says," *Toronto Star*, March 16, 2002, p. 3.

66. Katherine Macklem, "The view from the center," *Maclean's*, April 8, 2002, p. 53.

67. "Managing editor quits over editorial policies," *St. John's Telegram*, April 25, 2002. p. A9.

68. Peter Worthington, "Casting Asper-sions," *Toronto Sun*, January 13, 2002. Downloaded from http://64.34.111.53/articles/TorontoSun/torsun020113.htm.

69. Bill Schiller "Aspers' latest move spurs concern over media control: Worthington column cut from Southam's *Windsor Star*," *Toronto Star*, January 16, 2002, p. 14.

70. "Nfld columnist quits," *Broadcast News*, January 29, 2002.

71. Bill Schiller, "Axing of column sparks controversy," *Toronto Star*, January 12, 2002, p. A27.

72. http://www.dougcuthand.com/Columns300/Column378_dougcuthand.html Downloaded March 25, 2006.

73. Murdoch Davis, "The facts of the matter," *National Post*, January 31, 2002, p. A18.

74. Lyle Steward, "Sterile germ plasm dominates Canadian debate," *The Guild Reporter*, July 12, 2002, p. 3.

75. Keith Damsell, "CanWest scales back policy," *Globe and Mail*, February 12, 2002, p. B8.

76. Haroon Siddiqui, "CanWest censorship shameful," *Toronto Star*, March 10, 2002, p. A13.

77. Tim Harper, "Reporters protest at Asper paper: Journalists complain of 'censorship' after story rewritten," *Toronto Star*, March 7, 2002, p. A2.

78. Ibid.

79. Cal Millar, "Reporters penalized over protest," *Toronto Star*, March 10, 2002, p. A3.

80. Macklem, "The view from the center."

81. Keith Damsell, "CanWest accused of more meddling," *Globe and Mail*, March 8, 2002, p. B3.

82. See Rasha Mourtada, "Haroon and the sea of opinions," *Ryerson Review of Journalism*, Spring 2002.

83. Damsell, "CanWest accused of more meddling."

84. Millar, "Reporters penalized over protest."

85. Murdoch Davis, "Southam takes issue with *Star* editor," *Toronto Star*, March 13, 2002, p. A25.

86. Quoted in Canadian Journalists for Free Expression, "Not in the newsroom!: CanWest Global, chain editorials and freedom of expression in

Canada," April 2002. Availaible online at: http://www.cjfe.org/specials/canwest/canwintro.html

87. Warren Allmand, et al., "Let's press for press freedom," *Globe and Mail*, April 19, 2002 , p. A15.

88. Antonia Zerbisias, "Inquiry sought into media ownership," *Toronto Star*, April 14, 2002, p. A3.

89. Robert Fulford, "'Globe 53' don't know the meaning of freedom," *National Post*, April 22, 2002, p. A18.

90. Jane Gadd, "Journalists appeal to CanWest for truce," *Globe and Mail*, April 15, 2002, p. A2.

91. Canadian Journalists for Free Expression, "Not in the newsroom!"

92. Campbell Clark, "Senators push for media hearings," *Globe and Mail*, April 24, 2002, p. A4.

93. Keith Damsell and John Partridge, "CanWest's editorial problem regrettable, Black says," *Globe and Mail*, March 1, 2002, p. B3.

94. Ken Goldstein, CanWest defends itself," *Globe and Mail*, April 26, 2002, p. A19.

95. Raymond Heard, "Setting the facts straight: Critics of CanWest continue to overestimate the newspaper chain's size and influence," *Montreal Gazette*, April 25, 2002. p. B3.

Chapter 8: L'Affaire Russell Mills

1. Graham N. Green, "Double standard," *Ottawa Citizen*, June 1, 2002, p. B1.

2. "Time To Go: Chrétien's own behaviour inspired government rot. He must leave," *Ottawa Citizen*, June 1, 2002, p. B6.

3. "Journalistic responsibility," *Ottawa Citizen*, June 3, 2002, p. A14.

4. "Leadership and succession," *National Post*, June 4, 2002, p. A18.

5. "Vision and issues, please, not ego and mudslinging," *Vancouver Sun*, June 4, 2002, p. A14.

6. "He had no choice," *National Post*, June 4, 2002, p. A18.

7. Clark Davey, "A handy reader's guide to the politics of editorial writing: CanWest's promise of independence has not been kept," *Ottawa Citizen*, June 5, 2002, p. A17.

8. Nicholas Keung, "Ad critical of CanWest policy: 'Diversity of voices' lost, it says," *Toronto Star*, June 6, 2002, p. A3.

9. William E. Ardell, et. al., "Is freedom of the press being lost, one newsroom at a time? [advertisement]," *Globe and Mail*, June 6, 2002, p. A8.

10. Keith Damsell, "CanWest editorial policy blasted in ad," *Globe and Mail*, June 6, 2002, p. A6.

11. Keung, "Ad critical of CanWest policy."

12. Norman Webster, "An issue of public trust," *Montreal Gazette*, June 29, 2002, p. B7.

13. Russell Mills, "'Our job is to report the news, not make it,'" *Vancouver Sun*, June 15, 2002, p. A23.

14. "Group demands Mills' reinstatement: Rally of support outside Citizen offices," *Ottawa Citizen*, June 18, 2002, p. A6.

15. "Mills leaves Citizen," *Ottawa Citizen*, June 17, 2002, p. A2.

16. "Group demands Mills' reinstatement."

17. Tim Harper and Graham Fraser, "Publisher's ouster sparks an uproar: *Citizen* publisher says he was fired for editorial," *Toronto Star*, June 18, 2002, p. A1.

18. Keith Damsell, "Aspers feel the ire of Southam readers," *Globe and Mail*, June 22, 2002, p. A2.

19. Keith Damsell, "CanWest announces corporate shakeup," *Globe and Mail*, June 18, 2002, p. A4.

20. Gordon Pitts, "CanWest posts loss on ad woes: Results also reflect full inclusion of Post," *Globe and Mail*, April 26, 2002, p. B2.

21. Keith Damsell, "CanWest Global stock hits a six-year low: Controversial firing sparks investor fears," *Globe and Mail*, June 20, 2002, p. B2.

22. Keith Damsell, "CanWest takes a beating: A public relations nightmare, rising debt and a labour dispute have hurt the shares," *Globe and Mail*, July 10, 2002, p. B13.

23. Fabrice Taylor, "Cruel and unusual CanWest financing," *Globe and Mail*, July 10, 2002, p. B10.

24. Rob Ferguson, "CanWest Global sells east coast newspapers — Pressing need for capital overrides convergence strategy," *Toronto Star*, July 11, 2002, p. C12.

25. Chris Powell, "Putting it all together: Canada's media moguls are learning to assemble their diverse properties in advertiser-pleasing ways," *Marketing*, February 25, 2002, p. 15.

26. Leonard Asper, "Convergence game is on [letter to the editor]," *Marketing*, March 18, 2002, p. 23.

27. Shawn McCarthy and Jeff Sallot, "Publisher's firing sets off storm," *Globe and Mail*, June 18, 2002, p. A1.

28. Tonda MacCharles, "Chrétien shrugs off fears of media chill: Concentration of ownership not a concern, PM says," *Toronto Star*, June 19, 2002, p. A2.

29. Shawn McCarthy, "Government urged to expand mandate of CRTC to print," *Globe and Mail*, June 19, 2002, p. A4.

30. Graham Fraser and Tonda MacCharles, "Senators want media holdings inquiry: Publisher's firing by CanWest fuels diversity concern," *Toronto Star*, June 20, 2002, p. A6.

31. Webster, "An issue of public trust."

32. Kim Lunman and Shawn McCarthy, "Sacking of Ottawa publisher draws international criticism," *Globe and Mail*, June 19, 2002, p. A1.

33. Jeffrey Simpson, "What the Mills affair reveals about the PM, Aspers," *Globe and Mail*, June 18, 2002, p. A17.

34. Richard Gwyn, "Asper damages papers' credibility," *Toronto Star*, June 19, 2002, p. A25.

35. Margaret Wente, "Out of their depth, and sinking like a stone," *Globe and Mail*, June 20, 2002, p. A17.

36. Andrew Coyne, "Red herrings and red flags," *National Post*, June 20, 2002, p. A23.

37. Russell Mills, "Under the Asper thumb," *Globe and Mail*, June 19, 2002, p. A17.

38. Ibid.

39. Dirk Meissner, "CanWest issues memo pledging editorial integrity: Says journalists need not feel intimidated following Mills' firing," *Toronto Star*, June 22, 2002, p. A14.

40. Barbara Shecter, "CEO denies Chrétien had role in firing *Ottawa Citizen* publisher: 'The last people we listen to is politicians,' Leonard Asper says," *National Post*, June 21, 2002, p. A2.

41. Jennifer Wells, "Asper denies publisher's firing linked to Chrétien: CanWest CEO cites newspaper's lack of 'diversity,'" *Toronto Star*, June 21, 2002, p. A6.
42. Ibid.
43. "Asper promises a free press: Citizen's reporting violated journalistic principles, CanWest president says," *Ottawa Citizen*, June 21, 2002. p. A1.
44. "An interview with Leonard Asper," *Ottawa Citizen*. June 21, 2002, p. A19.
45. "Asper promises a free press."
46. Caroline Mallan, "Ex-publisher says Aspers libelled him: Fired *Citizen* chief wants apology and retraction from CanWest or he'll sue," *Toronto Star*, June 26, 2002, p. A8.
47. Russell Mills, "Pressing for freedom," *Time* (Canadian edition), July 1, 2002, p. 47.
48. Ibid.
49. Mills, "Pressing for freedom."
50. David Crane, "Media concentration requires public probing," *Toronto Star*, September 19, 2000, p. E2.
51. David Beers, "Caught in the Sun," *Vancouver*, June 2002, p. 22.
52. Canada, Standing Senate Committee on Transportations and Communications, "Interim report on the Canadian news media," April 2004, p. 37.
53. See http://www.communitymediacanada.ca/calculator/
54. See Marc Edge, *Pacific Press: The Unauthorized Story of Vancouver's Newspaper Monopoly.* (Vancouver: New Star Books), 2001, p. 365.
55. Canada, Competition Bureau, "The Competition Bureau's work in media industries: Background for the Senate Committee on Transport and Communications," April 2, 2004, pp. 16-17.
56. Terry Glavin, "CanWest's agenda should worry Canadians," *Winnipeg Free Press*, June 16, 2002, p. B4.
57. Ibid.
58. Paul Orlowski, "Educating in an era of Orwellian spin: Critical media literacy in the classroom," *Canadian Journal of Education*, 2006, p. 192.
59. Michael Smyth, e-mail to the author, May 10, 2007.
60. Quoted in Noel Hulsman, "What makes the *Vancouver Sun* tick?" *BC Business*, November 2003, p. 40.
61. Kevin Groves, Cecilia Jamasmie, Jock Paul and Lisa Johnson, "Still can't catch corporate crooks," *UBC Thunderbird Online Journalism Review*, January 2003. Available online at http://www.journalism.ubc.ca/thunderbird/archives/2002.12/alert.html.
62. See Richard Littlemore, "The most hated man in business," *BC Business*, July 2005, pp. 45-55.
63. Quoted in Hulsman, "What makes the *Vancouver Sun* tick?"
64. Gordon Gibson, "The prime minister will be brought down," *National Post*, June 3, 2002, p. A14.
65. Gordon Gibson, "'The damned thing just doesn't work,'" *Vancouver Sun*, June 5, 2002, p. A15.
66. Lunman and McCarthy, "Sacking of Ottawa publisher draws international criticism."
67. Damsell, "Aspers feel the ire of Southam readers."
68. Allan Fotheringham, "Confusion and despair at the top," *Globe and*

Mail, July 27, 2002, p. A2.

69. Rafe Mair, "Rafe Mair's June 27th editorial on Canwest Global," www.cknw.com. Downloaded August 5, 2002.

70. Ibid.

71. Charlie Smith, "Mair fights CanWest," *Georgia Straight*, July 4, 2002, p. 11.

72. Rafe Mair, "Rafe Mair's open letter to the editor in chief of the *Vancouver Sun*," www.cknw.com. Downloaded August 5, 2002.

73. Donna Logan, Interview with Rafe Mair, CKNW Radio, June 28, 2002.

74. Ibid.

75. Ibid.

76. See Marc Edge, "Balancing academic and corporate interests in Canadian journalism education," *Journalism and Mass Communication Educator,* Summer 2004, pp. 172-185.

77. Andrea Lobo, "J-school on Firm Ground," *Ubyssey*, November 21, 2000.

78. Natasha Norbjerg, "Journalism Students Start Student Union: School Re-naming a Factor," *Ubyssey*, February 16, 2001.

79. Marc Edge, "Balancing academic and corporate interests in Canadian journalism education," pp. 172–185.

80. Claire Hoy, "CRTC's Appointed Power Hacks," *Hill Times*, May 14, 2001, p. 6.

Chapter 9: Dishonest Reporting

1. Robert Bertuzzi "Mac honours media censor," *The Silhouette*, November 21, 2002.

2. Paul Lungen, "Izzy Asper: Philanthropist, renaissance man," *Canadian Jewish News*, October 16, 2003, p. 1.

3. Myron Love, "Human rights museum downplays funding rift," *Canadian Jewish News*, August 19, 2004, p. 28.

4. John Saunders, "CanWest campaign against CBC condemned," *Globe and Mail*, February 21, 2002, p. A8.

5. Antonia Zerbisias, "CanWest's anti-CBC campaign gets louder," *Toronto Star*, February 22, 2002, p. 10.

6. I. H. Asper , "So who's watching the CBC?: Nothing 'essential' about it," *National Post*, February 27, 2002, p. A17.

7. Izzy Asper, "CBC: No news, none of the time," *Ottawa Citizen*, May 22, 1999, p. B7.

8. Peter Mansbridge, "Izzy: He was never shy," *Maclean's*, October 20, 2003, p. 18.

9. Jennifer Wells, "Izzy's dream," *Maclean's*, February 19, 1996, p. 40.

10. Oakland Ross, "Asper's school slogan: 'Izzy clever . . . Izzy ever': Media giant had what it takes to win even as a teen," *Toronto Star*, August 5, 2000, p. A1.

11. Lungen, "Izzy Asper: philanthropist, renaissance man."

12. I.H. Asper, "Canada's pro-Palestinian, anti-Israel policy is shameful," *National Post*, July 5, 2001, p. 7.

13. Israel Asper, "Business school to encourage Israeli prosperity," *Calgary Herald*, June 23, 2001, p. OS8.

14. Kim Lunman, "Asper compares protesters to Nazis," *Globe and Mail*, September 11, 2002, p. A14.

15. John Doyle, "Global doc a misfire in tin-pot 'media war,'" *Globe and*

Mail, July 2, 2003. p. R2.
16. Antonia Zerbisias, "Global documentary seems unfair," *Toronto Star*, June 29, 2003, p. D10.
17. I.H. Asper, "We must end media bias against Israel," *National Post*, October 31, 2002, p. A22.
18. Ibid.
19. Ibid.
20. Israel Asper, "Media integrity lost in battle between Palestinians and Israelis," *Canadian Speeches*, January/February 2003, p. 28.
21. Ibid.
22. Doug Saunders, "Asper's charges of media bias 'bizarre,'" *Globe and Mail*, November 1, 2002, p. A12.
23. Tony Burman, "Asper should cover Israel, not lecture," *Globe and Mail*, November 9, 2002, p. A21.
24. Mike Gasher, "Where's the evidence, Mr. Asper?" *Globe and Mail*, November 7, 2002, p. A23.
25. Hugh Winsor, "We'd like to hear Russell Mills do the talking," *Globe and Mail*, December 18, 2002, p. A6.
26. David Estok, "Aspers' views won't disappear, but just take on subtle guise,' *Globe and Mail*, October 2, 2002, p. B11.
27. Keith Damsell, "CanWest setting up news centre in Winnipeg," *Globe and Mail*, November 22, 2002, p. B2.
28. Keith Damsell, "CanWest set to launch news hub," *Globe and Mail*, January 20, 2003, p. B2.
29. Stephen Kimber "Canada's lax rules enable CanWest domination," *Adbusters*, September/October 2003, p. 16.
30. Russell Mills, "CanWest plan a threat to local independence," *Globe and Mail*, January 22, 2003, p. B11.
31. Ibid.
32. Gordon Fisher, "Russell Mills, rest assured there is no CanWest bogey-man," *Globe and Mail*, January 29, 2003, p. B16.
33. Rob Ferguson "Two top executives leave CanWest — Newspapers drop Southam name," *Toronto Star*, January 29, 2003, p. E1.
34. "CanWest Global shares fall 16% to 52-week low," *Toronto Star*, October 5, 2002, p. E2.
35. David Paddon, "CanWest sells 4 small papers for $194 million," *Toronto Star*, January 28, 2003, p. C1.
36. Antonia Zerbisias, "CanWest slashing critics' beats — Film, TV, rock to be consolidated: Writers fear other positions next," *Toronto Star*, March 19, 2003, p. A4.
37. Tracy Tjaden, "*Vancouver Sun* cuts create concerns," *Business in Vancouver*, June 10, 2003, p. 3.
38. Keith Damsell, "Ads' placement in CanWest papers raises ire," *Globe and Mail*, April 4, 2003, p. B10.
39. Noel Hulsman, "What makes the *Vancouver Sun* tick?" *BC Business*, November 2003, p. 40.
40. Quoted in Tjaden, "*Vancouver Sun* cuts create concerns."
41. Quoted in Hulsman, "What makes the *Vancouver Sun* tick?"
42. Hulsman, "What makes the *Vancouver Sun* tick?"
43. Richard Littlemore, "The most hated man in business," *BC Business*, July 2005, p. 45.

44. Rod Mickleburgh, "Radio legend Rafe Mair fired," *Globe and Mail,* June 10, 2003, p. A1.

45. Charlie Smith, "Mair Crosses Corporate Boss," *Georgia Straight,* June 12, 2003, p. 15.

46. Graham Fraser, "Media expert criticizes cross-ownership: Prohibit convergence of TV, newspapers, Tom Kent tells Senate," *Toronto Star,* April 30, 2003, p. A19.

47. Russell Mills, "Reflections on the state of Canadian media," *Canadian Parliamentary Review,* Winter 2003, p. 27.

48. Terence Corcoran, "Media owners and freedom of the press," *National Post,* June 8, 2002. p. FP11.

49. Canada, Royal Commission on Newspapers, *Report.* (Ottawa: Queen's Printer), 1981, p. 1.

50. Terence Corcoran, "Media circus III," *National Post,* April 29, 2003, p. FP15.

51. Ibid.

52. Matthew Fraser, "Izzy Asper's Love Boat still cruising: CRTC protection has helped Global sail to profits," *National Post,* February 2, 1999, p. C7.

53. Ibid.

54. Bruce Leslie, "We play by the rules, says Global [letter to the editor]," *National Post,* February 8, 1999, p. C5.

55. Matthew Fraser, "Telemedia's conversion to continental culture," *National Post,* October 15, 1999, p. C7.

56. Leonard Asper, "CanWest isn't Ottawa's champion [letter to the editor]," *National Post,* October 26, 1999, p. C7.

57. Matthew Fraser, *Free For All: The Struggle for Dominance on the Digital Frontier.* (Toronto: Stoddart), 1999, p. 161.

58. Adrienne Macintosh, "Which way did he go?" *Ryerson Review of Journalism,* Spring 2004, pp. 19-20.

59. Matthew Fraser, "Beyond name-calling," *National Post,* May 15, 2001, p. A14.

60. Macintosh, "Which way did he go?" p. 20.

61. Antonia Zerbisias, "Missed opportunities in Asper interview: CanWest mogul is not sufficiently challenged," *Toronto Star,* February 13, 2002, p. F7.

62. Quoted in Macintosh, "Which way did he go?" p. 20.

63. Quoted in Ibid.

64. Matthew Fraser, "A man of big appetites, bigger heart," *National Post,* October 8, 2003, p. A9.

65. Ibid.

66. Rob Ferguson and Madhavi Acharya-Tom Yew, "Top editors, publisher ousted at *National Post*: Asper labels opposition 'axis of snivel,'" *Toronto Star,* May 2, 2003. p. A1.

67. Barbara Shecter, "CanWest commits to Post, replaces top managers," *National Post,* May 2, 2003, p. A2.

68. Antonia Zerbisias, "I'm happy for you, Matthew," *Toronto Star,* May 4, 2003, p. D10.

69. Matthew Fraser, "The *National Post* is here to stay: Owners, new editor have three-year plan for profit," *National Post,* May 2, 2003, A1.

70. Cobb, *Ego and Ink,* p. 323.

71. Ibid., p. 320.
72. Macintosh, "Which way did he go?" p. 21.
73. Kate Jaimet, "TV protectionists score win at Commons heritage committee: Keep content restrictions," *National Post*, June 10, 2003, p. FP6.
74. Ian Jack, "Give Canadian TV more cash, MPs say: Ottawa told to make decision on cross-media ownership," *National Post*, June 11, 2003, p. FP5.
75. David Taras, "Does Canadian TV have a future?" *Winnipeg Free Press*, July 6, 2003, p. B4.
76. Canada, Standing Committee on Canadian Heritage, *Our cultural sovereignty: The second century of Canadian broadcasting.* (Ottawa: Queen's Printer), 2003, p. 405.
77. Ibid., p. 411.
78. Antonia Zerbisias, "Heritage report buried," *Toronto Star*, July 10, 2003, p. A23.
79. Sean Davidson, "Report? What report? Critics warn that media concentration is threatening debate and broadcast reform," *Playback*, January 5, 2004, p. 1.
80. Arnold Amber, "Heritage Committee asks for freeze on additional cross-media ownership," *TNG Canada*, June 12, 2003. Available online at http://www.tngcanada.org/EN/news/2003/030612_heritage_report. shtml.
81. Patricia Graham, e-mail to the author, March 27, 2004.
82. I.H. Asper, "What we can do for Israel," *Ottawa Citizen*, July 30, 2003, p. A17.
83. Mazen Chouaib, "The problem: Israeli occupation," *National Post*, August 18, 2003, p. A12.
84. Israel Asper, "The problem: Arab intolerance," *National Post*, August 18, 2003, p. A12.
85. Melissa Radler, "Mogul with a message," *Jerusalem Post*, August 15, 2003, p. 15.
86. Russell Mills, "Judging Mr. Asper," *Media*, Fall/Winter 2003, p. 8.
87. Jonathon Gatehouse, "The risk taker," *Maclean's*, October 20, 2003, p. 40.
88. John Bugailiskis, "Izzy Asper: broadcaster, visionary, philanthropist, Canadian," *Broadcaster*, October 2003, p. 24.

Chapter 10. Like Father, Like Children

1. Jennifer Wells, "Izzy's dream," *Maclean's*, February 19, 1996, p. 40.
2. Katherine Macklem, "Can the Aspers do it?" *Maclean's*, April 8, 2002, p. 48.
3. Ibid.
4. Ric Dolphin, "Giants in a town they love to call home: The Asper family — Winnipeg's biggest boosters," *Calgary Herald*, May 10, 2002, p. A3.
5. Naomi Lakritz, "Leonard Asper wants to make a difference: The boss of 149 newspapers learned lessons about business with a lemonade stand when he was four," *Calgary Herald*, September 23, 2000, p. O8.
6. Trevor Cole, "Lenny: A mogul in the making," *Report on Business* Magazine, October 2000, p. 69.
7. Anthony Wilson-Smith, "Power moves from Toronto," *Maclean's*, August 21, 2000, p. 11.
8. John Geddes, "The Izzy and Leonard show," *Maclean's*, August 14, 2000, p. 14.

9. Ibid.
10. Cole, "Lenny," p. 70.
11. Ibid.
12. Geddes, "The Izzy and Leonard show."
13. David Berman, "Life with father," *Canadian Business*, September 1995, p. 50.
14. Dolphin, "Giants in a town they love to call home."
15. Gordon Pitts, *Kings of Convergence: The Fight for Control of Canada's Media*. (Toronto: Doubleday), 2002, p. 261.
16. Wells, "Izzy's dream."
17. David Olive, "Why Asper is waging a fake war with his scribes," *Toronto Star*, June 19, 2002, p. E1.
18. Pitts, *Kings of Convergence*, p. 258.
19. Tony Atherton, "Izzy Asper's national dream," *Ottawa Citizen*, July 13, 1996, p. E1.
20. Pitts, *Kings of Convergence*, p. 261.
21. Trevor Cole, "Revenge of the ratings king: How did CanWest Global evolve into the powerhouse that's eating CTV's lunch?" *Report on Business* Magazine, May 1996, p. 58.
22. Pitts, *Kings of Convergence*, p. 266.
23. Cole, "Lenny," p. 70.
24. Pitts, *Kings of Convergence*, p. 263.
25. Kevin Dawson, "The TV empire of Israel Asper," *Dublin Sunday Business Post*, January 3, 1999, p. 18.
26. Gay Abbate, "Former Post reporter sues Asper over firing," *Globe and Mail*, April 2, 2004, p. A8.
27. "Asper quits Bombers board," *Regina Leader-Post*, September 16, 2005, p. C2.
28. David Berman, "Channel changer: Izzy Asper has quietly built Canada's most ambitious multinational broadcasting company. The question is: who's preparing to take over from Asper?" *Canadian Business*, September 1995, p. 46.
29. Ed Pearce, "'Izzy' Asper: Going Global," *Ivey Business Journal*, November 2001, p. 26.
30. Ibid.
31. Berman, "Channel changer."
32. Rod McQueen, "Izzy Asper steps away from the fray: The business legend reflects on knowing when it's time to quit," *National Post*, April 15, 2000, p. D2.
33. David Olive, "Media moguls: It's in the genes," *Toronto Star*, February 22, 2003, p. F1.
34. John Geddes, "No asper-sions . . .," *Maclean's*, August 21, 2000, p. 8.
35. Ibid.
36. Ibid.
37. Gail Asper, "Charitable giving," *Globe and Mail*, July 2, 1997, p. B2.
38. Dolphin, "Giants in a town they love to call home."
39. Myron Love, "Community should respond to anti-Israel reporting: Asper," *Canadian Jewish News*, February 21, 2002, p. 31.
40. Donald Mcarthur, "Justice system tarnished by reporters' spin: Asper," *Ottawa Citizen*, June 12, 2005, p. A5.
41. David Asper, "This is not Conrad Black's last stand," *National Post*, Sep-

tember 16, 2003, p. A17.

42. Richard Reynolds, "How CanWest was won," *Sydney Morning Herald*, February 26, 2005. Downloaded from http://www.smh.com.au/news/Business/How-CanWest-was-won/2005/02/25/1109180109483.html.

43. John Gray, "The best & worst boards in Canada," *Canadian Business*, August 19, 2002, p. 30.

44. Pitts, *Kings of Convergence*, p. 249.

45. Olive, "Media moguls."

46. Krista Foss, "CanWest presses Ottawa on media legislation," *Globe and Mail*, March 2, 2002, p. B4.

47. Rob Ferguson and Madhavi Acharya-Tom Yew, "Top editors, publisher ousted at *National Post*: Asper labels opposition 'axis of snivel,'" *Toronto Star*, May 2, 2003. p. A1.

48. Leonard Asper, "Canada could do much better," *National Post*, January 17, 2001, p. A.16.

49. Hugh Winsor, "We'd like to hear Russell Mills do the talking," *Globe and Mail*, December 18, 2002, p. A6.

50. Maria Babbage, "Asper attacks CanWest critics," *Kingston Whig-Standard*, December 18, 2002, p. 20.

51. Graham Fraser, "Asper predicts 'advertainment,' user fees in the media of future: Says 'free ride'of reading papers on Internet will end," *Toronto Star*, December 18, 2002. p. A30.

52. James Baxter, "CanWest says size matters," *Calgary Herald*, December 18, 2002, p. C1.

53. Ibid.

54. Antonia Zerbisias, "Hard to feel sorry for broadcasters," *Toronto Star*, February 5, 2004, p. A19.

55. Grant Robertson, "Broadcasters score victory over Ottawa: Feds may have to refund $790-million after judge rules licence fees are illegal," *Globe and Mail*, December 15, 2006, p. B1.

56. Ian Jack, "CanWest CEO urges end to regulatory walls: CRTC licensing process is fragmenting audience, reducing revenues, Asper says," *National Post*, December 18, 2002, p. FP8 .

57. Richard Blackwell, "Asper wants Martin to back foreign control," *Globe and Mail*, November 14, 2003, p. B3.

58. Joe Paraskevas, "CBC should get out of local news, sports, Asper says: Stick to money-losing culture and drama, CanWest boss says," *Vancouver Sun*, March 18, 2003, p. C4 .

59. Leonard Asper, "Media bias and the Middle East," *National Post*, October 1, 2003, p. A16.

60. Ibid.

61. Ibid.

62. Ibid.

63. "CBC fails to get point, Asper says," *National Post*, October 3, 2003, p. A12.

64. The *National Post* editorial was published on June 5, not May 16. See "Neil Macdonald never quits," *National Post*, June 5, 2003, p. A19.

65. Neil Macdonald, "Mr. Asper, you owe me an apology," *Globe and Mail*, October 3, 2003, p. A29.

66. Ibid.

67. Antonia Zerbisias, "Curious silence greets discredited Hezbollah tale,"

Toronto Star, December 13, 2002, p. A37.
68. Christopher Dornan, "Ideology and idiocy," *Globe and Mail*, October 4, 2003, p. A23.
69. John Saunders, "Leonard Asper now biggest wheel," *Globe and Mail*. October 8, 2003, p. B7.
70. Pitts, *Kings of Convergence*, p. 250.
71. Paul Lungen, "Leonard Asper defends CanWest vision," *Canadian Jewish News*, January 3, 2002, p. 6.
72. See Hugh Miles, *Al-Jazeera: How Arab TV News Challenged the World*. (London: Abacus), 2005.
73. Colby Cosh, "Canada's cultural hypocrisy," *National Post*, July 16, 2004, p. A1.
74. James Gordon, "Canadians demand access to controversial, right-wing Fox News," *Vancouver Sun*, September 4, 2004, p. A4.
75. Antonia Zerbisias, "Fox News finally free of regulatory snares: Misinformation station OK'd at last," *Toronto Star*, November 19, 2004, p. C13.
76. James Gordon, "Feisty Fox News allowed in Canada," *Calgary Herald*, November 19, 2004, p. A4.
77. Antonia Zerbisias, "Who will protect us from Fox?: Censoring Al-Jazeera a double standard," *Toronto Star*, Jul 20, 2004, p. D5.
78. John Miller, "Which do you trust less: Al-Jazeera or Fox News?" *Globe and Mail*, March 26, 2005, p. D11.
79. David Olive, "Media moguls: It's in the genes," *Toronto Star*, February 22, 2003, p. F1.
80. David Dias, "Leonard Asper's master plan for Global domination," *Ryerson Review of Journalism*, Summer 2001, p. 42.
81. Tamara Gignac, "CanWest plans merger of newspaper, broadcast content," *Vancouver Sun*, February 5, 2004, p. F10.
82. Michelle DaCruz, "CanWest reaffirms convergence plan: Horizontal strategy key, Camilleri says," *National Post*, March 25, 2004, p. FP7.
83. Aaron Wherry, "Global's got him: Mike Bullard leaves CTV after six seasons," *National Post*, August 7, 2003, p. AL1.
84. Aaron Wherry, "Global gasses Mike Bullard for poor ratings," *Edmonton Journal*, March 13, 2004, p. E1.
85. Tony Van Alphen, "CanWest restructures in convergence push: Hires group of top U.S. executives," *Toronto Star*, October 5, 2004, p. D1.
86. Richard Blackwell, "CanWest shakes up its ranks with U.S. recruits," *Globe and Mail*, October 5, 2004, p. B1.
87. Paul Brent, "CanWest names Viner head of media unit," *National Post*, June 2, 2005, p. FP2.

Chapter 11: Media Bias

1. Mark D. Watts, David Domke, Dhavan V. Shah, and David P. Fan, "The politics of conservative elites and the liberal media argument," *Journal of Communication*, 1999, p. 54.
2. David Niven, *Tilt? The Search for Media Bias*. (New York: Praeger), 2002, p. 5.
3. Rick Perlstein, "Eyes right: Conservatives are winning the media war. How do they do it?" *Columbia Journalism Review*, March/April 2003, p. 52.
4. See Eric Alterman, *What Liberal Media? The Truth About Bias and the News*. (New York: Basic Books), 2003; Brian C. Anderson, *South Park Con-*

servatives: The Revolt Against Liberal Media Bias. (Washington: Regnery), 2005; Eric Boehlert, *Lapdogs: How the Press Rolled Over for Bush.* (New York: Free Press), 2006; L. Brent Bozell III, *Weapons of Mass Distortion: The Coming Meltdown of the Liberal Media.* (New York: Crown), 2004; David Brock, *The Republican Noise Machine: Right-Wing Media and How It Corrupts Democracy.* (New York: Crown), 2004; Jim A. Kuypers, *Bush's War: Media Bias and Justifications for War in a Terrorist Age.* (Lanham, MD: Rowman & Littlefield), 2006; Jim A. Kuypers, *Press Bias and Politics: How the Media Frame Controversial Issues.* (New York: Praeger), 2002.

5. Bernard Goldberg, *Bias: A CBS Insider Exposes How the Media Distort the News.* (New York: Regnery), 2001, pp. 77-78.

6. Martin Arnold, "Best seller lists get a new job," *New York Times,* January 10, 2002, p. E3.

7. See John W. C. Johnstone, Edward J. Salawski, and William W. Bowman, *The News People: A Sociological Portrait of American Journalists and Their Work.* (Urbana: University of Illinois Press), 1971; David H. Weaver and G. Cleveland Wilhoit, *The American Journalist: A Portrait of US News People and Their Work.* (Bloomington: Indiana University Press), 1991.

8. See Leo C. Rosten, *The Washington Correspondents.* (New York: Harcourt Brace), 1937.

9. See William Rivers, "The Correspondents after 25 years," *Columbia Journalism Review,* Summer 1962, pp. 4-10.

10. See Edith Efron, *The News Twisters.* (New York: Manor), 1972.

11. See Robert L. Stevenson, et al. "Untwisting 'The news twisters': A replication of Efron's study," *Journalism Quarterly,* Summer 1973, pp. 211-19.

12. See S. Robert Lichter and Stanley Rothman, "Media and business elites," *Public Opinion,* 1981, pp. 42-46, 59-60; S. Robert Lichter, Stanley Rothman, and Linda S. Lichter, *The Media Elite: America's New Power-brokers.* (New York: Hastings House), 1986.

13. Herbert J. Gans, "Are U.S. journalists dangerously liberal?" *Columbia Journalism Review,* November/December 1985, p. 31.

14. See J. Herbert Altschull, *Agents of Power: The Role of the News Media In Human Affairs.* (New York: Longman), 1984; Edward S. Herman and Noam Chomsky, *Manufacturing Consent: The Political Economy of the Mass Media.* (New York: Pantheon), 1988.

15. See Peter Phillips and Project Censored, *Censored 2007: The Top 25 Censored Stories.* 30th Anniversary Ed. (New York: Seven Stories Press), 2006; Robert A. Hackett and Scott Uzelman, "Tracing corporate influences on press content: A summary of recent NewsWatch Canada research," *Journalism Studies,* 2003, pp. 331-346.

16. See Pamela J. Shoemaker and Stephen D. Reese, *Mediating the Message: Theories of Influences on Mass Media Content.* 2nd ed. (New York: Longman), 1996.

17. See R.P. Vallone, L. Ross, &, M. R. Lepper, "The hostile media phenomenon: Biased perception and perceptions of media bias in coverage of the 'Beirut massacre,'" *Journal of Personality and Social Psychology,* 1985, pp. 577-585.

18. See Albert C. Gunther and Kathleen Schmitt, "Mapping boundaries of the hostile media effect," *Journal of Communication,* March 2004, pp. 55-70.

19. See Joseph Turow, *Breaking up America: Advertisers and the New Media*

World. (Chicago: University of Chicago Press), 1997.

20. Paul Farhi, "Good news for 'McPaper': After 15 years it climbs from money pit to profitability," *American Editor*, September 1997, p. 8.

21. Kathleen Cross, "Off balance: How the Fraser Institute slants its news-monitoring studies," *Canadian Forum*, October 1997, p. 10.

22. Tony Van Alphen, "Globe, CBC portray free trade in negative light, institute says," *Toronto Star*, October 13, 1988, p. D6.

23. Antonia Zerbisias, "News watchdog learns how to turn a blind eye," *Toronto Star*, January 28, 1994, p. A23.

24. Lydia Miljan and Barry Cooper, *Hidden Agendas: How Journalists Influence the News.* (Vancouver: UBC Press), 2003, p. 171.

25. Ibid., p. 175.

26. Ibid.

27. Ibid., p. 25.

28. Ibid., p. 10.

29. Ibid., p. 176.

30. Walter C. Soderlund and Walter I. Romanow, "The firing of Russell Mills, round two of national editorials, and the CanWest News Service initiative," in Soderlund and Hildebrandt, eds., *Canadian Newspaper Ownership in the Era of Convergence: Rediscovering Social Responsibility.* (Edmonton: University of Alberta Press), 2005, p. 130.

31. "Clear choice on Iraq," *Edmonton Journal*, March 11, 2003, p. A12.

32. Patricia Pearson, "See no evil, no more," *Globe and Mail*, April 19, 2003, p. A19.

33. Ibid.

34. Lawrence Martin, "It's not Canadians who've gone to the right, just their media," *Globe and Mail*, January 23, 2003, p. A19.

35. James Adams, "CanWest editing questioned: Reuters wants reporters' names removed when the word 'terrorist' added to stories," *Globe and Mail*, September 18, 2004, p. A9.

36. Hamza Hendawi, "U.S. attacks fail to weaken terrorists' grip on Fallujah," *Ottawa Citizen*, September 9, 2004, p. A9.

37. Steven Moore, "One man's terrorist . . . Re: U.S. attacks fail to weaken terrorists' grip on Fallujah, Sept. 9 [letter to the editor]," *Ottawa Citizen*, September 17, 2004, p. A15.

38. "Corrections," *Ottawa Citizen*, September 17, 2004, p. A2.

39. Ian Austen, "Reutrers asks a chain to remove its bylines," *New York Times*, September 20, 2004, p. C9.

40. Mazen Chouaib, "CanWest: Don't vilify Muslims," *Globe and Mail*, September 22, 2004, p. A17.

41. Scott Anderson, "One man's terrorist . . .," *Globe and Mail*, September 23, 2004, p. A18.

42. "Calling a terrorist a terrorist," *National Post*, September 18, 2004, p. A19.

43. Kelly McParland, "Call them the terrorists they are: The CBC may have a problem with that, but I don't," *National Post*, September 25, 2004, p. A11.

44. Robert Fulford, "Too much politeness, too little candour," *National Post*, September 25, 2004, p. A18.

45. See Marc Edge, "Cross ownership," in Wolfgang Donsbach, ed. *International Encyclopedia of Communication.* (Oxford: Blackwell). Forthcoming.

46. Gillian Doyle, *Media ownership: The Economics and Politics of Convergence in the UK and European Media*. (London: Sage), 2002, p. 26.

47. Powell's father Colin was U.S. Secretary of State from 2001–05 and appeared before the United Nations security council in 2003 to urge the invasion of Iraq.

48. Charles C. Layton, "News blackout," *American Journalism Review*, December 2003, p. 18.

49. Gal Beckerman, "Tripping up big media," *Columbia Journalism Review*, November/December 2003, p. 15.

50. See Robert W. McChesney, *The Problem of the Media: US Communication Politics in the 21st Century*. (New York: Monthly Review Press), 2004.

51. Frank Ahrens, "FCC drops bid to relax media rules," *Washington Post*, January 28, 2005, p. A1.

52. Les Brown, "Small is beautiful is the new mantra," *Television Business International*, April 2005, p. 1.

53. Steve Maich, "Better off without you: Convergence was just a tune the media giants hummed for awhile," *Maclean's*, May 16, 2005, p. 33.

54. Deb Halpern Wenger, "The road to convergence and back again," *Quill*, September 2005, p. 40.

55. Mary Lynn Young, "Extra! Extra! Newspaper war coming to Vancouver," *Globe and Mail*, January 27, 2005, p. B2.

56. Charlie Smith, "CanWest plans fourth daily," *Georgia Straight*, January 29, 2005, p. 12.

57. Michael McCullough, "CanWest buys share in new Metro daily," *Vancouver Sun*, March 15, 2005, p. D5.

58. Mason Wright, "Yipee Tyee," *This Magazine*, November/December 2005, p. 6.

59. Canada, Standing Senate Committee on Transportations and Communications, "Evidence — Unrevised," Vancouver, B.C., Monday, January 31, 2005.

60. Ibid.

61. Ibid.

62. Matthew Burrows, "CanWest controls too much media, Senate hearing told," *Westender*, February 7, 2005, p. 12.

63. Canada, Standing Senate Committee on Transportations and Communications, "Evidence — Unrevised," Vancouver, B.C., Tuesday, February 1, 2005.

54. Ibid.

65. Ibid.

66. Available online at http://www.cmrcccrm.ca/english/research.html.

67. Judy Monchuk, "Canadians believe media easily influenced: Survey," *Kingston Whig-Standard*, June 15, 2004, p. 11.

68. Joel Connelly, "Canada, U.S. can learn from other's press freedom," *Seattle Post-Intelligencer*, May 16, 2005, p. A2.

69. Mitchell Anderson, "Trust us, we're media: The global-warming debate ended long ago, but you wouldn't know it by reading Canadian newspapers," *Georgia Straight*, January 25, 2007, p. 33.

70. Quoted in Richard Warnica, "Campbell 'getting away with murder' on global warming: Tory pundit Spector lashes media for bias," *The Tyee*, February 13, 2007. Available online at http://thetyee.ca/Mediacheck/2007/02/13/GlobalWarming/

71. Commission on Freedom of the Press, *A Free and Responsible Press.* (Chicago: University of Chicago Press), 1947, p. 68.

72. Ibid., p. 51.

73. See Stephen D. Reese and Jane Ballinger, "The roots of a sociology of news," *Journalism and Mass Communication Quarterly*, Winter 2001, p. 649.

74. Warren Breed, "Social control in the news room," *Social Forces*, May 1955, p. 326.

75. Pamela J. Shoemaker and Stephen D. Reese, *Mediating the Message: Theories of Influences on Mass Media Content*, 2nd ed. (White Plains, NY: Longman), 1996, p. 170.

76. Herbert J. Gans, *Deciding What's News: A Study of CBS Evening News, NBC Nightly News, Newsweek and Time.* (New York, Random House), 1979, p. 94.

77. Charles Reina, "The memo," in Kristina Borjessin, ed., *Into the Buzzsaw: Leading Journalists Expose the Myth of a Free Press.* (Amherst, NY: Prometheus), 2002, p. 45.

78. Anthony Wilson-Smith, "Would we lie to you?" *Maclean's*, October 2, 2000, p. 15.

79. Vivian Smith, "In dear Victoria, the best is often free," *Victoria Times Colonist*, July 2, 2006, p. D3.

80. Sean Holman, "Writing coach written out," *Public Eye*, July 7, 2006. Available online at http://www.publiceyeonline.com/archives/001604. html

81. Roy Greenslade, "Odd tale of Canadian columnist's firing," July 16, 2006. Available online at http://blogs.guardian.co.uk/greenslade/2006/07/odd_tale_of_canadian_columnist.html

82. Sean Holman, "Sympathy pains," *Public Eye*, July 20, 2006. Available online at http://www.publiceyeonline.com/archives/001638. html#more

83. "CAJ opposes advertising's growing influence on news coverage," July 23, 2006. Available online at http://micro.newswire.ca/release.cgi?rkey=1407248862&view=42015-0&Start=0

84. Dennis Skulsky, "CanWest MediaWorks statement on the separation of editorial and advertising content," July 24, 2006. Available online at http://www.canwestmediaworks.com/newsroom/viewNews.asp?NewsroomID=349

85. Shannon Rupp, "Veteran journalist fired, rehired by *Times Colonist*," *The Tyee*, July 25, 2006. Available online at http://thetyee.ca/Mediacheck/2006/07/25/VivianSmith/

86. Sean Holman, "Publisher apologizes for putting Smith column on hiatus," *Public Eye*, July 24, 2006. Available online at http://www.publiceyeonline.com/archives/001643.html

87. Rob Wipond, "Advertisers make CanWest Global news go silent," *rabble.ca*, September 14, 2006. Available online at http://www.rabble.ca/arts_media.shtml?sh_itm=a9554214ad4512fb3a91f4b95d304d0c&rXn=1&

88. Andrew Kohut, "Self-censorship: Counting the ways," *Columbia Journalism Review*, May/June 2000, p. 41.

89. Stuart Soroka and Patrick Fournier, "With media ownership, size does matter," *Globe and Mail*, February 12, 2003, p. A17.

90. "Guild serves formal notice against Gazette order," *Globe and Mail*, February 6, 2002, p. A8.

91. Ingrid Peretz, "Arbitrator sides with reporters over byline use," *Globe and Mail*, October 28, 2003, p. A8.

92. Antonia Zerbisias, "Gazette gag pulled from mouths," *Toronto Star*, February 29, 2004, p. D10.

Chapter 12: The Press We Deserve

1. Antonia Zerbisias, "Campaign silent on media issues," *Toronto Star*, June 6, 2004, p. D10.

2. Keith Davey, "Preface," in Canada, *The Uncertain Mirror: Report of the Special Senate Committee on Mass Media, Vol. I.* (Ottawa: Information Canada, 1970), p. vii.

3. Ben Bagdikian, *The Media Monopoly.* (Boston: Beacon Press), 1983, p. 99.

4. *Canada, The Uncertain Mirror: Report of the Special Senate Committee on Mass Media, Vol. I.* (Ottawa: Information Canada), 1970, p. 63.

5. Ibid., p. 47.

6. Ibid., p. 50.

7. Ibid., p. 60.

8. Ben Bagdikian, *The Media Monopoly.* (Boston: Beacon Press), 1983, p. 11.

9. Canada, *The Uncertain Mirror*, p. 4.

10. Keith Davey, *The Rainmaker: A Passion for Politics.* (Toronto: Stoddart), 1986, p. 153.

11. Andy Hoffman and Andrew Willis, "Trust chill forces CanWest to cut its IPO," *Globe and Mail*, October 4, 2005, p. B1.

12. Juliet O'Neill, "RCMP probe exonerates Goodale," *Ottawa Citizen*, February 16, 2007, p. A4.

13. Grant Robertson, "CanWest pays former executive $2.9-million," *Globe and Mail*, December 1, 2005, p. B6.

14. Grant Robertson, "CanWest says rising costs may force cuts," *Globe and Mail*, January 12, 2006, p. B7.

15. Steve Maich, "Bleeding newspapers dry," *Maclean's*, January 30, 2006, p. 22.

16. See Gilbert Cranberg, Randall Bezanson, and John Soloski, *Taking Stock: Journalism and the Publicly Traded Newspaper Company.* (Ames: Iowa State University Press), 2001.

17. Maich, "Bleeding newspapers dry."

18. Grant Robertson, "TV revenue slide hits CanWest profit," *Globe and Mail*, January 13, 2006, p. B3.

19. Grant Robertson, "Analyst questions papers' viability," *Globe and Mail*, January 14, 2006, p. B3.

20. "CanWest to shop its 45% stake in Ireland's TV3," *National Post*, January 17, 2006. p. FP4.

21. Barbara Shecter, "CanWest stands behind Post, Viner says: Indicators looking good," *National Post*, February 8, 2006, p. FP5.

22. Grant Robertson, "CanWest stops publishing money-losing freebie Dose," *Globe and Mail*, May 18, 2006, p. B3.

23. Barbara Shecter, "CanWest sells Irish network," *National Post*, May 20, 2006, p. FP4.

24. Wojtek Dabrowski, "Advertising doldrums deepen, cutting CanWest profit by 75%," *Toronto Star*, July 7, 2006, p. F3.

25. Grant Robertson, "Irish sale pushes CanWest to profit: Deal main factor in avoiding quarterly loss," *Globe and Mail*, November 3, 2006, p. B4.
26. "Media firm's rating under review," *National Post*, July 8, 2006, p. FP4.
27. "DBRS puts CanWest debt under review," *Toronto Star*, July 12, 2006, p. F3.
28. Romina Maurino, "Some skeptical about CanWest's Turkish radio moves," *Montreal Gazette*, April 18, 2006, p. B5.
29. "It's far from quiet on the CanWest front," *Television Business International*, August 1, 2006, p. 1.
30. Richard Blackwell, "CanWest buys *Jerusalem Post* stake: Prized Hollinger property will be half owned by Aspers," *Globe and Mail*, November 17, 2004. p. B4.
31. Hadar Horesh, "'It was a mistake to do business with Eli Azur,'" *Ha'aretz*, March 7, 2005. Downloaded May 12, 2006 from www.haaretz.com/hasen/spages/548663.html.
32. Barbara Shecter, "*Jerusalem Post* ruling," *National Post*, June 3, 2006, p. FP4.
33. Robertson, "TV revenue slide hits CanWest profit."
34. Barbara Shecter, "CanWest slams Ottawa over handling of trusts: Goodale's musings cost us $150M, chairman says," *National Post*, January 13, 2006, p. FP1.
35. Roy MacGregor, "Newsman turned candidate throws challenge to media: Check your bias at the door," *Globe and Mail*, September 26, 2005, p. A2.
36. Shecter, "CanWest slams Ottawa over handling of trusts."
37. David Asper, "The Asper Agenda: 15 points for Canada," *Calgary Herald*, April 24, 2001, p. A15.
38. Christopher Dornan, "Should Asper have joined Harper on stage?" *Globe and Mail*, January 20, 2006, p. A9.
39. Colin Campbell, "All in the family: The line between the media business and politics is getting fuzzy," *Maclean's*, October 25, 2006, p. 40.
40. Ibid.
41. Canada, Standing Senate Committee on Transportations and Communications, "Final report on the Canadian news media," Volume 1, June 2006, p. 24.
42. Ibid., Volume 2, p. 49.
43. Canada, Standing Senate Committee on Transportations and Communications, "Interim report on the Canadian news media," April 2004, p. 39.
44. Charles Gordon, "Senate meddles with media, but someone's got to do it," *Ottawa Citizen*, June 25, 2006, p. A12.
45. "Ottawa must not control the press," *Toronto Star*, June 23, 2006, p. A22.
46. Don Martin, "CBC drifts into PM's agenda," *National Post*, June 22, 2006, p. A5.
47. Antonia Zerbisias, "Media do a poor job on media report," *Toronto Star*, July 10, 2006, p. C2.
48. Ibid.
49. "CanWest gives notice it may withdraw from Canadian Press," *National Post*, June 29, 2006, p. FP6.
50. Jeff Sallott, "Copps asks committee to consider future of CP," *Globe and Mail*, October 30, 1999, p. B3.

51. Jennifer Ditchburn, "Oda cancels fundraiser over ethical questions," *Toronto Star*, November 8, 2006, p. A22.

52. Michael Geist, "Shameful lobbying no way to make public policy," *Toronto Star*, November 13, 2006, p. F3.

53. Barbara Shecter, "Broadcast rules hinder TV industry, Asper says," *National Post*, June 16, 2006, p. FP7.

54. Grant Robertson, "Let media consolidate, Asper urges," *Globe and Mail*, June 16, 2006, p. B3.

55. Bill Curry, "Minister cancels bash over lobbyist's role," *Globe and Mail*, November 8, 2006, p. A4.

56. Paul Samyn, "Feds might run rights museum: Tories ready to pay operating cost by making it a national institution," *Winnipeg Free Press*, December 19, 2006, p. A1.

57. Val Ross, "Coming to a town near you: National museums," *Globe and Mail*, December 20, 2006, p. R1.

58. Alexander Panetta, "Journalists stage boycott of PM's press conference," *Toronto Star*, May 24, 2006, p. A4.

59. Kady O'Malley, "How did CanWest get its scoop? By going on PM's list," *Hill Times*, August 28, 2006, p. 1.

60. Antonia Zerbisias, "Star and Globe? You're all invited to my funeral," *Toronto Star*, December 6, 2005, p. D5.

61. Madhavi Acharya-Tom Yew, "Torstar chief sees more co-operation between Toronto's daily newspapers: Joint distribution likely in 10 years," *Toronto Star*, September 13, 2006, p. D4.

62. Eric Reguly, "CanWest left standing in media musical chairs," *Globe and Mail*, December 3, 2005, p. B2.

63. Lori McLeod and Barbara Shecter, "Torstar deal about television: Prichard," *National Post*, December 3, 2005, p. FP3.

64. Grant Robertson and Jacquie McNish, "BGM grabs CHUM for $1.4-billion," *Globe and Mail*, July 13, 2006, p. B1.

65. Simon Tuck, "A test for Tories' free-market mantra," *Globe and Mail*, July 13, 2006, p. B5.

66. Ibid.

67. Rick Westhead, "BCE media deal stokes controversy," *Toronto Star*, May 18, 2006, p. C6.

68. Barbara Shecter, "Bell Globemedia transfer approved," *National Post*, July 22, 2006, p. FP1.

69. Antonia Zerbisias, "Minister brushes off warnings on media: Oda downplays Senate report on ownership," *Toronto Star*, December 2, 2006, p. A17.

70. Barbara Shecter, "Atlantis deal worth the risk, Asper says," *Ottawa Citizen*, January 12, 2007, p. FP1.

71. Grant Robertson, Andrew Willis and Eric Reguly, "Alliance deal could produce TV spinoff," *Globe and Mail*, January 12, 2007, p. B1.

72. Barbara Shecter, "Broadcasting opportunity," *National Post*, January 13, 2007, p. FP3.

73. Barbara Shecter, "CanWest's new Alliance," *National Post*, January 11, 2007, p. A1.

74. Etan Vlessing, "Alliance deal rests on 'control': Goldman Sachs involvement will draw CRTC scrutiny," *Hollywood Reporter*, January 12, 2007. Downloaded January 28, 2007 from http://www.hollywoodre-

porter.com/hr/search/article_display.jsp?vnu_content_id=1003531822

75. Antonia Zerbisias, "Alliance deal a test of control," *Toronto Star*, January 12, 2007, p. C8.

76. Shecter, "Atlantis deal worth the risk, Asper says."

Conclusion: The Pictures in Our Heads

1. "Liberalism's last prayer," *Wilson Quarterly*, Spring 2005, p. 84.

2. Walter Lippmann, *Public Opinion*. (New York: MacMillan), 1922, p. 47.

3. Ibid., p. 43.

4. Ibid., p. 158.

5. Ibid., p. 358.

6. See David Carr, "A comeback overshadowed by a blog," *New York Times*, September 11, 2006, p. C1.

7. Jack Shafer, "The perfervid Peretz," *Washington City Paper*, April 12, 1991. Available online at http://www.slate.com/id/2134011/

8. Peter Dykstra, "Windbags, tyrants, and chickens," *Progressive*, October 1991, p. 14.

9. Hilton Kramer, "Life after liberalism," *New Criterion*, September 1994, p. 4.

10. "New Republic's library and web site cited as CanWest buys stake," *National Post*, January 20, 2006, p. FP4.

11. Patricia Best, "New Republic deal a lesson in networking," *Globe and Mail*, March 1, 2007, p. B2.

12. Katharine Q. Sellye, "New Republic cuts back, but bulks up its image," *New York Times*, February 24, 2007, p. B7.

13. See Marc Edge, "Balancing academic and corporate interests in Canadian journalism education," *Journalism and Mass Communication Educator*, Summer 2004, pp. 172-185.

14. Peter Desbarats, "Who's on the barricades?" *Globe and Mail*, June 3, 1998, p. A21.

15. Ibid.

16. Erin Steuter, "He who pays the piper calls the tune: Investigation of a Canadian media monopoly," *Web Journal of Mass Communication Research*, September 2004. Available online at http://www.scripps.ohiou.edu/wjmcr/vol07/7-4a.html.

17. Jennifer Lewington, "Journalism school saved by one vote," *Globe and Mail*, October 30, 1993, p. A3.

18. See Lee B. Becker, Tudor Vlad, Maria Tucker, and Renée Pelton, "2005 enrollment report: Enrollment growth continues, but at reduced rate," *Journalism and Mass Communication Educator*, Autumn 2006, p. 299; Marc Edge, "Balancing academic and corporate interests in Canadian journalism education," *Journalism and Mass Communication Educator*, Summer 2004, pp. 172-185.

19. Peter Desbarats, "News about the history of news," *Globe and Mail*, June 3, 1989, p. C16.

20. See Steven H. Chaffee and Everett M. Rogers, eds., *The Beginnings of Communication Study in America*. (Thousand Oaks: Sage), 1997.

21. See Garth Jowett, Ian Jarvie and Kathryn H. Fuller, *Children and the Movies: Media Influence and the Payne Fund Controversy*. (New York: Cambridge University Press), 1996.

22. See Stanley J. Baran and Dennis K. Davis, *Mass Communication*

Theory: Foundations, Ferment, and Future. 3rd ed. (Belmont, CA Wadsworth), 2003, pp. 188-189.

23. John Eggerton, "Missing FCC study sparks inquiry," *Broadcasting and Cable*, September 18, 2006, p. 14.

24. John Eggerton "Missing in D.C.," *Broadcasting and Cable*, September 25, 2006, p. 16.

25. Mark Fitzgerald, "Stung for alleged cover up, FCC posts staff cross-ownership studies," *Editor and Publisher*, January 9, 2007. Downloaded from editorandpublisher.com/eandp/departments/business/article_display.jsp?vnu_content_id=1003522909.

26. Richard Linnett and Ira Teinowitz, "Media merge, marketers pay," *Advertising Age*, February 17, 2003, p. 1.

27. "FCC studies find more competition among various media forms," *Television Digest*, October 7, 2002.

28. David Pritchard, "A tale of three cities: 'Diverse and antagonistic' information in situations of local newspaper/broadcast cross-ownership," *Federal Communications Law Journal*, December 2001, p. 51.

29. "Mass media," *Communications Daily*, October 4, 2002.

30. Lee Bacchus, "Man who would be king of CKVU impatient to get his dream started," *Vancouver Sun*, February 23, 1987, p. F5.

31. Canada, Standing Senate Committee on Transportations and Communications, "Final report on the Canadian news media," Volume 2, June 2006, p. 67.

32. Canada, Competition Bureau, "The Competition Bureau's work in media industries: Background for the Senate Committee on Transport and Communications," April 2, 2004, p. 12.

33. Canada, Standing Senate Committee on Transportations and Communications, "Final report on the Canadian news media," Volume 2, June 2006, p. 67.

34. Ibid., p. 50.

35. Norman Spector, "Senate communications report fails to look outside the news box," *Globe and Mail*, July 3, 2006, p. S1.

36. "Media report misses most pressing issues," *Montreal Gazette*, June 30, 2006, p. B6.

37. Canada, Standing Senate Committee on Transportations and Communications, "Final report on the Canadian news media," Volume 2, p. 65.

38. Ibid.

39. David L. Altheide and Robert P. Snow, *Media Worlds in the Postjournalism Era.* (Hawthorne, NY: Aldine de Gruyter), 1991, p. 3.

40. Ibid., p. 4.

41. James W. Carey, *Communication as Culture: Essays on Media and Society.* (New York: Routledge), 1989, p. 20.

42. David Taras, *Power and Betrayal in the Canadian Media.* 2nd ed. (Peterborough, Ont: Broadview Press), 2001, p. 4.

43. Ibid., p. 235.

44. See Ian Austen, "Does bigger mean better?" *Calgary Herald*, September 14, 1994, p. C4.

45. Catherine Murray, "Wellsprings of knowledge: Beyond the CBC policy trap," *Canadian Journal of Communication*, 2001, p. 48, fn. 12.

46. "BCE Inc. further re CTV acquisition.," *Regulatory News Service*, July 18, 2000.

47. See Paul Orlowski, "Educating in an era of Orwellian spin: Critical media literacy in the classroom," *Canadian Journal of Education*, 2006, p. 176-198; Leah R. Vande Berg, Lawrence A. Wenner, and Bruce E. Gronbeck, "Media literacy and television criticism: Enabling an informed and engaged citizenry," *American Behavioral Scientist*, 2004, pp. 219-228.

48. I.H. Asper, letter to Conrad Back, August 9, 2000. Available online at http://www.usdoj.gov/usao/iln/hot/us_v_black_exhibits/canwest_121.pdf.

49. I.H. Asper, "Re: Various matters arising from memoranda exchanges [memo to Conrad Back],'" June 26, 2000, p. 3. Available online at http://www.usdoj.gov/usao/iln/hot/us_v_black_exhibits/canwest_108.pdf.

50. See John Gray, "How the press was won," *Canadian Business*, December 10, 2001, p. 61.

51. I.H. Asper, letter to Conrad Back, August 9, 2000. Available online at http://www.usdoj.gov/usao/iln/hot/us_v_black_exhibits/canwest_121.pdf.

52. "A Canterbury Tale: Concept Document," May 31, 2000, p. 17. Available online at http://www.usdoj.gov/usao/iln/hot/us_v_black_exhibits/canwest_006.pdf

53. Canada, Royal Commission on Newspapers, *Report*. (Ottawa: Minister of Supply and Services), 1981, p. 219.

54. Walter C. Soderlund and Kai Hildebrandt "Preface," in Soderlund and Hildebrandt, eds., *Canadian Newspaper Ownership in the Era of Convergence: Rediscovering Social Responsibility*. (Edmonton: University of Alberta Press), 2005, p. xiii.

55. Walter C. Soderlund, Ronald H. Wagenberg, Kai Hildebrandt, and Walter I. Romanow, "Ownership rights vs. social responsibility: Defining an appropriate role for newspaper ownership," in ibid., p. 149.

56. Ibid., p. 145.

57. Stephen Kimber, "Canada's lax rules enable CanWest domination," *Adbusters*, September/October 2003, p. 16.

58. Taras, *Power and Betrayal in the Canadian Media*, p. 1.

59. John Doyle, "Offensive? How about fair criticism?" *Globe and Mail*, December 18, 2006, p. R3.

60. Dennis Earl, "Zerbisias won't be replaced at *The Star*," Fading to Black [online blog], June 12, 2007. Downloaded July 16, 2007 from http://mediafade.blogspot.com/2007/06/zerbisias-wont-be-replaced-at-star.html.

61. Brian Brennan, "Canadian media de-Zerbified," *Back of the book.ca*, July 3, 2007. Downloaded July 6, 2007 from http://backofthebook.ca/media/2007/07/canadian-media-de-zerbified.html

62. See Daniel Yergin and Joseph Stanislaw, *Commanding Heights: The Battle for the World Economy*. (New York: Simon & Schuster), 2002.

63. Ben Bagdikian, *The Media Monopoly*. (Boston: Beacon Press), 1983, p. 237.

Appendix: The Asper Agenda

1. David Asper, "The Asper Agenda: 15 points for Canada," *Calgary Herald*, April 24, 2001, p. A15.

BIBLIOGRAPHY

Donald E. Abelson, *Do Think Tanks Matter? Assessing the Impact of Public Policy Institutes.* (Montreal: McGill-Queen's University Press), 2002.

Nurith C. Aizenman, "The man behind the curtain: Richard Mellon-Scaife and $200 million of his money is the man behind the conservative revolution," *Washington Monthly,* July/August 1997, pp. 28-34.

Eric Alterman, *What Liberal Media? The Truth About Bias and the News.* (New York: Basic Books), 2003.

David L. Altheide and Robert P. Snow, *Media Worlds in the Postjournalism Era.* (Hawthorne, NY: Aldine de Gruyter), 1991.

David L. Altheide and Jennifer N. Grimes, "War programming: The propaganda project and the Iraq war," *Sociological Quarterly,* September 2005, pp. 617-643.

J. Herbert Altschull, *Agents of Power: The Role of the News Media in Human Affairs.* (New York: Longman), 1984.

Arnold Amber, "Heritage Committee asks for freeze on additional cross-media ownership," *TNG Canada,* June 12, 2003. Available online at http://www.tngcanada.org/EN/news/2003/030612_heritage_report. shtml.

Brian C. Anderson, *South Park Conservatives: The Revolt Against Liberal Media Bias.* (Washington: Regnery), 2005.

Mitchell Anderson, "Trust us, we're media: The global-warming debate ended long ago, but you wouldn't know it by reading Canadian newspapers," *Georgia Straight,* January 25, 2007, p. 33-34.

I.H. Asper, *The Benson Iceberg: A Critical Analysis of the White Paper on Tax Reform In Canada.* (Toronto: Clarke, Irwin), 1970.

Israel Asper, "Media integrity lost in battle between Palestinians and Israelis," *Canadian Speeches,* January/February 2003, p. 28.

Leonard Asper, "Observations on the media, Canada and Winnipeg," *Canadian Speeches,* January/February 2001, p. 54.

Benoit Aubin, "Back to the future," *Maclean's,* July 21, 2003, p. 32.

Valerie Aubourg, "Organizing Atlanticism: the Bilderbeg Group and the Atlantic Institute, 1952-63," *Intelligence and National Security* 2003, pp. 92-105.

Australia, Senate Environment, Communications, Information Technology and the Arts Legislation Committee, "Report on the Broadcasting Services Amendment (Media Ownership) Bill 2002," June 2002.

Australian Broadcasting Authority, "Investigation Into Control: CanWest

Global Communications Corporation/The Ten Group Ltd.," November 1995.

Australian Broadcasting Authority, "Investigation Into Control: CanWest Global Communications Corporation/The Ten Group Ltd., Second investigation," April 1997.

Ben Bagdikian, *The Media Monopoly*. (Boston: Beacon Press), 1983.

James Bamford, *A Pretext for War: 9/11, Iraq, and the Abuse of America's Intelligence Agencies*. (New York: Doubleday), 2004.

Stanley J. Baran and Dennis K. Davis, *Mass Communication Theory: Foundations, Ferment, and Future*. 3rd ed. (Belmont, CA: Wadsworth), 2003.

Allan Bartley, "The regulation of cross-media ownership: The life and short times of PCO 2294," *Canadian Journal of Communication*, Summer 1988, pp. 45-59.

Lee B. Becker, Tudor Vlad, Maria Tucker, and Renée Pelton, "2005 enrollment report: Enrollment growth continues, but at reduced rate," *Journalism and Mass Communication Educator*, Autumn 2006, pp. 297–327.

Gal Beckerman, "Tripping up big media," *Columbia Journalism Review*, November/December 2003, p. 15.

David Beers, "Caught in the Sun," *Vancouver*, June 2002, pp. 21–25.

Julian Beltrame, "Whose news?" *Maclean's*, May 28, 2001, p. 34.

Barry Berlin, *The American Trojan Horse: US Television Confronts Canadian Economic and Cultural Nationalism*. (New York: Greenwood), 1990.

David Berman, "Channel changer: Izzy Asper has quietly built Canada's most ambitious multinational broadcasting company. The question is: who's preparing to take over from Asper?" *Canadian Business*, September 1995, pp. 46–53.

David Berman, "Life with father," *Canadian Business*, September 1995, p. 50.

David Berman, "Life without father," *Canadian Business*, November 1995, p. 98.

Rosalie Bertell, "An elite pow wow," *Peace Magazine*, July/August 1996, p. 6.

Conrad Black, *A Life in Progress*. (Toronto: Key Porter), 1993.

Conrad Black, *Franklin Delano Roosevelt: Champion of Freedom*. (New York: Public Affairs), 2003.

Eric Boehlert, *Lapdogs: How the Press Rolled Over for Bush*. (New York: Free Press), 2006.

Kristina Borjessin, ed., *Into the Buzzsaw: Leading Journalists Expose the Myth of a Free Press*. (Amherst, NY: Prometheus), 2002.

L. Brent Bozell III, *Weapons of Mass Distortion: The Coming Meltdown of the Liberal Media*. (New York: Crown), 2004.

Warren Breed, "Social control in the news room," *Social Forces*, May 1955, pp. 326–355.

David Brock, *The Republican Noise Machine: Right-Wing Media and How it Corrupts Democracy*. (New York: Crown), 2004.

Allan Brown, "Newspaper ownership in Australia." *Journal of Media Economics*, Fall 1993, pp. 49–64.

Les Brown, "Small is beautiful is the new mantra," *Television Business International*, April 2005, p. 1.

Peter A. Brown, "Bias among the media," *Nieman Reports*, Spring 2002, p. 87.

Steven D. Brown and James Delodder, "When is a creationist not a creationist? Appreciating the miracles of public opinion polling," *Canadian Journal of Communication*, 2003, pp. 111-119.

John Bugailiskis, "Izzy Asper: broadcaster, visionary, philanthropist, Canadian," *Broadcaster*, October 2003, pp. 24-26.

Jonah Butovsky, "Phony populism: The misuse of opinion polls in the *National Post*," *Canadian Journal of Communication*, 2007, pp. 91-102.

Colin Campbell, "All in the family: The line between the media business and politics is getting fuzzy," *Maclean's*, October 25, 2006, p. 40.

Canada, Competition Bureau, "The Competition Bureau's work in media industries: Background for the Senate Committee on Transport and Communications," April 2, 2004.

Canada, Royal Commission on Newspapers, *Report*. (Ottawa: Minister of Supply and Services), 1981.

Canada, Standing Committee on Canadian Heritage, *Our cultural sovereignty: The second century of Canadian broadcasting*. (Ottawa: Queen's Printer), 2003.

Canada, Standing Senate Committee on Transport and Communication, "Interim report on the Canadian news media," April 2004.

Canada, Standing Senate Committee on Transportations and Communications, "Final report on the Canadian news media," Vol. 1 & 2, June 2006.

Canada, *The Uncertain Mirror: Report of the Special Senate Committee on Mass Media, Vol. I*. (Ottawa: Information Canada), 1970.

"Canadian chain editorials called bad for business," *Editor and Publisher*, January 7, 2002, p. 20.

Canadian Journalists for Free Expression, "Not in the newsroom! CanWest Global, chain editorials and freedom of expression in Canada," April 2002.

James W. Carey, *Communication as Culture: Essays on Media and Society*. (New York: Routledge), 1989.

Steven H. Chaffee and Everett M. Rogers, eds., *The Beginnings of Communication Study in America*. (Thousand Oaks: Sage), 1997.

Paul Chard, "Too many gringos," *Canadian Business*, September 1995, p. 48.

Patricia Chisholm, "Tycoon of the tube," *Maclean's*, November 27, 1995, p. 36.

Reese Cleghorn, "The press shifts to the right, but slowly: Words have moved the center as journalists followed the pols," *American Journalism Review*, December 1994, p. 4.

Chris Cobb, *Ego and Ink: The Inside Story of Canada's National Newspaper War*. (Toronto: McClelland & Stewart), 2004.

Trevor Cole, "Lenny: A mogul in the making," *Report on Business* Magazine, October 2000, pp. 66-75.

Trevor Cole, "Revenge of the ratings king: How did CanWest Global evolve into the powerhouse that's eating CTV's lunch?" *Report on Business* Magazine, May 1996, pp. 58-61.

Commission on Freedom of the Press, *A Free and Responsible Press*. (Chicago: University of Chicago Press), 1947.

Farrel Corcoran, *RTÉ and the Globalisation of Irish Television*. (Bristol: Intellect Books), 2004.

Gilbert Cranberg, Randall Bezanson, and John Soloski, *Taking Stock: Journalism and the Publicly Traded Newspaper Company*. (Ames: Iowa State University Press), 2001.

Kathleen Cross, "Off balance: How the Fraser Institute slants its news-monitoring studies," *Canadian Forum*, October 1997, pp. 10-11.

Daniel J. Czitrom, *Media and the American Mind: From Morse to McLuhan*. (Chapel Hill, NC: University of North Carolina Press), 1982.

Keith Davey, *The Rainmaker: A Passion for Politics*. (Toronto: Stoddart), 1986.

Sean Davidson, "Report? What report? Critics warn that media concentration is threatening debate and broadcast reform," *Playback*, January 5, 2004, p. 1.

Stefano DellaVigna and Ethan Kaplan, "The Fox News effect: Media bias and voting," National Bureau of Economic Research Working Paper No. 12169, March 30, 2006. Available online at http://elsa.berkeley.edu/~sdellavi/wp/foxvote06-03-30.pdf

David Dias, "Leonard Asper's master plan for Global domination," *Ryerson Review of Journalism*, Summer 2001, pp. 42-48.

Gillian Doyle, "Convergence: 'A unique opportunity to evolve in previously unthought-of ways' or a hoax?" in Christopher T. Marsden and Stefaan G. Verhulst, eds., *Convergence in European Digital TV Regulation*. (London: Blackstone), 1999, pp. 141-154.

Gillian Doyle, Media ownership: *The Economics and Politics of Convergence in the UK and European Media*. (London: Sage), 2002.

Peter J. S. Dunnett, *The World Newspaper Industry*. (London: Croom Helm), 1988.

"Duopoly," *The Economist*, April 28, 2001, p. 38.

Peter Dykstra, "Windbags, tyrants, and chickens," *Progressive*, October 1991, p. 14.

Marc Edge, "Balancing academic and corporate interests in Canadian journalism education," *Journalism and Mass Communication Educator*, Summer 2004, pp. 172–185.

Marc Edge, "Byline wars at the *Vancouver Province*: The battle was not the only problem at the paper," *Media*, Spring 2001, pp. 6–7.

Marc Edge, "Convergence and the 'Black News Hole': Canadian Newspaper Coverage of the 2003 Lincoln Report," *Canadian Journal of Media Studies*, April 2007. Available online at http://cjms.fims.uwo.ca/issues/02-01/index/html.

Marc Edge, "Cross ownership," in Wolfgang Donsbach, ed. *International Encyclopedia of Communication*. (Oxford: Blackwell). Forthcoming.

Marc Edge, *Pacific Press: The Unauthorized Story of Vancouver's Newspaper Monopoly*. (Vancouver: New Star Books), 2001.

Marc Edge, "The good, the bad, and the ugly: Financial markets and the demise of Canada's Southam Newspapers," *International Journal on Media Management*, Winter 2003, pp. 180-189.

Marc Edge, "The pain of the obdurate rump: Conrad Black and the flouting of corporate governance," in Robert G. Picard, ed. *Corporate Governance of Media Companies*. (Jönköping, Sweden: Jönköping International Business School), 2005, pp. 165-182.

Marc Edge, "The press we deserve: A legacy of unheeded warnings," *Textual Studies in Canada*, Fall 2002, pp. 1-11.

Edith Efron, *The News Twisters*. (New York: Manor Books), 1972.

John Eggerton, "Missing FCC study sparks inquiry," *Broadcasting and Cable*, September 18, 2006, p. 14.

John Eggerton "Missing in D.C.," *Broadcasting and Cable*, September 25, 2006, p. 16.

Paul Farhi, "Good news for 'McPaper': After 15 years it climbs from money pit to profitability," *American Editor*, September 1997, p. 8.

Gordon Fisher, "Have the journalists lost control?" *Canadian Issues*, August 2003, pp. 10–11.

Mark Fitzgerald, "Stung for alleged cover up, FCC posts staff cross-ownership studies," *Editor and Publisher*, January 9, 2007. Downloaded from editorandpublisher.com/eandp/departments/business/article_display. jsp?vnu_content_id=1003522909.

Matthew Fraser, *Free For All: The Struggle for Dominance on the Digital Frontier*. (Toronto: Stoddart), 1999.

Erwin Frenkel, *The Press and Politics in Israel: The Jerusalem Post from 1932 to the Present*. (Westport, CT: Greenwood Press), 1994.

Ritchie Gage, "Asper vision," *Manitoba Business*, December 1995, p. 5.

Herbert J. Gans, "Are U.S. journalists dangerously liberal?" *Columbia Journalism Review*, November/December 1985, pp. 29–33.

Herbert J. Gans, *Deciding What's News: A Study of CBS Evening News, NBC Nightly News, Newsweek and Time*. (New York, Random House), 1979.

Mike Gasher, "Does CanWest know what all the fuss is about?" *Media*, Winter 2002, p. 8.

Jonathon Gatehouse, "The risk taker," *Maclean's*, October 20, 2003, p. 40.

John Geddes, "No asper-sions . . .," *Maclean's*, August 21, 2000, p. 8.

John Geddes, "The Izzy and Leonard show," *Maclean's*, August 14, 2000, p. 14.

Susan Gittins, *CTV: The Television Wars*. (Toronto: Stoddart), 1999.

Bernard Goldberg, *Bias: A CBS Insider Exposes How the Media Distort the News*. (New York: Regnery), 2001.

John Gray, "The best & worst boards in Canada," *Canadian Business*, August 19, 2002, p. 30.

John Gray, "How the press was won," *Canadian Business*, December 10, 2001, pp. 61-63.

Edward Greenspon, "Bicker, bicker: The three-way partnership at Global TV is acrimonious, to say the least," *Report on Business*, March 1988, pp. 24-36.

Edward Greenspon, "The acquisitor," *Report on Business*, October 1987, p. 27.

Kevin Groves, Cecilia Jamasmie, Jock Paul and Lisa Johnson, "Still can't catch corporate crooks," *UBC Thunderbird Online Journalism Review*, January 2003. Available online at http://www.journalism.ubc.ca/thunderbird/archives/2002.12/alert.html.

Albert C. Gunther and Kathleen Schmitt, "Mapping boundaries of the Hostile Media Effect," *Journal of Communication*, March 2004, pp. 55-70.

Donald Gutstein, "Who funds the Fraser Institute?" *Straight Goods*, October 28, 2004. Downloaded from http://www.yourmedia.ca/library_articles/041028_gutstein_fraser.html.

Donald Gutstein, "Harperstein," *Georgia Straight*, July 6, 2006, p. 39.

Donald Gutstein, "US-style media monitor comes north, *The Tyee*,

February 10, 2005. Available online at http://thetyee.ca/Media-check/2005/02/10/FraserInstituteSpinCycle/.

Robert A. Hackett and Scott Uzelman, "Tracing corporate influences on press content: A summary of recent NewsWatch Canada research," *Journalism Studies*, 2003, pp. 331-346.

Hershel Hardin, *Closed Circuits: The Sellout of Canadian Television*. (Vancouver: Douglas & McIntyre), 1985.

Edward S. Herman and Noam Chomsky, *Manufacturing Consent: The Political Economy of the Mass Media*. (New York: Pantheon), 1988.

Seymour Hersh, "Selective intelligence," *New Yorker*, May 12, 2003, p. 44.

Hollinger International Inc., "Report of investigation by the special committee of the board of directors," August 30, 2004, p. 8. Available online at http://www.sec.gov/Archives/edgar/data/868512/000095012304010413/y01437exv99w2.htm

Noel Hulsman, "What makes the *Vancouver Sun* tick?" *BC Business*, November 2003, p. 40.

Jennifer Hunter, "Plotting a takeover: CanWest Global closes in on Griffiths family's control of WIC," *Maclean's*, January 19, 1998, p. 46.

"It's far from quiet on the CanWest front," *Television Business International*, August 2006, p. 1.

Shanto Iyengar and Donald R. Kinder, *News That Matters: Television and American Opinion*. (Chicago: University of Chicago Press), 1987.

John W. C. Johnstone, Edward J. Salawski, and William W. Bowman, *The News People: A Sociological Portrait of American Journalists and Their Work*. (Urbana: University of Illinois Press), 1971.

Garth Jowett, Ian Jarvie and Kathryn H. Fuller, *Children and the Movies: Media Influence and the Payne Fund Controversy*. (New York: Cambridge University Press), 1996.

Chaim Kaufmann, "Threat inflation and the failure of the marketplace of ideas: The selling of the Iraq war," *International Security*, Summer 2004, pp. 5-48.

Tasha Kheiriddin and Adam Daifallah, *Rescuing Canada's Right: Blueprint for a Conservative Revolution*. (Mississauga, ON: J. Wiley and Sons Canada), 2005.

"Kimber on tour," *Maclean's*, May 27, 2002, p. 9.

Stephen Kimber "Canada's lax rules enable CanWest domination," *Adbusters*, September/October 2003, p. 16.

Stephen Kimber, "In the wonderful world of Iz, it's 1984 all over again," in David Dadge, ed., *Silenced: International Journalists Expose Media Censorship*. (Amherst, NY: Prometheus Books), 2005, pp. 43-60.

Alexandra Kitty, *Outfoxed: Rupert Murdoch's War on Journalism*. (New York: Disinformation), 2005.

Alec Klein, *Stealing Time: Steve Case, Jerry Levin, and the collapse of AOL Time Warner*. (New York: Simon & Shuster), 2003.

Andrew Kohut, "Self-censorship: Counting the ways," *Columbia Journalism Review*, May/June 2000, p. 41.

Hilton Kramer, "Life after liberalism," *New Criterion*, September 1994, pp. 4-11.

Steven Kull, Clay Ramsay, and Evan Lewis, "Misperceptions, the media, and the Iraq war," *Political Science Quarterly*, Vol. 118, No. 4, 2003–2004, pp. 569–598.

Adam B. Kushner, "The bogus editorial voice in Canada," *Masthead*, Summer 2002, p. 22.

Jim A. Kuypers, *Bush's war: Media Bias and Justifications for War in a Terrorist Age*. (Lanham, MD: Rowman & Littlefield), 2006;.

Jim A. Kuypers, *Press Bias and Politics: How the Media Frame Controversial Issues*. (New York: Praeger), 2002.

Charles C. Layton, "News blackout," *American Journalism Review*, December 2003, pp. 18-31.

Eve Lazarus,"Third network? No thanks!" *Marketing*, March 9, 1998, p. 28.

"Liberalism's last prayer," *Wilson Quarterly*, Spring 2005, pp. 84-85.

S. Robert Lichter and Stanley Rothman, "Media and business elites," *Public Opinion*, 1981, pp. 42-46, 59-60.

S. Robert Lichter, Stanley Rothman, and Linda S. Lichter, *The Media Elite: America's New Powerbrokers*. (New York: Hastings House), 1986.

Trudy Lieberman, *Slanting the Story: The Forces That Shape the News*. (New York: New Press), 2000.

Richard Linnett and Ira Teinowitz, "Media merge, marketers pay," *Advertising Age*, February 17, 2003, p. 1.

Walter Lippmann, *Public Opinion*. (New York: MacMillan), 1922.

Richard Littlemore, "The most hated man in business," *BC Business*, July 2005, pp. 45-55.

Donna Logan, "Opportunities and challenges," *Media*, Fall 2000, p. 15.

Myron Love, "Community should respond to anti-Israel reporting: Asper," *Canadian Jewish News*, February 21, 2002, p. 31.

Myron Love, "Human rights museum downplays funding rift," *Canadian Jewish News*, August 19, 2004, p. 28.

Paul Lungen, "Izzy Asper: philanthropist, renaissance man," *Canadian Jewish News*, October 16, 2003, p. 1.

Paul Lungen "Leonard Asper defends CanWest vision," *Canadian Jewish News*, January 3, 2002, p. 6.

John R. MacArthur, "The lies we bought: The unchallenged 'evidence' for war," *Columbia Journalism Review*, May/June 2003, p. 62.

Adrienne Macintosh, "Which way did he go?" *Ryerson Review of Journalism*, Spring 2004, pp. 18-21.

Katherine Macklem, "Can the Aspers do it?" *Maclean's*, April 8, 2002, pp. 48-53.

Katherine Macklem, "The view from the center," *Maclean's*, April 8, 2002, p. 53.

Steve Maich, "Better off without you: Convergence was just a tune the media giants hummed for awhile," *Maclean's*, May 16, 2005, p. 33.

Steve Maich, "Bleeding newspapers dry," *Maclean's*, January 30, 2006, p. 22.

Andrea Mandel-Campbell, "Television South American-style," *Manitoba Business*, May 1995, p. 30.

Peter Mansbridge, "Izzy: He was never shy," *Maclean's*, October 20, 2003, p. 18.

Robert W. McChesney, *The Problem of the Media: US Communication Politics in the 21st Century*. (New York: Monthly Review Press), 2004.

Wendy McLellan, "Study paints journalists as left-wing, biased messengers," *Media*, Winter 2003, p. 31.

Linda McQuaig, *Behind Closed Doors: How the Rich Won Control of Cana-*

da's Tax System — And Ended Up Richer. (Markham, Ont.: Viking), 1987.

Hugh Miles, *Al-Jazeera: How Arab TV News Challenged the World*. (London: Abacus), 2005.

Lydia Miljan and Barry Cooper, *Hidden Agendas: How Journalists Influence the News*, (Vancouver: UBC Press), 2003.

John Miller, *Yesterday's News: Why Canada's Daily Newspapers Are Failing Us*. (Halifax: Fernwood), 1998.

Russell Mills, "Judging Mr. Asper," *Media*, Fall/Winter 2003, p. 8.

Russell Mills, "Pressing for freedom," *Time* (Canadian edition), July 1, 2002, p. 47.

Russell Mills, "Reflections on the state of Canadian media," *Canadian Parliamentary Review*, Winter 2004. p. 27.

Aaron J. Moore, "A chill in Canada," *Columbia Journalism Review*, March/April 2002, p. 40.

Rasha Mourtada, "Haroon and the sea of opinions," *Ryerson Review of Journalism*, Spring 2002, pp. 12-19.

Nina Munk, *Fools Rush In: Steve Case, Jerry Levin, and the Unmaking of AOL Time Warner*. (New York: Harper Business), 2004.

Catherine Murray, "Wellsprings of knowledge: Beyond the CBC policy trap," *Canadian Journal of Communication*, 2001, pp. 31-52.

Peter C. Newman, "The inexorable spread of the Black Empire," *Maclean's*, February 3, 1992, p. 68.

Peter C. Newman, "Thumbs down to Izzy Asper's national dream," *Maclean's*, December 23, 1996, p. 40.

Peter C. Newman, *Titans: How the New Canadian Establishment Seized Power*. (Toronto: Viking), 1998.

David Niven, *Tilt? The Search for Media Bias*. (New York: Praeger), 2002.

Kady O'Malley, "How did CanWest get its scoop? By going on PM's list," *Hill Times*, August 28, 2006, p. 1.

Paul Orlowski, "Educating in an era of Orwellian spin: Critical media literacy in the classroom," *Canadian Journal of Education*, 2006, pp. 176-198.

Larry Patriquin, *Inventing Tax Rage: Misinformation in the National Post*. (Halifax: Fernwood), 2004.

Ed Pearce, "'Izzy' Asper: Going Global," *Ivey Business Journal*, November 2001, p. 26.

Rick Perlstein, "Eyes right: Conservatives are winning the media war. How do they do it?" *Columbia Journalism Review*, March/April 2003, p. 52.

Peter Phillips and Project Censored, *Censored 2007: The top 25 Censored Stories*. 30th Anniversary Ed. (New York: Seven Stories Press), 2006.

Gordon Pitts, *Kings of Convergence: The Fight for Control of Canada's Media*. (Toronto: Doubleday), 2002.

Chris Powell, "Putting it all together: Canada's media moguls are learning to assemble their diverse properties in advertiser-pleasing ways," *Marketing*, February 25, 2002, p. 15.

Chris Powell, "Royal buys into Global convergence," *Marketing*, June 25, 2001, p. 4.

David Pritchard, "A tale of three cities: 'Diverse and antagonistic' information in situations of local newspaper/broadcast cross-ownership," *Federal Communications Law Journal*, December 2001, pp. 31-51.

Robin Raizel, "Mixed-up media," *Canadian Business*, October 15, 2001, p. 65.

Sheldon Rampton and John Stauber, *Weapons of Mass Deception: The Uses of Propaganda in Bush's War on Iraq*. (New York: Tarcher/Penguin), 2003.

Krishna Rau, "A million for your thoughts," *Canadian Forum*, July/August, 1996, pp. 11-17.

Stephen D. Reese and Jane Ballinger, "The roots of a sociology of news," *Journalism and Mass Communication Quarterly*, Winter 2001, pp. 641-648.

Andrew Rich and R. Kent Weaver, "Think tanks in the U.S. media," *Harvard International Journal of Press/Politics*, Fall 2000, pp. 81-103.

William Rivers, "The correspondents after 25 years," *Columbia Journalism Review*, Summer 1962, pp. 4-10.

Piers Robinson, *The CNN Effect: The Myth of News Media, Foreign Policy and Intervention*. (New York: Routledge), 2002.

Leo C. Rosten, *The Washington Correspondents*. (New York: Harcourt Brace), 1937.

Werner J. Severin and James W. Tankard Jr., *Communication Theories: Origins, Methods, and Uses in the Mass Media*. 4th ed. (New York: Longman), 1997.

James Shanahan and Michael Morgan, *Television and its Viewers: Cultivation Theory and Research*. (Cambridge: Cambridge University Press), 1999.

Donald L. Shaw and Maxwell E. McCombs, *The Emergence of American Political Issues: The Agenda-setting Function of the Press*. (St. Paul, MN: West Publishing), 1977.

Robert Sheppard and Patricia Chisholm, "A strategic retreat," *Maclean's*, May 8, 1999, p. 30.

Pamela J. Shoemaker and Stephen D. Reese, *Mediating the Message: Theories of Influences on Mass Media Content*. 2nd ed. (New York: Longman), 1996.

Richard Siklos, *Shades of Black: Conrad Black and the World's Fastest Growing Press Empire*. (Toronto: Minerva), 1996.

Charlie Smith, "CanWest plans fourth daily," *Georgia Straight*, January 29, 2005, p. 12.

Charlie Smith, "Left-wing shtick: The *Vancouver Province* is criticized for removing a professor's comments from a story," *Media*, Spring 2000, p. 15.

Charlie Smith, "Mair crosses corporate boss," *Georgia Straight*, June 12, 2003, p. 15.

Charlie Smith, "Mair fights CanWest," *Georgia Straight*, July 4, 2002, p. 11.

Dallas Smythe, "On the political economy of communications." *Journalism Quarterly*, 1960, pp. 563–572.

Dallas W. Smythe and Tran Van Dinh, "On critical and administrative research: A new critical analysis" *Journal of Communication*, September 1983, pp. 117–127.

Walter C. Soderlund and Kai Hildebrandt, eds., *Canadian Newspaper Ownership in the Era of Convergence: Rediscovering Social Responsibility*. (Edmonton: University of Alberta Press), 2005.

Robert Sparks, Mary Lynn Young, and Simon Darnell, "Convergence, corporate restructuring, and Canadian online news, 2000-2003," *Canadian Journal of Communication*, 2006, pp. 391-423.

John Stackhouse, "Izzyvision," *Report on Business*, May 1990, pp. 76-83.

Erin Steuter, "He who pays the piper calls the tune: Investigation of a Canadian media monopoly," *Web Journal of Mass Communication Research*, September 2004. Available online at http://www.scripps. ohiou.edu/wjmcr/vol07/7-4a.html.

Robert L. Stevenson, et al. "Untwisting 'The news twisters': A replication of Efron's study," *Journalism Quarterly*, Summer 1973, pp. 211-19.

Diane Stone, *Capturing the Public Imagination: Think Tanks and the Policy Process*. (Portland, OR: Frank Cass), 1996.

Scott Stossel, "The man who counts the killings," *Atlantic Monthly*, May 1997, pp. 86-104.

Kara Swisher with Lisa Dickey, *There Must Be a Pony In Here Somewhere: The AOL Time Warner Debacle and the Quest for a Digital Future*. (New York: Crown Business), 2003.

David Taras, *Power and Betrayal in the Canadian Media*. 2nd ed. (Peterborough, Ont: Broadview Press), 2001.

David Taras, "The winds of right-wing change in Canadian journalism," *Canadian Journal of Communication*, 1996, pp. 485-495.

Paul Taylor, "Third service, third network: The CanWest Global system," *Canadian Journal of Communication*, 1993, pp. 469-477.

"The Gazette Intifada," *Media*, Winter 2002, p. 6.

Clive Thompson, "Fraser attacks! A leaked document reveals the right-wing Fraser Institute's five-year plan for media domination," *This Magazine*, May/June 1997, p. 18

Tracy Tjaden, "*Vancouver Sun* cuts create concerns," *Business in Vancouver*, June 10, 2003, p. 3.

Joseph Turow, *Breaking Up America: Advertisers and the New Media World*. (Chicago: University of Chicago Press), 1997.

R.P. Vallone, L. Ross, &, M. R. Lepper, "The hostile media phenomenon: Biased perception and perceptions of media bias in coverage of the 'Beirut massacre,'" *Journal of Personality and Social Psychology*, 1985, pp. 577–585.

Leah R. Vande Berg, Lawrence A. Wenner, and Bruce E. Gronbeck, "Media literacy and television criticism: Enabling an informed and engaged citizenry," *American Behavioral Scientist*, 2004, pp. 219–228.

Mary Vipond, *The Mass Media in Canada*. (Toronto: Lorimer), 1989.

Mark D. Watts, David Domke, Dhavan V. Shah, and David P. Fan, "The politics of conservative elites and the liberal media argument," *Journal of Communication*, 1999, pp. 35–58.

David H. Weaver and G. Cleveland Wilhoit, *The American Journalist: A Portrait of US News People and Their Work*. (Bloomington: Indiana University Press), 1991.

Jennifer Wells, "He styles himself as the editorial savior of papers, but critics label Conrad Black an editorial storm trooper," *Maclean's*, November 11, 1996, p. 56.

Jennifer Wells, "Izzy's dream," *Maclean's*, February 19, 1996, p. 40.

Deb Halpern Wenger, "The road to convergence and back again," *Quill*, September 2005, p. 40.

Anthony Westell, "Journalism foundation launched," *content* [Centre for Investigative Journalism], November/December 1990, p. 10.

Hugh Wilford, "Calling the tune? The CIA, the British left, and the Cold

War, 1945–60," *Intelligence and National Security*, 2003, pp. 41–50.

Anthony Wilson-Smith, "Power moves from Toronto," *Maclean's*, August 21, 2000, p. 11.

Anthony Wilson-Smith, "War of words: The gloves are off over circulation claims in the newspaper industry," *Maclean's*, February 8, 1999, p. 48.

Anthony Wilson-Smith, "Would we lie to you?" *Maclean's*, October 2, 2000, p. 15.

Mason Wright, "Yipee Tyee," *This Magazine*, November/December 2005, p. 6.

Daniel Yergin and Joseph Stanislaw, *Commanding Heights: The Battle for the World Economy.* (New York: Simon & Schuster), 2002.

INDEX

Alliance Atlantis, xii, 7, 243, 244, 245

Asper, Ruth "Babs", xv, 12

Asper, David, 12, 56, 64, 136-138, 141, 167, 186, 190, 192-194; and David Milgaard, 193; and Mike Bullard, 207; and the *Montreal Gazette*, 136-137; Asper agenda, speech, 125, 265-267; CanWest: Channel 5 (UK) purchase, 64-67; programming strategy, 38; Torstar alliance, 240; Fraser Institute, 107; in New Zealand, 193; Israel Asper defence, 113; legal battles, 162, 194; Liberal defence, ix, 6, 112, 113, 153-154; personality, 141, 193; political affiliations, 234; REM reference, 6, 137; Russell Mills firing, 7, 156, 159; Winnipeg Blue Bombers, xii, 159; views, 196

"Asper Disaster" (2002), 3, 165, 262

Asper, Gail (*see also* Asper Foundation),12, 24, 190, 192, 194, 195; fundraising, 239

Asper, Israel "Izzy", 49. 63. 67, 106, 141, 159; and author, xii-xiii; and Conrad Black, 75, 82, 113-114, 258; and convergence, 3, 206; and Jean Chrétien, 156; and Robert Lantos, 183, 243; and Thomson, 258; background, 4, 9-13; *The Benson Iceberg*, 4, 14, 198; Bilderberg meetings, 84; biography (*see also* Levine,

Allan), xiv; business sense, 10, 53; Canadian Museum for Human Rights, 171, 238-239; Channel 5 (UK) bid, 66-67; CBC criticism, 21, 171-173; children, 190-207; Conservative Party support of, 7; compassion, 170, 171, 183, 185; criticism, 184; death, 189, 204, 235, 260; employee relations, 51-52; Global TV buyout, 30-31; heart attack, 35; Jewish background, 25, 52; legal battles, xiii, 35-36, 37, 41, 45, 47, 60, 170, 183; Leon Ausereper AKA Leon Asper, 10-11; mafias: "Canadian and Jewish", 52; "Winnipeg", xiv-xv; on taxes, 4, 8, 12-15, 22, 23, 24, 99, 180, 186, 265; Order of Canada, 41-42; personality, 9-10, 185; philanthropy (*see also* Asper Foundation), 170-171, 173, 190; political career, 4, 9, 12, 15-26, 27, 28; political position, 4, 5, 8, 228, 265; succession, 194-195; *Titans*, xv, 123; views, 44, 174-177, 186, 188, 189, 196; Western alienation claims, 16, 18-19, 20, 21-23, 253, 256; *Winnipeg Free Press*, 82, 258; Work ethic, 10

Asper, Leonard, 12, 47, 48, 68, 107, 118, 123, 148, 157, 160, 167, 179, 183-184, 186, 189, 190, 193, 194, 195, 233, 244, 245, 250; CBC, 199-204, 205; convergence vision, 116, 117, 123, 126, 157-158, 206, 207;

Matthew Fraser, 183, 184; media concentration, 197, 198, 238; Network TEN, 244; on anti-Israel bias, 199-204; personality, 193, 197; public benefits requirement, 238, 257; Southam name change, 179; staff cuts, *National Post*, 129; UBC visit, 122; views, 198, 233

Asper Foundation, 170, 174, 190, 192, 195

Australia; and Izzy Asper, 74; Australian Broadcasting Authority (ABA), 74; convergence experience, 127-128; cross ownership, 60, 111; foreign ownership, 52-53, 59, 60, 128; Network TEN, 52-61; Paul Keating, prime minister, 73, 74, 111

Bell Globemedia, xv, 112; Bell Canada Enterprises (BCE) sale, 2005, 239; CHUM buy, 241; layoffs, 241; *Globe and Mail* merger, 112

Black, Conrad, 33, 70-85, 150, 155; *A Life in Progress*, 73; American Publishing Co., 71; and Israel Asper, 75, 82, 112-114, 150, 258-259; acquisitions: Australia, 54, 73-74; *Chicago Sun-Times*, 70, 71-72; *Jerusalem Post*, 70, 233; staff cuts, 72-73; Southam, 3, 186; sale, 69, 83-84, 159, 213; takeover, ix, 70, 75-79, 80-81, 259; *Telegraph* (UK), 65, 70, 81; profits, 70-71; staff cuts, 70; Barbara Amiel Black, x, 106, 259; on flogging, 74; Bilderberg Group, 83-84; conservative stance, 70, 81, 95-99, 102, 211; convergence, 81; corporate governance, 78; David Radler, ix, 71-73, 78, 106-107; "human chain saw", 71; effect on journalism, 70; employee relations, 71, 72-73, 77; Hollinger International, x, 70, 73, 74, 77,

78-79, 80, 81, 84, 85, 106, 196-197, 233, 237, 254; hyperbole, 74, 78, 79; legal battles, 73, 81-82; fraud trial, 2007, xii, 113, 196, 258; background, x, 5, 190-197; Liberal antipathy, 79, 80, 81, 96-97; Lordship, 81-82; *National Post*, 70, 80-81, 85, 95, 103, 112, 114, 186, 196, 211, 213, 214; *Globe and Mail* merger, proposed, 258-259; sale, 5, 125;

Black Wednesday, 1980, 109, 262

Bush, George W., 84, 94, 95, 142, 214

Canada Development Corporation (CDC), 33, 34

Canadian Broadcasting Corporation (CBC), 17, 28, 33; "back to basics", 236; bias, alleged, 5, 176, 188, 199-204, 211; criticism of, 5, 21, 184, 171-173, 186-187, 188-189, 216; Friends of Canadian Broadcasting, 40-41, 47-48, 118, 123-124; funding increases, 115, 187, 188; move to Winnipeg, proposed, 115; US programming, 36

Canadian Journalism Foundation, 120-122, 124

Canadian Radio and Telecommunications Commission (CRTC), 5, 29, 35, 37, 39: 40, 54, 110, 145, 158, 162, 236, 255; al-Jazeera application, 205; Alliance Atlantis deal, xii, 245; Bell Globemedia, 241; broadcasting profits, 230; broadcasting review, 238; diversity, 2007, xii; Canadian content (Cancon), 38, 41, 42, 44; rules, 31. 44; CanWest licence renewal, 2001, 112, 117, 119, 123, 132, 169, 254; convergence, 5, 112, 118-119, 261; fees, 199; Fox News Network application, 204-205; Global takeover, 30; ownership oversight, 240, 241-242, 244,

258, 259, 263; public benefits program, 48, 115, 121, 251, 256-257, 262; firewall provision, 6, 115-117; journalism school funding, 250; Robert Lantos, 41-42; simultaneous substitution rule, 1973, 31, 35, 37; Western International Communications (WIC), 46, 47, 48

CanWest Capital Corp., 32-33

CanWest Global, xi, 28; acquisitions: Alliance Atlantis, xii, 7, 243, 244, 245; attempted, NetStar, 45; Australia, Network TEN, 5, 52-61; employee relations, 232; profits, 57, 60-61; sale, 244; Chile, La Red, 5, 61; employee relations, 61; sale, 62; *National Post*, 125; *New Republic*, xii, 250; New Zealand, 5, 233; TV3, 51-52, 68-69; employee relations, 51-52, 57, 69; programming, 69; Southam, ix, x, 3, 5, 31, 57-58, 61, 68, 69, 84-85, 111-112, 115, 122, 157, 159, 163, 171, 213, 254, 256, 258; Turkey, 233; UK, Channel 5, 64-67, 193; advertising, 129, 157, 238; advertorials, 180; and Conservative Party, 8, 228, 235, 237, 239, 245; and Hollinger, 254; and *National Post*, 118; Bev Oda, 7, 237-238, 239, 242; Canadian Association of Broadcasters, 47; Canadian content (Cancon), 31, 39, 43-44; Canadian Press pullout, 8, 237; *Confrontation at Concordia*, film, 174; company editorials, 6, 7, 131-143, 145, 147, 153, 154, 155, 162, 178, 185, 260; no-rebuttal order, 132; contributions, 170; journalism schools, 122-123, 169; political parties, 4-5, 165, 229, 234; convergence, 85, 116, 157, 171, 206-207, 218, 253-255, 256, 258; corporate governance, 197; criticism, 143, 150, 158, 161 (*see*

also Global Television); debt, 128, 241, 242; *Dose*, 219, 232, 233; Gordon Fisher, news president, 179; history (*see also* Levine, Allan), xiv, xvi, 28, 57; income trust, 231, 234; legal battles, 38, 56-57, 199; licence renewal hearings, 2001, 117, 123, 132, 254; MediaWorks name change, 207; MediaWorks Income Fund, 231; partnerships, Goldman Sachs, 243-244; Peter Viner, 52, 55; Prime TV, 43; profits, 5, 34, 44, 235, 243-244; programming, US, 31, 36; protests: against, 143, 149; response, 150; at, 6; by, 113; Rafe Mair, 166-169, 181; Richard Camilleri, 157, 206, 207; share prices, 50, 157, 179, 207; Torstar, 76, 240; views, 185, 213-214, 215, 234; anti-global warming, 222-223; anti-Palestinian, alleged, 215-216; pro-Liberal Party, 222-223, 234; Western International Communications (WIC), 45-46, 47, 48

Chrétien, Jean, 40, 80, 146, 152-`53, 163, 167, 186, 189; and the Aspers, 5, 39, 156; David Asper defence, ix, 6, 112, 113, 114, 161; Russell Mills firing, 7, 158, 161; Shawinigate, 96-97, 115

concentration, media *see* media concentration

convergence (*see also* media, convergence), xi, 3, 5, 83, 109-112, 112-113, 115-119, 125-129, 212, 216; and Friends of Canadian Broadcasting, 124; Bev Oda, 242; CanWest, 85, 116, 206-207, 218, 253-255, 256, 258; CRTC oversight, 261; informed public's view, 217; Israel Asper view, 206; Russell Mills, 182; Torstar, 240; UBC conference, 123; UK views, 126-127

cross ownership, xi, 85, 109-

110, 154, 182; ban, 1982, 220;
in Australia, 217, 244, 261; in
Europe, 217; in UK, 216-217, 261;
in US, 217-218, 261; Leonard
Asper, 198; Lincoln Committee
recommendations, 187;
maximums, proposed, 261
Desmarais, Paul, 76-78

Fox News Network, 204; bias,
alleged, 94-95; against
Canadians, 205; "Fox effect",
94; pro-American, 206, 225; in
Canada, 204-206
Fraser Institute, 98-99, 103, 104-
106, 107; and David Asper, 107;
bias, alleged, 213; CanWest
connections, 107; five-year-plan,
leaked, 104-105; history, 105;
media monitoring, 211
Fraser, Matthew; *Free for All*, 184;
Israel Asper criticism, sea change,
184, 185

Global Television, 26, 29, 33, 157,
161; advertising, 38-39, 245; and
the CBC, 5; BCTV, 220; CanWest
takeover, 33; "cash machine",
5, 35-37, 184; history, 29-31;
"idea free zone", 183; "*Love Boat*
network", 5; Peter Kent, editor,
and Conservative candidate,
228, 234; profits, 232, 245;
programming, American, 5, 31,
37, 183, 184; blunders, 30, 207;
Canadian, 37, 40, 43, 245
Globe and Mail, The, xv, 7, 17, 20, 23,
24, 28, 34, 47, 51, 54, 58, 78, 106,
107, 117, 119, 138, 139, 142, 148,
149, 150, 156-157, 162, 165, 167,
168, 174, 178, 204, 215, 216, 250,
262; and Canadian Journalism
Foundation, 121; "axis of snivel",
197; Bell Globemedia merger, 112;
bias, alleged, 80, 176, 196, 211, 212;
CanWest-Torstar alliance, 240;
colour printing, 81; convergence,

83, 112; Gordon Fisher, 179;
National Post merger, proposed,
258-259; Neil Macdonald, 203;
Patricia Pearson, 213-214; Rafe
Mair, 181; Russell Mills, 159-160,
178; *Toronto Star* link, 240
Harper, Stephen; and CanWest,
239; and David Asper, 7, 234-235;
election, 241

journalism: ethics, 122; schools;
Brigham Young University,
convergence experience, 218-219;
call for monitoring, 234; Canada
vs US, 251; CanWest donation,
122; Carleton University,
Director Christopher Dornan,
204; Professor Roger Bird, 178;
Concordia University, Professor
Mike Gasher, 140, 177; corporate
funding, x, 250-251, 257, 263;
Ryerson, Chair Vince Carlin, 141;
Ethics Chair Peter Desbarats, 29,
117, 120, 139-140, 250-251,
257; Professor John·Miller, 139;
Professor Matthew Fraser,
183; Simon Fraser University,
NewsWatch Canada, 257;
University of British Columbia,
Director Donna Logan, 116-
117; 122, 168-169, 221; Director
Stephen Ward, xiv; University
of King's College, Instructor
Stephen Kimber, 142-144;
University of Western Ontario,
Professor David Estok, 178

labour relations *See* employee
relations
Levine, Allan, 28, 34, 62; Asper
biography, proposed, xiv; *From
Winnipeg to the World*, xiv, 51, 52,
57, 64

media; advertising rules, effect,
252; broadcast, 90; advertising,
238; Canadian; Americanization

of, 32, 35, 183, 184; CBC, 87; CTV, 87; "flexibility", 242; origins, 87; profits, 230; concentration, US, 253; cross ownership, US, 252; effects, 86-108; limited, 87, 252; powerful, 86, 89, 90, 93-94, 95; propaganda, 87, 247-248; *National Post*, 97, 130; foreign ownership, in Australia, 52-53; Harold Innis, 89, 90, 248; Marshall McLuhan, 89-90; *New Republic*, pro-Israel bias, alleged, 249-250; news; bias, alleged, xiv, 92-93, 101-102, 208, 221-223; conservative, 101-104, 208, 210, 212, 213, 222; liberal, 208-213; owners', x, 223-227; reporters', x, 175-176, 212; Canadian; bias, alleged, 102-107; concentration, 3, 6, 8, 79, 80, 85, 107, 111, 112, 118, 122, 129, 196, 197, 212, 227, 228, 229, 230-231, 235, 236, 238, 239-; 245, 254, 255, 259-260, 262, 263, 264; criticism, 88, 90; cross ownership, 154, 182, 188-199, 235, 261; foreign ownership, 7, 82, 187, 199, 228, 243, 244; profits, 229-230, 236; regulation, 229; in New Zealand, 51; regulators; Australian Broadcasting Authority (ABA), 74; CRTC, Canada, *see* Canadian Radio and Telecommunications; Commission; Federal Communications Commission (FCC), US, 88, 111, 143, 217-; 218, 252, 253; (IRTC), Ireland, 68; "selling crack", 253; symbolism, 94; think-tanks, 99-100, 102-103

Mills, Russell, 178, 182, 235; firing, ix, 7, 155-163, 177-178, 227; Israel Asper obituary, 189

Montréal Gazette, The, 131-139, 141-142, 150, 161, 167, 235; byline strike, 2001, ix, 6, 134-138, 141, 148, 227; "Gazette intifada", 135, 141; ownership influence, 131-143, 227

Murdoch, Rupert, 65, 111, 126, 128; Channel 5 (UK) bid, 65, 66; Fox News Network, 94, 206; political stance, 94-95; Network TEN, 244

National Post, The, 4, 102, 122, 149, 159, 164, 174, 177, 184, 197, 203, 219; advertising, 211; advertorials, 180; bias, alleged, 80, 95-99, 129-130, 131, 211, 213-215, 216; CanWest Global, purchase, 125; rebuttal mechanism, 118; CBC, 236; David Asper column, 112; response, 113, 114, 117; Friends of Canadian Broadcasting, 124; income trust exemption, 231; Jean Chrétien, ix, 5, 6, 96, 153, 186; Ken Whyte, editor, 114, 125; launch, 1998, 3, 80-81, 103; Leonard Asper, 198; losses, 128-129, 185, 199, 207; Matthew Fraser, 124; editor, 42, 43, 183, 185; Peter C. Newman, 123; printing, 80, 240; staff cuts, 128-129; Terence Concoran, 129, 182, 211, 232;

Newman, Peter C., 24, 25, 35, 39; and Friends of Canadian Broadcasting, 123-124; Asper biography, proposed, xiv, xv; *National Post* column, 123; *Titans* book, xv, 123

9/11, 94, 95, 144, 200, 214, 218

Oda, Bev, 7, 237-238, 239, 242

Ottawa Citizen, The, 152-156, 161, 172, 238; byline strike, 156; ownership influence, 160, 226; "terrorist" labelling, 215, 216

Pacific Press (*see also Vancouver Sun, Province*), ix, xiv; CanWest, 163, 164-165, 168, 234; potential breakup, 261; profit pooling arrangement, 229, 240

Pacific Press: The Unauthorized Story of Vancouver's Newspaper Monopoly, ix, xiii

Province, The, ix, xii, 107, 132, 157, 167, 219 ; bias, alleged, 107, 166
public policy institutes ("think tanks"), 99-100; in Canada, 102-108

Royal Commission on Newspapers, 1981, 259

Senate of Canada; Davey committee, 236; hearings into news media, 2003, 182, 221; 2005, 219; Lincoln committee, 2002, 138, 141, 150, 187-188, 197, 198, 228, 242, 260; media inquiry, 7, 162; report, 1970, 109, 263; 2006, 228, 235-237, 241, 242, 254
simulcasting (*see also* CRTC, simultaneous substitution rule), 35, 38
Sinclair Jr., Gordon, *Izzy Asper: The Maverick, the Mogul and All That Jazz,* xiv, xv
Southam, 4, 80, 114, 124, 125, 138, 148, 154, 178; Asper control, 3, 6, 112, 132, 139; name change, 7, 179; CanWest purchase, ix, x, 3, 5, 31, 57-58, 61, 68, 69, 84-85, 111-112, 115, 122, 157, 159, 163, 171, 213, 254, 256, 258; income trust, 232; Conrad Black takeover, 1996, 70, 75-79, 80-81, 150, 259; Harvey, 76; *News and the Southams,* xiv; Torstar, 76, 79

Television (*see also* media, broadcast); demographics; La Red, 61; Network TEN, 57-58; Prime, 43; effects, 92; ideology, 92; origins, 87; political view, 92; power, 90, 91; violence, 91-92;
Toronto Star, The, 6, 44, 116, 123, 134, 138-139, 144, 145, 147-148, 155, 199, 262; "axis of snivel", 197; David Asper, 137, 161, 192; *Globe and Mail* link, 240; ownership influence, 227
Trudeau, Pierre, 21, 26

Vancouver Sun, The, ix, xiii, 153, 157, 163-164, 168, 219, 238; advertorials, 220; bias, alleged, 222; pro B.C. Liberal, 165, 181, 220; Donna Logan interview, 122; Lincoln committee report, 188; ownership influence, 132, 227; staff cuts, 179, 180-181, 220; Southam sale, 109
Victoria Times-Colonist, The, 80; Vivian Smith firing, 225-226

Zerbisias, Antonia, 44, 47, 116, 118, 123, 184, 186, 187-188, 199, 203, 227, 229, 240, 242, 245, 262